T0271916

Habitable Exoplanets for Extra-Terrestrials

This book explores the questions of What, Why, When, How and Where we might find Extra-Terrestrials (a.k.a. Aliens) and their habitats throughout the Universe – and Who might they be?

Starting from ourselves and the Earth and eventually speculating about life-forms that might span multiple Universes, it provides an accessible introduction to extra-terrestrial life, the search for extra-terrestrial intelligence and exoplanets. It will enable readers to appreciate, follow and sometimes answer questions on life and planets outside Earth. It details these exciting topics by pondering what comprises an alien life form and what suitable habitats might exist for them inside and out of our solar system. The book also looks ahead to the future and the realities of finding alien life and the possibilities of mankind leaving Earth and living on another planet.

This guide is accessible to those without a formal scientific or mathematical background. It will also be of interest to students of astrobiology, astronomy, astrophysics, planets throughout the Universe, the origin and evolution of life-forms beyond the Earth and (perhaps) of the ultimate limits attainable by life in the Universe, who are looking to learn more about those same topics, but which are outside their own specialisms.

Key Features:

- Combines the exciting topics of extra-terrestrial life, the search for life outside Earth and exoplanets into one accessible guide.
- Contains no complex mathematical formulae or content.
- Authored by a professional educator and a professional *and* amateur astronomer, with a life-long interest in getting everyone and anyone as fascinated by astronomy and its related subjects as he himself has been, since discovering the subject in early secondary school.

Chris Kitchin is currently a Professor Emeritus at the University of Hertfordshire and a freelance writer of astrophysics books. From 1987 to 2001, he was director of the University's Observatory and from 1996 to 2001, he was also head of the Division of Physics and Astronomy. He took early retirement in 2001 in order to concentrate on his writing. Kitchin has written 14 single-author books, with several being translated into Chinese, German, Japanese and Polish and some having multiple editions. He has also contributed to another dozen or so books, and he has written hundreds of articles ranging from popular astronomy to specialist research. In 1997, he was awarded the title of Professor of the Public Understanding of Astronomy.

Habitable Exoplanets for Extra-Terrestrials

C.R. Kitchin

CRC Press
Taylor & Francis Group
Boca Raton London New York

CRC Press is an imprint of the
Taylor & Francis Group, an **informa** business

Designed cover image: An artist's impression of the possible view from a spacecraft close to the TRAPPIST-1 exoplanet system. The seven currently known exoplanets are indicated by arrows. (Reproduced by kind permission of ESO/N. Bartmann/spaceengine.org).

First edition published 2025
by CRC Press
2385 NW Executive Center Drive, Suite 320, Boca Raton FL 33431

and by CRC Press
4 Park Square, Milton Park, Abingdon, Oxon, OX14 4RN

CRC Press is an imprint of Taylor & Francis Group, LLC

Library of Congress Cataloging-in-Publication Data
Names: Kitchin, C. R. (Christopher R.), author.
Title: Habitable exoplanets for extra-terrestrials / C.R. Kitchin.
Description: First edition. | Boca Raton, FL: CRC Press, 2025. | Includes bibliographical references and index. | Summary: "This book explores the questions of What, Why, When, How and Where might we might find extra-Terrestrials (a.k.a. Aliens) and their habitats throughout the Universe- and Who might they be? Starting from ourselves and the Earth and eventually speculating about life forms that might span multiple Universes, it provides an accessible introduction to extra-terrestrial life, the search for extra-terrestrial intelligence, and exoplanets. It will enable readers to appreciate, follow and sometimes answer questions on life and planets outside Earth. It details these exciting topics by pondering what comprises an alien life form and what suitable habitats might exist for them inside and out of our solar system. The book also looks ahead to the future and the realities of finding alien life and the possibilities of mankind leaving Earth and living on another planet. This guide is accessible to those without a formal scientific or mathematical background. It will also be of interest to students of astrobiology, astronomy, astrophysics, planets throughout the Universe, the origin and evolution of life-forms beyond the Earth and (perhaps) of the ultimate limits attainable by life in the Universe, who are looking to learn more about those of the same topics which are outside their own specialisms"--Provided by publisher.
Identifiers: LCCN 2024011954 | ISBN 9781032159485 (hbk) | ISBN 9781032152813 (pbk) | ISBN 9781003246459 (ebk)
Subjects: LCSH: Habitable planets. | Life on other planets.
Classification: LCC QB820 .K58 2025 | DDC 576.8/39--dc23/eng/20240617
LC record available at https://lccn.loc.gov/2024011954

ISBN: 9781032159485 (hbk)
ISBN: 9781032152813 (pbk)
ISBN: 9781003246459 (ebk)

DOI: 10.1201/9781003246459

Typeset in Times
by Deanta Global Publishing Services, Chennai, India

I take great pleasure in dedicating this book to the eight ETIs
(Extraordinary Terrestrial Intelligences - a.k.a. Border Collies):

Honey, Jess, Misty, Pip, Rowan, Spruce, Willow and Wills.

whose lives, characters, friendliness, love, joie de vivre and independence
I have been privileged to share over many decades and
still continue to share and to remember.

I just hope that when Extra Terrestrial Intelligences are found they live
up to such high standards as those which my ETIs have set for them.

Contents

THEME 2   ETs and ETIs (Plus HUMANKIND)

THEME 3   Suppose There Really Are Aliens Somewhere 'Out There' – What Do We Do About It?

Frontispiece – The Earth seen from 67,000 km out in 1969 from Apollo 10. (Reproduced by kind permission of NASA.)

Is this the only place in this entire Universe of about

100,000,000,000,000,000,000,000 stars

and perhaps

1,000,000,000,000,000,000,000,000 planets

where life may be found?

It could be so.

In which case

We had better look after it very carefully.

On the other hand, if even only one planet in a million has life on it – that is still

1,000,000,000,000,000,000 inhabited planets

in the Universe.

Perhaps, by the time you've read this book, you'll have a clearer idea of which of these possibilities might be the case.

But even if inhabited planets are common, the Earth is the only one which *we* have got.

So,

We had *still* better look after it very carefully.

Acknowledgements

My thanks are due to Stephen H. Dole for the idea for this book's theme and title. His book *Habitable Planets for Man*[1] analysed the chances of finding planets within the solar system and beyond which might be colonisable by humankind. Sixty years later and we know of many thousands of planets outside the solar system, and our understanding of life-forms and their requirements has improved a hundred-fold. The time therefore seems ripe to extend his speculations out into the wider universe and to cover more than just humankind's needs. Sadly, though, Stephen died in the year 2000 so I cannot express my gratitude to him in person.

A special individual 'thank you' is also due to Christine Danes for checking through the biological and astrobiological sections of this book for any mistakes made through my more workaday knowledge of those subjects.

Finally, a myriad thanks are due to the tens of thousands of astrobiologists, astronomers, astrophysicists, biologists, chemists, geologists, mathematicians, meteorologists, physicists, space scientists and (some) SF authors whose investigations, discoveries, ideas and results underlie the material discussed throughout this book. Although space and the requirements of readability preclude me from naming them all,[2] the credit for our present and future understanding of these aspects of the Universe is theirs alone.

The responsibility for *any* mistakes remaining in the book is all mine and I can only echo Leonardo Fibonacci's plea for forgiveness in his 1202 CE book *Liber Abaci*[3]:

> "If, by chance, something less or more proper or necessary I omitted, your
> indulgence for me is entreated, as there is no one who is without fault, and in all
> things is altogether circumspect."

---

[1] Blaisdell Publishing Co. 1964.

[2] In a few cases, where I felt that readers might be likely to wish to pursue the topic, or that it was especially significant, a source reference has been given in a footnote.

[3] *Fibonacci's Liber Abaci: A Translation into Modern English of Leonardo Pisano's Book of Calculation*. Sigler, L. E., Springer, 2002. N.B., Leonardo Pisano was Leonardo Fibonacci's original name and it was this book which introduced Europeans to the Hindu-Arabic base-10 numbering system.

Preface

Raison d'-etre

The Universe is BIG.

If something CAN happen,

then it very probably WILL.[4]

AIMS

My aim for this book is to introduce readers from many backgrounds, interests and ages to a question and its ramifications, the answer to which has long thwarted humankind, but which may soon be answered.

This Most Human of Questions[5] is:

Is there life elsewhere in the Universe?

Trying to see what is involved with this question and what will be needed to look for its answer will require us to get to grips with the cutting edges of many aspects of many modern sciences including:

Astronomy, Biology, Chemistry, Geology and Physics

in particular.

But WORRY-NOT – no-one can claim to be an expert in all of those topics. Indeed, the scope of modern science is now such that it is only the optimist (or braggart) who can claim to be an expert in even more than *one* sub-topic of *one* of those main sciences.

All the topics needed, therefore, will be described and will have sufficient background material for you to be able to follow the arguments in this book through to their conclusions.

So, be you a

professional scientist,
a citizen scientist,
a student,
a well-read lay person,
a simple enquirer after knowledge

or

someone for whom science is currently a totally closed subject

and,

be you young or old,

---

[4] The author's own and preferred version of Murphy's Law (a.k.a. Sod's Law, Finagle's law).
[5] MHoQs.

then, with a bit of concentration at times, you should be able to find your way successfully towards understanding what we currently know about this MHoQs and why it may shortly be answered.

The main three sciences concerned in addressing the question and so also in the material presented in this book are thus:

- Exobiology
- Exoplanets

and

- SETI.[6]

We tackle these in the order of their current progress towards achieving concrete results. Thus, at the time of writing:

- Thousands of actual examples of Exoplanets are known.
- No non-terrestrial-types life-forms have been found, but the study of terrestrial life-forms is very advanced now at the atomic and molecular scales, and we have found abiogenetically produced molecular building blocks for life throughout the inter-stellar medium and in many other situations.
- SETI has yet to detect any clearly non-natural signals. New instruments now coming on stream and other new developments, however, will soon increase the search areas, frequencies covered and signal patterns sought, by orders of magnitude over the total of everything which has been attempted to date. So, if they are there to be found, the probability of making detections is increasing apace.

The basic structure of the book thus follows the above order, but, of course, there are frequent cross references and the final chapters synthesise the earlier discussions.

It is likely that individual scientists working in the scientific areas covered within this book will think that I have given insufficient weight to the *vital* importance which *their* particular specialisms have to my themes, but, if the book is not to have the mass of a large galaxy, then limits have to be drawn to its contents – and I have drawn them at the points where I think the reader will have sufficient information to follow the descriptions and arguments then being presented. Numerous references are given throughout the book for readers who would like to research into some particular topic further than may be covered here.

Nonetheless, I hope that *all* my readers find this book interesting, informative and mind-expanding and that it provides explanations and solutions for at least some questions which your curiosity may have prompted about the MHoQs.

Chris Kitchin
Hertford, March 2024

---

[6] Search for Extraterrestrial Intelligence. In common usage, 'SETI', has two slightly differing connotations. The first is as a general term for all activities related to this topic, as is the case here. The second is as a short-hand for the 'SETI Institute', a not-for-profit organisation set up four decades ago, which supports, promotes, studies, researches, publicises and generally does all that it can do to advance all aspects of the Search for Extraterrestrial Intelligence. The SETI institute is largely supported by philanthropic gifts. For example, it has just, at the time of writing, received $200 million from one of its benefactor's estates. It has only minor funding from 'official' sources.

These two usages will be distinguished in this book as just 'SETI' when it is generic term for the whole subject and 'SETI Institute' when the institute itself is intended.

Preparatory notes: The Molehills in Our Path Which Never Become Mountains

MATHEMATICS

Many potential readers who *are* prepared to work through a written argument, even if it has graphs and illustrations, give up at the sight of a single equation. So be comforted – equations and higher forms of mathematics will *not* be encountered.

Simple numbers and simple arithmetic (plus, minus, multiply, divide, squares and square roots) *will* be encountered. However, because astronomy, chemistry, physics, etc., are involved, those simple numbers can be very, very large and/or very, very small. We have unavoidably therefore, at times, also used index numbers such as:

$$10^9 \text{ and } 10^{-3}$$

(these two examples are of the 'written-out' numbers, one billion and one thousandth, or 1,000,000,000 and 0.001), in order to avoid the typesetters from running out of zeros.

If you are rusty about index notation or have never come across it at all, then Appendix A provides a guide. Alternatively, an internet search for, say, 'a simple guide to the index notation of numbers' will provide hundreds of easy tutorials – and don't be afraid – it is *just* a convenient *notation* – it is not proper mathematics that is involved.

STYLE AND RELATED MATTERS

This is a 'serious' book, which is to say, I have tried to make it as complete, thorough, accurate and up-to-date (to the end of March 2024) as possible. However, I hope that it is not serious in the sense of being unnecessarily difficult, dull or tedious. If you read through the main text, without diverting into the Boxes and Footnotes, then you should get a straightforward account of the main topics.

Most of the boxes then explain material which may not be familiar to all readers or may be quite technical and which can be missed if you are happy just to accept what is in the main text without knowing some of the underlying 'whys' and 'wherefores'. In one or two cases the boxes outline some background/related material which I just thought might be of interest.

The footnotes vary rather more: some just give the details of the sources of quotations, translations and the like, a few are internet links to pages which some readers may find of interest, many footnotes will list the words making up an acronym or abbreviation, others cover a difficult/technical subject which is in the main text, but which can be explained briefly, and a lot more are just amusing things related to the topic(s) in the text and which I find diverting and which I hope appeal to you as well.

The Appendices are either short (then usually just listing data) or long (for when an explanation being given is too extensive for a Box).

Almost all quotations are taken from their original sources and, where applicable, are given in their original forms (however, translations *are* provided). Where the original source could not be traced, then a later and, I hope reliable, reference is provided.

I have tried to keep the style light, readable and, in some places, conversational (i.e., I have an imaginary conversation with you, the reader, during which, astonishingly, your responses are always exactly as I had desired). I hope that you find that this approach suits you, but if not, I can only apologise and refer you to my Leonardo Fibonacci quote in the Acknowledgements (above).

In support of the hoped-for readable nature of this book, as already mentioned, equations have been avoided. However, sometimes I have done the calculations for you and just give the results. If you try to check those calculations (and I hope some readers will do so, since I am not infallible – see Fibonacci quote again), then you may get slightly different results. In some cases, though I hope that they are few or none, I may have got something wrong. Most of the time, however, the difference will be due to the rounding up or down of numbers.

Thus, the well-known number relating a circle's radius and circumference and usually given the name, 'π',[7] has the numerical value:

$$3.141\ 592\ 653\ 589\ 79\ \ldots\ \text{(it goes on forever).}$$

Now if, for example, the radius of a star is known to lie between 500,000 km and 1,000,000 km (and such levels of uncertainty are not at all unusual for the subjects dealt with in this book), then it would be pointless to use a value of π accurate to 14 decimal places to calculate its circumference. In fact, in this case, I would take a value for the radius of 750,000 km and multiply it by 3.1, getting a value of 2,325,000 km for the circumference and then quote the calculated value of the circumference in this book as ~2,000,000 km (you will encounter the symbol, '~', meaning 'approximately', many, many more times throughout this book).

ALERTS

This is a book which will be speculating about things which are beyond our current levels of knowledge and understanding and so some imagination will be needed at times. As far as possible, though, these speculative journeys will keep within the bounds of what we currently know or which we can sensibly extrapolate from that current knowledge. I will indicate when a discussion is speculative and give some indication of the level of probability/fantasy using the following in-text symbols;

® Reality – not speculation at all, but proven knowledge. Usually, this symbol will only be used where the author feels that the reader may perhaps be in some doubt that what she/he is reading is *not* fictional.

For example, there are thousands of exoplanets, of many different types and conditions, many of which are relatively close to the solar system.

❀ Quite possible, even probable, given today's scientific and technical knowledge.

For example, sending micro-spacecraft to nearby exoplanet systems.

❀❀ Possible, given a few decades of likely developments from today's scientific and technical knowledge.

For example, establishing permanent, but wholly enclosed, human colonies and industries in orbit or on the Moon, Mars, asteroids, etc.

❀❀❀ Quite speculative, but probably not ruled-out completely by today's scientific and technical knowledge.

---

[7] A Greek letter. The complete list of Greek letters, which are often used in naming things in astrophysics (e.g., the star Sirius, is also called α CMa – see below), is given in Appendix C.

For example, the terraforming (Chapter 18) of solar system planets and satellites.

𝕾𝕾𝕾𝕾 Unlikely and will probably require unforeseeable developments of today's scientific and technical knowledge.

For example, cannibalising the giant solar system planets to construct Dyson swarms (Chapter 18).

𝕾𝕾𝕾𝕾𝕾 Ruled-out by today's and foreseeable scientific and technical knowledge.

For example, traveling faster than light.

"Ah!", I'm sure you will say after this last point "but what about tomorrow's, or the day-after's scientific and technical knowledge" – Well, that's fair comment – it's just that it's beyond the remit of *this* book.

Having given the caveat in the last paragraph, it will undoubtedly transpire, at some future time(s), that some of the '𝕾' speculations go on to become abject flops, whilst some of the '𝕾𝕾𝕾𝕾𝕾' speculations proceed to triumphs presently inconceivable.

UNITS AND MEASUREMENTS

The SI[8] system of units is generally used throughout this book.

One exception to this rule, though, is for the distances to astronomical objects. For this purpose, light-years (ly, a non-SI unit) will be used, rather than astronomers' usually preferred units of parsecs (pc[9]). This is because light years enable the travel times to distant objects to be calculated very easily. The speed of light is usually given the symbol 'c' and has a value very close[10] to 300,000,000 m/s or 300,000 km/s.

Light is just one of many forms of electromagnetic radiation (e-m radiation) and that also includes: radio waves, microwaves, infrared, ultra-violet, x-rays and γ rays – and the speed of light is the *maximum* speed which *any* and *all* of them may attain.

The time taken for a radio signal (say), travelling at the speed of light, sent from an inhabited exoplanet *10 ly* away from us, to arrive at the Earth is thus *10 years*.[11] If those same ETIs[12] were to launch a spacecraft to visit us at 1% of the speed of light, that would take 1,000 years (10 ly/divided by 0.01 c) to get here.

If you are rusty about SI units or have never come across them at all, then Appendix B provides an introductory guide. Alternatively, an internet search for, say, 'a simple guide to the International System of Units' will again provide hundreds of easy introductions and tutorials.

The SI system is widely used because it has been developed to be a self-consistent set of measurements building on just seven basic types of measurements/quantities. These basic quantities are:

- Time (SI unit; the second, symbol 's').
- Length (SI unit; the metre, symbol 'm').

---

[8] System International. In full; International System of Units.

[9] 1 ly = 0.3066 pc = 9,460,730,472,580,800 m (exactly) = 9.460 730 472 580 800 x 10^{15} m = 9.460 730 472 580 800 x 10^{12} km.Useful approximations; 1 ly ≈ ten trillion kilometres ≈ 10^{13} km ≈ 6,200,000,000,000 miles ≈ 6 trillion miles ≈ 63,000 AU (the astronomical unit, AU, is the distance between the Earth and the Sun) and 1 pc = 3.262 ly. The pc is also a non-SI unit. Useful approximation; 1 pc ≈ 3 1/3 ly ≈ 33 trillion kilometres ≈ 20 trillion miles ≈ 200,000 AU.

[10] Exact value; c = 299,793,458 m/s = 2.997 934 58 x 10^8 m/s = 299,792. 458 km/s ≈ 670,618,866 miles per hour. The symbol 'c' for the speed of light comes from the word 'Celerity'.

[11] The terms 'Year' and 'Day', as used in this book, refer to terrestrial years and terrestrial days; 1 year = 365 days or 365.25 days, etc. For other planets the term 'Orbital period' will generally be used for 'year' and 'Rotational period' or 'Sol' for 'day', although some exceptions may be encountered; such as "the Martian 'year' = 686.98 days = 669.6 Martian Sols".

[12] In this book I use the convention of ETs (Extraterrestrials) for non-terrestrial life in any form, although mostly for when it is non-intelligent and ETIs (Extraterrestrial Intelligences) for non-terrestrial *Intelligent* life-forms.

- Mass (SI unit; the kilogram, symbol 'kg').
- Electric current (SI unit; the ampere, symbol 'A').
- Temperature (SI unit; the kelvin, symbol 'K').
- Amount of substance (SI unit; the mole, symbol 'mol').
- Luminous intensity (SI unit; the candela, symbol 'cd').

If you are already about to give up, DO NOT WORRY. Some of these units, such as seconds and metres, you probably *are* acquainted with already and the last two (mol and cd) are not to be found in this book anyway.

As already mentioned, a slightly more detailed guide to the SI is to be found in Appendix B, along with some conversions which you may find useful, such as:

$$1 \text{ kg} = 2.20462 \text{ pounds} = 35.274 \text{ ounces}$$

and there are many other sources of information about SI on the internet and in almost any book on physics (or science more generally).

There are, though, a few inconsistencies in the SI system and one or two other quirks which you may find it useful know about now and these are:

- Prefixes. Small and large quantities of the basic units are indicated by prefixes, some of which you are very likely to have encountered already, such as:
  - microsecond (symbol; μs, one-millionth of a second)
  - kilometre (symbol; km, one thousand metres).
  A complete list of the prefixes is given in Appendix B.
- Mass (i). Very unfortunately, the name chosen for the basic SI unit of mass is the kilogram. This has great potential for confusion and many people think (quite logically) that the basic SI unit of mass is therefore the gram.
  This is made even worse by the continuing use of 'gram' as though it *were* the base unit, thus a millionth of a kg is called a milligram not a μkg (microkilogram) as should be the case.
  I am sorry about this (I would have done it differently), but you just have to remember this inconsistency.
- Mass (ii). On the plus side for the SI system,[13] there are different units for mass and weight (a.k.a. force). Thus, mass has units of kg, while weight, a force on that mass due to the local gravitational field, has units of newtons[14] (symbol 'N'). Sometimes the term kilogram-force (kgf) is used for weights, with 1 kgf = 9.80665 N and 1 N = 0.101972 kgf.
- Temperature. The Kelvin scale (a.k.a. absolute temperature) is based upon the lowest temperature that theoretically could ever be reached: zero degrees kelvin (0 K). To date, the lowest temperature achieved in practice stands at 0.000 000 000 038 K.
  The Kelvin scale is related to the, perhaps more familiar Celsius (Centigrade) and Fahrenheit temperature scales, by the relationships
  $$0 \text{ K}^{15} = -273.15 \text{ °C}$$
  $$= -459.67 \text{ °F}$$

---

[13] This is *good* because mass and weight (or force) are *different* quantities. Thus, when you buy a one-pound bag of sugar in a supermarket on Earth, that bag of sugar will weigh one pound (the *force* of the *Earth*'s gravity on the sugar) and it will also have a *mass* of one pound.
However, in some future supermarket on the Moon, the same one-pound bag or sugar would weigh only about 2⅔ ounces (= 0.166 pounds – i.e., the *force* of the *Moon's* gravity on the sugar), but its *mass* would still be one pound.

[14] Force is not a basic quantity in the SI system, thus its unit, the newton, may also be written in the basic units as: kg m s⁻². It is therefore related to the unit of mass (kg) by the local force of gravity which has the same units as acceleration, i.e., metres per second per second or m s⁻². This is further explained in Appendix B.

[15] The convention for the Kelvin temperature is not to use the superscript 'o' as in °C etc.

When it seems more sensible (i.e., for values around 'room' temperature) °C will be used in this book and these may be converted to K by adding 273.15 to the °C value.

DATES

For events earlier than recorded history, the date is the *years ago* (yra) from the present day. For recorded historical events, the date is in the conventional notation of 'Before Common Era' (BCE) and 'Common Era' (CE). The '~' prefix indicates an approximate value for pre-recorded-history events, while c. (circa) also indicates an approximate value and is the customary practice for recorded events. For events far into the future, yrh (*years hence*) is used.

Complications such as the beginning of the year being March 25th or December 25th instead of January 1st and the Julian/Gregorian calendar changeovers have not been investigated. The pleasure (and hard work) of sorting out such details is left for readers, who may need such accuracy, to find for themselves. Dates in other calendars, such as the Chinese and Hebrew, may generally be converted to the Gregorian calendar and back by algorithms easily to be found on the internet.

NAMES AND LABELS

The naming or labelling of the constituents of the Universe has grown up over centuries and can only be described as a complete and utter mess.

A major centre for astronomical archives, the Centre de données astronomique de Strasbourg,[16] currently contains over 9,500 catalogues of stars, nebulae, galaxies, exoplanets, etc. and almost every single one uses a different means of labelling its entries – and there are a great many other systems in addition to those based upon catalogues.

Thus, the brightest star in the sky, Sirius, is also called:

> α CMa, 9 CMa, HD 48915, HR 2491, FKS 257, BD-16 1591, IRAS 06429-1639 …
> with at least 50 more alternatives

My advice is: 'don't worry about star's names; just accept them'.

Fortunately, at least for the time being, the naming of exoplanets has a bit more logic. The exoplanet is named by taking (one of) the host star's names and adding a lowercase letter to it. For the host star's first discovered exoplanet 'b' is used (the host star is considered 'a', but *that* name will not ever be used), while it is 'c', if a second exoplanet is found, then 'd', 'e', 'f' and so on.

Thus, the host star TRAPPIST-1 has seven currently known exoplanets (Cover Image) labelled:

TRAPPIST-1b
TRAPPIST-1c
TRAPPIST-1d
TRAPPIST-1e
TRAPPIST-1f
TRAPPIST-1g

and

TRAPPIST-1h

In this case the alphabetical/discovery order of the exoplanets also happens to be their order in terms of their distances from the host star, but this is not always the case. Thus τ Cet's exoplanets (three of which are currently unconfirmed) when placed in order of distance from the star are:

---

[16] Freely accessible by anyone at https://cds.u-strasbg.fr/.

τ Cet b
τ Cet g
τ Cet c
τ Cet h
τ Cet d
τ Cet e
τ Cet f

and

τ Cet i

A very few exoplanets now have proper names, thus 51Peg b[17] is also called Dimidium.

The IAU (International Astronomical Union) over the last decade or so has occasionally held competitions (see https://www.nameexoworlds.iau.org/) to enable a few exoplanets and their host stars to be given 'official' names. The process offers a selected host star and its exoplanet(s) to a participating country. That country then holds its own internal competition amongst its populace to suggest names.[18] Thus, the population of Morocco has recently chosen the names *Titawin, Saffar, Samh* and *Majriti* for the host star, υ Andromedae and its three exoplanets υ Andromedae b, c and d, respectively. There is not, however, much indication at the time of writing that such names have entered common usage.

The labellings in cases of exoplanets around one star in a binary or multiple star system or around both stars in a binary star system or free-floating in space all still have to be sorted out at the time of writing.

---

[17] The first 'normal' exoplanet to be discovered (Chapter 6).
[18] There is thus little chance of getting an exoplanet named after yourself.

1 A Question: Just *WHY* Are We Looking for Other Living Entities and Other Planets?

ANSWER:

Because evolution (Chapter 10) has made the human race curious about just about everything and anything and has given us enough brain power to follow up that curiosity effectively.

So – our search for Extra-terrestrials[19] and Exoplanets[20] is just 'Blue-Sky[21] Curiosity', right?

RIGHT!!

and so, it is a waste of time and money, right?

WRONG!!

Curiosity may kill off cats, but in evolutionary terms for humankind it has paid off in spades – and today we still retain all of that original nagging itch for knowledge, whether it be immediately 'useful' or not.

Almost all terrestrial life-forms exhibit the phenomenon which we call 'Curiosity' but which in a more general sense may be regarded as 'Reactions to Environmental Pressures'. For plants and other essentially immobile forms of life, the reaction may be as slight as growing towards the light. Basic cellular life, though, may be able to move away from danger or towards food using various abilities. Amoebae, for example, can change their shapes to extend projections called 'false feet' (pseudopodia) which then adhere to the surface, contract and so pull the organism towards its desired goal.

For both plant life and amoebic life, such reactions are survival tactics and evolution ensures that, amongst those entities, those which are best in such tactics will prosper over their less able rivals – in due course evolving to different life-forms (i.e., becoming new species).

More complex life-forms will generally have more complex and more effective survival tactics, though not essentially any different from those of amoebae. For early humankind, such as Homo

---

[19] a.k.a., Aliens, BEMs (Bug-Eyed Monsters), ETs (Extra-terrestrials), ETIs (Extraterrestrial Intelligences), LGMs (Little Green Men), Outsiders and many other appellations.

[20] The term derives from 'extra-solar planet' – i.e., a planet which is *not* a part of the solar system

[21] This term is a synonym for 'basic curiosity', or in science for 'basic research' – i.e., research without a specific goal in mind at the time that it was/is undertaken. The term 'blue-sky' derives from John Tyndall's 1869 experiments which showed that the sky's blue colour arose from blue sunlight being preferentially scattered by molecules in the Earth's atmosphere. At the time, Tyndall's result had no practical use, but, like much other blue-sky research, it rapidly *did* find valuable applications.

Thus, Louis Pasteur, in 1862, had proposed that there was no such thing as spontaneous generation of microbes and bacteria, but could not produce samples sufficiently free from existing bacteria to prove that this was the case. Tyndall's blue-sky apparatus provided the means for measuring the purity of the air in Pasteur's experimental set-ups so that only those air samples, which were completely sterile, could be selected out – so enabling Pasteur to succeed in demonstrating his revolutionary hypothesis – and thus leading to the saving of millions of human lives through the technique of pasteurization and via the laying of the foundations of modern bacteriology.

DOI: 10.1201/9781003246459-1

habilis, the survival tactics, which by then could justifiably be called curiosity, might be phrased in today's terms as:

where are the ripe apples?

or

is that a lion following me? (Figure 1.1)

FIGURE 1.1   Panthera Leo, a more-than-adequate reason for humankind's development of curiosity? (Reproduced by kind permission of Pixabay / Gerhard.)

The reactions to such questions are identical to those of the amoeba; hurry to the food source, flee from the danger.

However, a few Homo habilis individuals might also have wasted their time thinking about idle blue-sky questions such as:

Why is that forest on fire following that big lightning strike?

or

What has caused that fragment of rock to have a sharp edge where it's broken?

and these, like many more blue-sky questions, paid for themselves later on and in ways which could never have been anticipated initially.

In the above two example thoughts, the pay-offs are obvious: the control of fire and the invention of stone tools. Campfires would have reduced the danger of lions by keeping them at a distance, while stone axes and stone spearheads turned the lions from being predators into being prey.

Early humankind's blue-sky enquiries have developed so far from their beginnings that few people now need to fear lions and even fewer need to eat them. Blue-sky curiosity though has continued to come across with the goods throughout the 2 million years since some Newtonian intellect amongst the early Homo habilis tribes started thinking her/his thoughts about the Universe.

I will only give two further examples, lest this book becomes in need of re-titling 'A History of Blue-Sky Enthusiasms' and these final examples are much more recent.

Many of you may be reading this book on the internet via your smartphones. Even if reading a hard copy, you are likely to be in a room illuminated by LED lights and with the TV news running softly in the background. None of these actions and a very great many other parts of our current life-styles would be possible without those blue-sky explorations which led to:

the discovery of electricity

and

the discovery of radio waves.

The practical values of both electricity and radio to modern humankind are unlikely to need extolling further to any readers of this book, yet both discoveries started as blue-sky investigations.

The first significant investigations of what we now call electricity (and also of magnetism) began with William Gilbert's simple blue-sky curiosity about the peculiar properties of amber. Amber, when rubbed with a piece of fur or cloth attracts small items, such as feather fragments and bits of dry leaf towards itself (Figure 1.2). These peculiar properties of amber had probably been known to Thales of Miletus around 600 BC and were certainly known to Aristotle some 250 or so years later. But Gilbert, via the experiments described in his book, *De Magnete*[22] (published in 1600), was able to show that the force generated by amber and now called static electric charge, was different from the force from a lodestone, which is now called magnetic attraction or repulsion. He was even led to invent a detector, called a 'Versorium', for the presence or absence of the force on a piece of amber. Four and a bit centuries onwards from Gilbert's small-scale, blue-sky foolings around, the all-pervading colossus which we call electricity[23] now dominates our lifestyles.

FIGURE 1.2   Left – A small specimen of Baltic amber lying next to some dried leaf fragments. After rubbing the amber against some wool, the resulting electrostatic charge on the amber attracts some of the fragments upwards to itself (right). (© C. R. Kitchin 2023. Reproduced by permission.)

Radio waves were discovered around 1886 by Heinrich Hertz whilst investigating the production of sparks using capacitors and induction coils. This was not intentionally true blue-sky research, since Hertz was attempting to verify some predictions of Maxwell's equations, nonetheless, Hertz reputedly regarded both Maxwell's work and his own as having no practical applications; it was

---

[22] Full title: *De Magnete, Magneticisque Corporibus, et de Magno Magnete Tellure* (On the Magnet and Magnetic Bodies, and on That Great Magnet the Earth).

[23] The word 'electricity' derives from the Greek name for Amber – 'ηλεκτρον' (pronounced 'ilektron', when you transpose the Greek letters into their Latin-letter-pronunciation-equivalents).

'pure science'. However, only nine years after Hertz's work, Guglielmo Marconi was demonstrating wireless telegraphy, using these Herztian waves[24] across distances of up to two miles and in 1902 made the first broadcast across the Atlantic – and the rest, as they say, "is history".

Thus, humankind's insatiable curiosity about all and everything, whether directed towards solving an immediate problem (lions) or with no apparent application (sparks) has always paid off sooner or later and very often in quite unanticipated ways. As Kipling put it in his poem, *I keep Six Honest Serving Men*[25]:

> " …
> I know a person small –
> She keeps ten million serving men,
> …
> One million Hows, two million Wheres,
> And, seven million Whys!"

To return to the start of this chapter therefore, I need to make no apology for writing about topics which are currently 'blue sky', nor, indeed, should you feel any necessity to explain why you have an interest in them. History, extending back through years, centuries, millennia, millions of years and billions of years justifies both of us. Probably and in some presently unknown fashion, research and investigation into exoplanets and extra-terrestrials *will* pay off in substantive terms at some future time.

But we do not even need that justification; humankind's, other animals' and other living entities' inquisitiveness, curiosity and survival tactics are, literally, inborn. For ourselves we articulate this innate instinct as wanting to know *all* the answers to; 'Who?', 'What?', 'When?', 'Where?', 'Why?' and 'How?' just *because* we possess avid and voracious enquiring minds as parts of our very natures. Humankind's thirst for knowledge, as Larry Niven and Edward Lerner write in *Juggler of Worlds*[26] is:

> " 'Ah, curiosity' – a very human trait."

and

> "The difficulty with curiosity (is), it (knows)[27] no bounds."

At the end of the day, curiosity is just a part all of us and no further discussion is needed, or as Carl Sagan[28] expressed it (far better than I):

> "The open road still softly calls … . We invest far-off places with a certain romance. The appeal … has been meticulously crafted by natural selection as an essential element in our survival. Long summers, mild winters, rich harvests, plentiful game – none of them lasts for-ever. It is beyond our powers to predict the future. Catastrophic events have a way of sneaking up on us, of catching us unaware. Your own life, or your band's, or even your species' might be owed to a restless few – drawn, by a craving they can hardly articulate or understand, to undiscovered lands and new worlds."

So, then, with all of *that* out of our way; *on with the motley.*

---

[24] Now called Radio Waves.
[25] Rudyard Kipling. *The Elephant's Child*. From *Just So Stories*. Macmillan. 1902.
[26] Tor Books. 2008.
[27] Author's amendments from the actual 'was' and 'knew', to make it more apt for *this* book.
[28] *Pale Blue Dot*. Random House. 1994

2 Setting the Scene

REALITY (#1)

This book is based around three main cutting-edge sub-topics within Astrophysics, Biology and Physics. In combination, they seek to solve a question[29] which has probably crossed the minds of much of humankind in some form or other and at some time or other. Viz:

Are

We

ALONE

in the Universe?

It is a concern which has been with us for millennia, at least – and the plethora of beliefs which humankind has developed, often involving Creators, Deities, Ghosts, Jujus, Magicians, Oracles, Spirits, Super-Natural Beings, Supreme Beings, Witches, UFOs and the like, mostly existing somewhere 'out-there', probably reflects an innate longing *not* to be *alone*.

Presently, the answer to this MHoQs question is:

YES

– which is to say; we *do* seem to be *alone* in the Universe,

an all-important caveat, however, must follow that answer:

… on the evidence which we have to date.

Today, in the early decades of the 21st century, we may reasonably hope that the answer, based upon firm evidence may soon change to:

NO

within the next minute, day, month, year, century[30] …

What has changed from the accepted idea over many millennia to enable such a hope now to be expressed? The answer to *that* question lies in the aforementioned 'cutting-edge sub-topics within Astrophysics, Biology and Physics', and those sub-topics are:

Exobiology, Exoplanets and SETI.

---

[29] A.k.a. MHoQs – see earlier discussions.
[30] Of course, the answer may actually *be* '**YES**', in which case the wait could be for an awfully long time.

These sub-topics have all recently emerged as the subjects of increasingly large numbers of news items and features, ranging from popular newspaper, radio, TV and internet articles and programmes to abstruse peer-reviewed research articles. This increase in their exposure, of course, reflects a wider awareness that the answer to the MHoQs may soon be within our grasp.

These same subjects have naturally also been the basis of much of SF (science fiction) and science fantasy writing and for a much longer period. It would appear, though, that the story-lines are becoming more credible with the passing of time; books/films such as *Contact*[31] and *Close Encounters of the Third Kind*[32] are a quantum leap improvement on such comic-magazinish earlier efforts such as the *Creature from the Black Lagoon*,[33] perhaps because they also may soon become (almost) non-fiction.

Whilst a lot of all these SF items are well-written, accurate and informative, there are some which may, perhaps most politely, be described as 'over-imaginative'. The terms 'extraterrestrial intelligences' and 'exobiology', in particular, are likely to awaken in some readers' minds images of gigantic apes, multi-legged spider-types, towering robots, tyrannosaurs and many disparate types of bug-eyed monsters, all intent on invading the Earth using various designs of zap guns whilst carrying off nubile young (human) ladies for unspecified purposes.

You may be relieved (or you may not) to learn that such entities will make no further appearances in this book.

Even the less extravagant SF novels, films, videos, etc. mostly assume that Earth-type life, even human-type life, will be encountered on far-distant exoplanets at much the same level of technological development as ourselves. Were the Universe to be infinite in its extent, then to be sure, such discoveries *would* be there to be made ®.

Indeed, in a (mathematically) infinite Universe there would not only be many planets *similar* to the current Earth in all respects, there would be an infinite number of planets which are *identical* to the current Earth in all respects ®. Furthermore, there would be an infinite number of Earths identical to our current Earth except that a century ago an emerging butterfly flew left upon leaving its chrysalis instead of straight ahead and there would be another infinite number of Earths wherein the butterfly flew right or upwards or … – and so on and so on … ® .

If it sounds as though the previous paragraph is even more extravagant imagining than the bug-eyed monster scenarios, then you may be astonished to learn that this is just the simple consequence of a mathematically infinite Universe, and mathematics students in their final school years ® could easily prove it to you.[34]

Fortunately, perhaps, the Universe is not infinite, at least in so far as we can tell and/or, not at the moment. Captain Kirk's[35] many planetary landings will therefore hardly ever be met by extraterrestrial beings looking remarkably like a couple of terrestrial B-Movie Stars in thick make-up.

What, then, would be a more realistic account of *Star Trek*'s Enterprise's travels, assuming that the faster-than-light-travel-problem 💲💲💲💲💲 had been solved? We will also assume that only stars known by other investigations to possess planetary systems are selected for visitations and that terrestrial-type (carbon, hydrogen and oxygen-based) life-forms are the only sort to be found.

Taking the first 1,000,000 such explorations, the statistics might resemble those in Table 2.1.

[31] Carl Sagan, Simon and Schuster, 1985.

[32] Steven Spielberg, EMI Films/Columbia Pictures, 1977.

[33] Maurice Zimm, Universal Pictures, 1954.

[34] Ask him/her about Euclid's prime number theorem. Beware, though, of becoming interested in infinity; there are lots of different ones and some are bigger than others. In fact, there are an infinite number of unique infinities!

[35] One of the Captains of the Star Ship Enterprise, from *Star Trek*, the long running science fiction set of programmes, created initially in 1966 by Gene Roddenberry.

TABLE 2.1

Author's (Probably-More-Realistic-than-Roddenberry's) Account of the Star Ship Enterprise's voyages

What was found during Enterprise's missions	Number of instances[36]
Planets currently inhabited by life-forms resembling terrestrial B-Movie Stars in thick make-up	0
Planets currently inhabited by intelligent life-forms	1[37]
Planets inhabited by intelligent life-forms at some time in the past	1
Planets potentially inhabitable by intelligent life-forms (i.e., probably colonisable by humankind)	1
Planets currently inhabited by life-forms up to animal-level (i.e., possibly colonisable by humankind)	10
Planets currently inhabited by life-forms up to plant-level (i.e., perhaps colonisable by humankind)	100
Planets currently inhabited by life-forms at the micro-organism-level	1,000
Planets with organic-type molecules[38] present on them which have been produced abiotically	100,000
Planets barren of all life or any evidence of past life	898,887
Total	1,000,000

NB: the term 'Planets', as used in this table, includes planets, satellites, asteroids, comets and the like, even possibly, pre- and proto-planetary circumstellar discs.

Of course, in the actual world, an Enterprise-type spacecraft's speed would not be many times faster than light but be limited to a few hundreds of kilometres per second at best, so it would take about the current age of the Universe[39] to make the 1,000,000 visits listed in Table 2.1 ®.

At this point I proffer apologies to readers whose expectations may have just been dashed. But howsoever science fiction writers may try to gloss over the fact:

the Universe is BIG.

Thus, Douglas Adams tells us that the introduction to the iconic *Hitchhikers Guide to the Galaxy*[40] begins this way:

"Space" *it says*, "is big. Really big. You just won't believe how vastly hugely mindbogglingly big it is. I mean you may think it's a long way down the road to the chemist's, but that's just peanuts to space."

---

[36] These are just my guesses. I hope that they are pessimistic, I fear they are optimistic.

[37] Despite such an apparently low probability estimate, this would still put the number of intelligent life-forms *currently* living in the Milky Way galaxy at ~10,000 or more and also suggest perhaps, that there may be 10,000,000,000,000,000 (10^{16}) ETIs *currently* within the visible Universe ®.

[38] Britannica defines an organic molecule (compound) as:

"**Organic compound**; any of a large class of chemical compounds in which one or more atoms of carbon are covalently linked to atoms of other elements, most commonly hydrogen, oxygen, or nitrogen. The few carbon-containing compounds not classified as organic include carbides, carbonates, and cyanides."

https://www.britannica.com/science/organic-compound.

[39] Estimated at around 13,800,000,000 years.

[40] First broadcast on BBC Radio 4, in March 1978. Now available in many follow-on books and other media formats.

But the *Guide* does not even *begin* to appreciate the extent of the problem – and the Universe contains all of what we call space – so it must be even bigger.

Frequently, when astronomers try to describe vast distances and the like, resort is made to human-type comparisons and analogies, such as:

"walking continuously, it would take you almost 9 years to cover the Earth-Moon distance".

I have never found such parallels to be of much help. The fact is humankind does not have the ability to appreciate and understand distances, sizes, masses, temperatures, forces and the like, which are very much beyond our own personal physical experiences. Thus, when it comes to going to the stars, never mind to the galaxies and beyond, we all just have to accept that

"we ain't going nowhere fast"[41]
(and we ain't sending out no messages much faster neither).

So, if your expectations were dashed a few moments ago, I'm sorry, but that is just the way it is.

REALITY (#2)

If you have read this far, then you will know that in searching for answers to the MHoQs, we are not going to be zooming around the universe physically looking for life. What, then, can we do? As mentioned in the preface, the approach in this book is to look first at the topic of exoplanets followed by exobiology and then by SETI. This is to say, in the order of going from our most well-established to our weakest knowledge/understanding of those topics. Here we will look at why each of these topics is important to answering the MHoQs before going on to more detail in the later chapters.

Executive Summary

An executive summary of what the main topics of this book involve could look as follows:

- **Exoplanets** – these are planets, i.e., coldish and smallish[42] masses of rocks, metals, liquids, gases, etc., forming some of the many components of the Universe but not constituting a part of the Solar System.[43] The term can also include other entities similar to those within the Solar System but which also lie outside it such as satellites (moons), asteroids, comets and the like (see note to Table 2.1). The latter, though, may also be labelled separately as exomoons, exocomets, etc.

    The usual expectation is that the exoplanets will 'belong to' (i.e., be orbiting) a star, but many may just wander independently through the galaxy and even between the galaxies (free-floating exoplanets).

    Like the planets within the Solar System, currently known exoplanets are found in two main classes: gas giants, such as Jupiter, and rocky planets, such as the Earth (Figure 2.1, Chapters 3 to 8).

---

[41] John Revitte. https://www.lyrics.com/lyric-lf/5834610/john+revitte/Going+Nowhere+Fast. Album 2021.

[42] On astronomical scales – i.e., the Earth should be considered as 'Quite Small' but not 'Tiny'.

[43] The planets and other bodies within the Solar System almost certainly have no fundamental differences from exoplanets (etc.). It is just that the first exoplanets were found only about three decades ago and so are still regarded as somewhat exotic entities. Where Solar System Planets and Exoplanets *do* differ is in that we know prodigiously more detail about the Solar System Planets than we do about any Exoplanet – though that difference will ameliorate somewhat over the next century or so.

FIGURE 2.1 An artist's impression of HD 40307, a cooler star than the Sun which has six exoplanets (some disputed). Only the three innermost exoplanets, which are Super-Earths (small, rocky exoplanets, but with masses a few times that of the Earth – see Chapter 7), are illustrated and have the names: HD 40307b, HD 40307c and HD 40307d. Even the third of these (HD 40307d) is only 7.5% of the Earth–Sun distance away from its host star (0.1321 A.U.) and its orbital period (year) is ~ 20½ days. (Reproduced by kind permission of ESO[44].)

The first 'normal' (Chapter 6) exoplanet was detected in 1995 and is known as 51 Peg b, whilst at the time of writing, some 5,580[45] exoplanets are known (and there are between ~10,000 and ~20,000 possible exoplanets whose reality has yet to be confirmed).

Exoplanets are important to answering the MHoQs because we know of only one set of examples of life in the Universe and that set inhabits the surface of the Earth. Whilst it may be possible for life to occupy other niches in the Universe, the first and simplest extrapolation from our current knowledge is to assume that an exoplanet will need a surface and have close-to-terrestrial climatic conditions, in order for extraterrestrial life to inhabit it (other possibilities will be looked at in Chapter 13).

- **Exobiology** – This is nominally a subset of biology. However, since exobiology is normally regarded as the 'Study of Life in the Universe', biology should, strictly speaking, be a subset of exobiology – and if our MHoQs does turn out to have the answer 'No' (see above), then perhaps that is the way it will soon become.

For the moment, though, it is exobiology which is the junior partner, if only because currently, just a few very specialised aspects of the whole of biology are involved in exobiology's development.

Exobiology is also known as astrobiology, and it is developing from terrestrial biology and many other sciences, since those are the only areas of knowledge which we have at the moment (Figure 2.2). Its essential purpose is to determine whether life exists elsewhere in

[44] European Southern Observatory.
[45] The author has used the '*NASA Exoplanet Archive*' (https://exoplanetarchive.ipac.caltech.edu) and '*The Extrasolar Planets Encyclopaedia*' (http://exoplanet.eu/) wherever possible, as the definitive sources for statistics and data such as numbers, names, etc. throughout this book. There is some hope that by 2030 to 2035, the vast majority of exoplanets within 500 to 600 ly of the Earth will have been discovered, but what that 'final' number will be is still anyone's guess. *My* guess would be that it will at least equal the number of stars within 600 ly of the Earth, i.e., ≥ 10 million exoplanets Ⓡ.

the Universe other than on the Earth (i.e., our MHoQs) and, if so, how it may have originated, what forms it may take, where it may be found and how we may detect it.[46] A subset of exobiology is concerned with *intelligent* life elsewhere in the Universe and how to
communicate it, and this takes us onto SETI (see below).

FIGURE 2.2   An image of the Earth, centred on Africa, obtained by the Deep Space Climate Observatory.
All that we can deduce, infer and guess about the way(s) in which life-forms may originate and what forms
they may take elsewhere in the Universe currently has to be based on information which we obtain from this
single source: the Earth. (Reproduced by kind permission of NASA.)

As remarked when outlining the subject of exoplanets and above, we only have one
example to work from so far: the Earth and its life-forms.

In mathematical terms, we have a graph with a single point or measurement (Figure 2.3)
on it and we are trying to predict where the next point might lie. Any mathematician would

---

[46] A more 'official' definition (NASA's *Astrobiology Road Map*, Des Marais D. J., *et al.* Astrobiology, 715, **8**, 2008) lists
the aims of the subject as

- Emergence of life on Earth,
- Earth-life's evolution and interaction with the environment,
- Evolution,
- Environmental limits,
- Habitable environments within the Solar System and the Universe,
- SETI

and

- Signatures of life.

Related discussions, definitions and material may also be found in later updates of NASA's road map and in ESA's
AstRoMap (Horneck G.N., *et al*, Astrobiology, 201, **16**, 2016).

tell you that the second point could be anywhere on the graph; a single point has no predictive value. A second point on the graph, however, changes the situation totally; joining the
two points with a straight line and extending it in both directions suggests, with a certain
level of probability, positions near where a third point might lie. Likewise, the discovery
of a second form of life within the Universe, which has developed independently of the

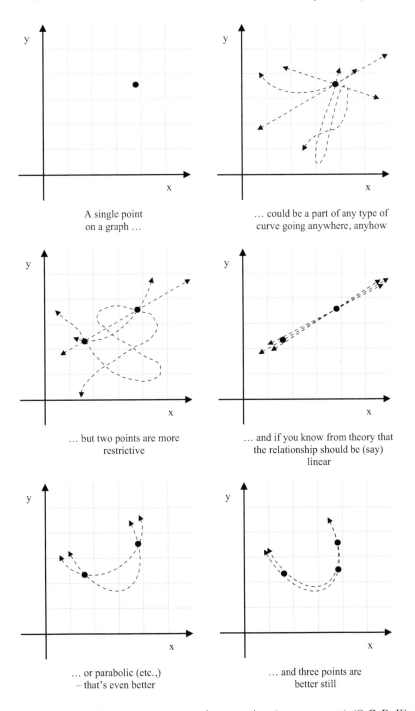

FIGURE 2.3 Extrapolations from one, two or more known points (measurements). (© C. R. Kitchin, 2022.
Reproduced by permission.)

Earth's life-forms, even if it were to be simpler than the simplest forms of life now on Earth, would expand our understanding of life in the Universe a trillion-fold.[47]

Of course, our knowledge of a single point on the graph (life on Earth) is not without its value. For a start, it tells us that life *is* possible.[48] This may seem a trivial point, but in the infinite Universe mentioned in Chapter 1, it would mean that terrestrial-type life would have originated *independently* an infinite number of times. In our probably finite Universe, though, a single life-origination event (Chapter 9) remains a possibility.

More profitably, our knowledge of life on Earth does allow for some theoretical and experimental predictions. For example, we know about many, if not all, of the biochemicals and processes which are necessary for our life-forms and we can conduct laboratory experiments to show that some of those biochemicals can be formed by non-living (abiotic) processes (Chapter 9). Similar biochemicals, also originating abiotically, have been detected in interstellar nebulae, the interstellar medium, in planetary atmospheres, and many other places away from the Earth, so that the likely pre-conditions for terrestrial-type life can be investigated.

Theoretical predictions may be further pursued by knowing the processes which may be crucial to terrestrial-type life and looking for analogous functions using other types of atoms, molecules, mixtures and compounds. In this context, silicon is often one candidate suggested to replace carbon's role in terrestrial-type life chemistry though it now seems that this example may be over-hopeful.[49] Similarly, ammonia, methyl alcohol and sulphuric acid[50] might act as solvents in place of water.

All these astrobiological studies and many others will also contribute to our ability to recognise life (Chapter 11) when we see it. This is not a given, even if the life is similar to our own; think of a tortoise in deep hibernation; it might fail to show any of the normal signs of living for weeks or months at a time, and Tardigrades (Chapter 9) can suspend their animation for years to the point where we currently do not have *any* means of telling that they are *not* dead, save by restoring them to their normal living conditions.

Terrestrial extremophiles (Chapter 12), by contrast, might be thought to be non-terrestrial-type life, even though they have evolved on Earth, since they can endure conditions that would be instantly fatal to most 'normal' Earth-life.

If ET life is based upon quite different biochemical processes from those on Earth, then it might be very difficult indeed to recognise it as being life (Chapter 11). Isaac Asimov's short SF story, *The Talking Stone*,[51] if you can find it, is salutary in this respect. Recently, an AI (artificial intelligence) approach to distinguishing between terrestrial molecular

---

[47] Probably a gross underestimate.

[48] If you want to initiate a discussion on the paradoxes involved in knowing that life isn't possible, then the best of luck – but don't involve me.

[49] Silicon is the next heavier atom after carbon in carbon's group in the periodic table and has a similar outer electron structure. Thus, it can form the equivalent of carbon's methane, a tetrahedral molecule with a central silicon atom surrounded by four hydrogen (or other) atoms (c.f. Figure 11.4). However, silicon's multiple bonds to other atoms are much weaker than those of carbon, so that some silicone equivalents of carbon-based organic molecules are structurally weak or may not even form at all.

Furthermore, silicon can only with difficulty be persuaded to form the equivalent of carbon's benzene rings – see Chapters 9 to 11. Silicon does not form long chain molecules in the same form as those of carbon, although analogous chain structures partially based upon silicon are possible. Thus, polydimethylsiloxane (a.k.a. silicon oil) has a 'backbone' of alternating silicon and oxygen atoms and is the nearest silicone equivalent to polyethylene (whose 'backbone' just comprises carbon atoms).

Silicon atoms *can*, though, also partially replace carbon atoms in some organic (Box 9.1) molecules, so perhaps hybrid carbo-silicate life might be possible?

[50] These are all toxic to human life-forms, but life-forms which have evolved to use sulphuric acid (say) as their organic solvent would probably regard water as a highly lethal liquid.

[51] *The Magazine of Fantasy and Science Fiction*. Mystery House. 1955. Also appears in *Asimov's Mysteries*. Doubleday. 1968

samples from abiotic, fossil and currently living sources has achieved a claimed 90% accuracy.[52] Such an approach may well be applicable to distinguishing between abiotic and biotic samples from other Solar System sources. It may even be able to identify life-forms which are not DNA-based, since it looks for patterns in molecular distributions arising from the requirements for life, not for specific molecule types, using pyrolytic gas chromatography.[53] Such an approach is in its infancy at the time of writing and may be expected rapidly to become much more sophisticated – so – Watch This Space.

- **SETI** – This acronym derives from 'Search for Extraterrestrial Intelligence' and that, really, says most of what it's all about.

    Until recently, the SETI has meant looking for radio emissions coming to us from the depths of space and for which no known natural production process could provide an explanation. In particular, a modulation of the signal in some way which enables it to contain information is sought. At a simple level this could be like our Morse code, wherein the distress signal 'SOS' is sent out via three short beeps, followed by three longer beeps and then three short beeps again. Much more sophisticated modulations are possible though (Chapter 15).

    Similar searches at shorter wavelengths have begun recently and/or are planned for the near future. At all the wavelengths currently being searched, new instruments with much wider fields of view, wider pass bands and/or multiple pass bands and faster responses are also starting to come into use or are being constructed. Within a few years, the SETI observational capacity will have increased many times over today's possibilities.

    Serious SETI work began around 1960, but to date, no clearly non-natural signal has been detected. A single rather puzzling signal *was* received in 1977 by the Ohio State University's 33 × 103 m radio telescope (the now dismantled 'Big Ear'). That signal is widely known as the Wow! signal from a note scribbled on the computer print-out by the excited discoverer. The Wow! signal is discussed further in Chapter 15, suffice it here to say that while no clear natural mechanism for the emission has been found, neither can it be clearly seen to be of ETI origin and so it remains a puzzle.

    Although SETI searches for intelligences, it is possible to imagine non-intelligent life-forms that could emit radio signals; terrestrial electric eels,[54] for example, generate voltages of up to 600 V, more than enough to produce large sparks in air and so produce radio waves. It seems unlikely that these radio signals could become powerful enough to be detectable many parsecs away, but if that were to happen, then I'm sure that the SETI researchers would be just as delighted to become successful SETnon-I researchers as they would to become successful SETI searchers.

    Radio signals going outwards from the Earth produced by intelligences (ourselves) have, of course, been happening since before the 1880s and with rapidly increasing strengths ever since that time. Outward-going radio signals from the Earth must thus now be filling a spherical region of space some 150 ly in radius and containing around 8,000 to 10,000 stars. Now, not all stars have exoplanets, but some stars have many exoplanets. So, as a very rough approximation, there may be 10,000 exoplanets or so, whose inhabitants, if any and if they have radio receivers, now know that we exist on the Earth.[55] Those extraterres-

---

[52] Cleaves H. J, *et al*. *A robust, agnostic molecular biosignature based on machine learning.* Proc. Nat. Acad Sci. 41, **120**, 2023.

[53] Details of this technique are beyond the scope of this book but may be found in the Authors' book, *Remote and Robotic Investigations of the Solar System*, 2018, Taylor & Francis, or via an internet search for 'gas chromatography'. Also, Box 9.3 discusses a related technique.

[54] Actually, a type of freshwater fish.

[55] This may be an over-estimate. Some recent work suggests that because of interstellar gas and dust, our current radio receivers would only be able to detect our current levels of radio emission out to a 'few ly'.

FIGURE 2.4 A modern radio and TV transmission mast and aerials. How Hertz and Marconi would have loved to have had this sort of equipment! – and how far out have signals sent from it travelled by now? (Reproduced by kind permission of Pixabay and Paul Brennan.)

trials' SETI programmes will therefore now have been/are being successful, but whether they will broadcast something back to us, though, is another matter entirely.

Apart from all our raucous noise from Wi-Fi, mobile phones, radios, TVs, GPS systems and communications with distant spacecraft, etc., a few signals have been sent out aimed deliberately at particular stars and/or known exoplanets. These events are essentially publicity stunts and, as such, have succeeded in that purpose. In no other way have they produced any results.

BRINGING IT ALL TOGETHER

With our 'bricks' of Exoplanets, Exobiology and SETI, we may now start on building our 'houses' of 'Habitable Exoplanets for Extraterrestrials'.

As already remarked, we are basing our speculations upon just one known example, Life on Earth – and that, therefore, is our starting point. We shall also continue close to home by looking at habitats within the Solar System, some of which we have already visited, others of which may be visited soon and all of which are known in much greater detail than any object beyond the Solar System.

We know that humankind can exist in space and that we can also land on and survive for several days on the surface of the Moon. Perhaps within the next decade we will know whether we can survive for weeks or months on the surface of Mars. In all these cases, though, our presence is made possible only by taking small simulacra of our terrestrial environment with us (such as spacecraft and space suits) and living inside them. Our next 'Small Step or Giant Leap'[56] will thus be to examine where else within the Solar System similar levels of exploration/usage may be possible. We shall also examine if environments away from the Earth may be modifiable (terraformed – Chapter 18) to allow humankind to exist with no more protection than we now use on Earth.

During this examination of potential habitable environments within the Solar System, we shall also be on the look-out for life-forms which may already have managed to find homes 'out there'. These may or may not be related in some way to terrestrial life-forms and may have migrated out from the Earth, or it may be that life originated elsewhere within the Solar System and then migrated to the Earth (see 'panspermia', Chapter 9).

Then, of course, the life-forms might have evolved or have otherwise originated quite independently of the Earth and so be quite different in all respects from terrestrial life. There might even be artefacts or other evidence of exolife visiting the Solar System from distant exoplanetary systems (see Fermi paradox, Chapter 16) – it is even, very remotely, possible that ETIs are there now, just waiting for us to join them. Then again and *much* more likely, there may be nothing at all to be found anywhere within the Solar System.

Our second 'Small Step or Giant Leap' will not be for humankind but will be to explore what range of terrestrial environments non-human but terrestrial life-forms[57] can occupy successfully and to see what other potential Solar System environments might be homes for such entities or their selectively bred/genetically modified descendants. This should also give us some clearer ideas about the types of exoplanets and the conditions on them where some form(s) of terrestrial life analogues might evolve.

Whilst within the Solar System we may be able to find niches for life-forms of some types, including humankind, or adapt those niches and/or life-forms until they do suit each other, the theme of this book is really to look outside the Solar System. Thus, having reached some understanding of the conditions needed for terrestrial type life-forms and perhaps also of the conditions required for it to originate from inanimate matter, we may start looking at what 'Des Res'[58] there may be amongst the exoplanets already discovered.

At the time of writing some 5,580 exoplanets are known to exist, with about another 10,000 to 20,000 candidates for possible exoplanets waiting to be studied in more detail. As we shall see in Chapter 13, exoplanet researchers call regions around exoplanet host stars where conditions might be suitable for terrestrial-type life 'Goldilocks Zones' or 'Habitable Zones'.[59] An exoplanet orbit-

---

[56] cf. Neil Armstrong's utterance on descending onto the surface of the Moon 'One small step for man, one giant leap for mankind'.

[57] Some terrestrial life-forms can survive in boiling water or in solutions as alkaline as caustic soda or as acidic as pure nitric acid – see Chapter 12.

[58] 'Desirable Residences' for those of you unfamiliar with the UK Estate Agents' abbreviations.

[59] The definition of a 'Habitable Zone' is essentially anthropocentric, and it is where there is some possibility of finding water present as a liquid. As NASA sometimes expresses it, we need to 'Follow the Water' in our search for life. Discussions of other possibilities and of the apparently unlikely places where liquid water might exist are covered in many places later in this book.

In any case, 'habitable zone' is probably somewhat of an exaggeration; the Earth clearly orbits within the Sun's habitable zone, yet if we take the Earth as a whole, only a few very thin surface layers, amounting to about 0.15% of the Earth's mass, actually *do* contain life-forms. Furthermore, the zone for which conditions are appropriate for the *origin* of life (Chapter 9) may well differ from the zone in which life, once it exists and has evolved somewhat, can *continue* to exist.

(Continued)

ing within one of those zones, which is also suitable in terms of its size, atmosphere, solid surface, composition and so on is classed as a 'Potentially Habitable Exoplanet' and one list at the time of writing contains 61 such candidates. Almost all of these, though, have host stars that are much less massive and much cooler than our Sun, so they may be 'res' but are probably not 'des', at least for humankind. Five of the host stars are somewhat more like the Sun, but their candidate planets are either much more massive than the Earth implying surface gravities (i.e., weights) twice or more their terrestrial equivalents, or of unknown mass. Birds would thus be unlikely to be able to fly and gazelles would have physiques like those of rhinoceroses on these exoplanets.

We are still in the early stages of having enough known exoplanets to try doing statistical analyses, but taking the rough figures given in the previous paragraph, we may see that around 1% of exoplanets are potentially habitable and around 0.1% might be not-very-good Earth analogues. Whether such exoplanets will actually be hosting life will depend upon a myriad of other factors – but – we have now made a start on such investigations.

My guesstimates in Table 2.1 would suggest that even if all 61 of the currently suspected potentially habitable exoplanets *do* bear life, it is likely to still be at a micro-organism level. Rather, therefore, than having the problem of communicating with Extraterrestrial *Intelligences*, we are much more likely to have the problem of how can we tell that there *is* life on an exoplanet, where it exists only in the form of a few organic molecules floating in a puddle somewhere. In future chapters and especially in Chapters 14 to 17, we shall therefore be as much concerned with this very basic question as with SETI itself.

Another intriguing consideration is whether we *can* send spacecraft to exoplanets – and also, perhaps, whether we *should* do so. Many science fiction stories envisage humankind visiting and exploring exoplanets and perhaps founding colonies 🌀🌀🌀🌀. Whilst various warp factors, hyperspaces, etc. are brought in by the SF writers to enable their heroes and heroines to journey to their chosen exoplanets in times ranging from a few minutes to a few tens of years, the reality is that visiting an exoplanet and returning to Earth within a human lifetime is just *not* within the present bounds of the possible.

Founding colonies after journeys lasting many human generations (see 'Space Arks', etc., Chapter 18) via what are essentially very small travelling planets might be possible 🌀🌀🌀🌀 and is the subject of quite a number of SF stories. However, apart from the enormous cost and the resources needed for such projects, there is the ethical question of whether or not we have any right to commit our descendants, centuries from now, to live their lives under such conditions and in such places without any consent on their part.

Another possibility might be to send the makings of humankind and terrestrial life to exoplanets in the forms of frozen embryos, spores, seeds, etc. together with automated machinery to reanimate those rudimentary beginnings 🌀🌀🌀🌀. The ethical objects to this, though, are orders of magnitude worse than for the 'Space Arks'. A related and more practicable possibility with today's knowledge and techniques but with somewhat different ethical considerations would be to send out basic terrestrial life-forms (bacteria, etc.) in large numbers of small, cheap spacecraft to be seeded (like sowing a field of wheat) into large numbers of exoplanetary systems 🌀🌀 / 🌀🌀🌀, there to

---

We should also consider that there may be planets which are even better than the Earth for the origin, evolution and development of life-forms. In fact, it would be prodigiously implausible that the Earth should be the uniquely Best-Planet-for-Life in a whole Universe containing, maybe, 1,000,000,000,000,000,000,000,000 planets Ⓡ. Planets, potentially better for life than the Earth, have been labelled as 'Superhabitable Planets'. The main problem with the concept is how to define what is *Better* in this context – and that I leave up to you to think about.

A number of AI approaches to defining and identifying habitable zones are currently under development but have yet to achieve widespread use. See Cobb-Douglas Habitability Score (CDHS), for example.

A galactic habitable zone has been proposed based upon the availability of elements heavier than helium and rates of occurrence of supernovae and the like. This might extend, possibly, from somewhere Galactic-Centre-ish to somewhere further out than is the Earth from the galactic centre. This seems, to the author, to be rather premature, and it is not clear that it would be very meaningful anyway.

'do their own thing' and evolve to suit the exoplanetary conditions available (see also panspermia, Chapter 9).

Sending completely automatic spacecraft to exoplanets, though, along the lines of the many Solar System explorers (e.g., the Viking and New Horizons spacecraft), to explore, but not to attempt to colonise exoplanets, is a distinct possibility within the next few decades 🎐🎐. The spacecraft would be tiny, maybe only a few tens of millimetres in size, launched from the Solar System at relativistic speeds[60] using powerful lasers (see light sails, Chapter 18) and in large numbers. They would simply fly through the exoplanet systems making such observations and measurements as might be required and then broadcast that data back to us. At 10% of the speed of light, we could have the results of such a mission to Proxima Centauri (Figure 2.5) about 47 years after launch, or even less if higher spacecraft speeds can be reached.[61]

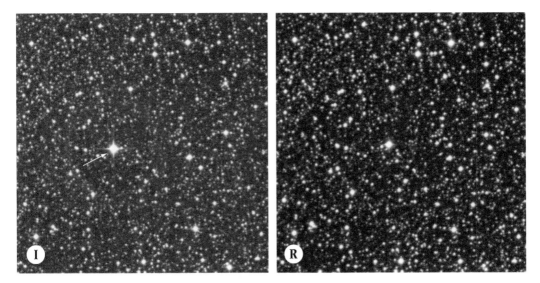

FIGURE 2.5   The currently nearest star to the Earth (other than the Sun), Proxima Centauri. The star field on the left is viewed in the NIR[62] that on the right is at red wavelengths and the bright stellar image just to the left of centre in both images is Proxima Centauri. The other objects to be seen are much more distant stars. Each image covers an area of the sky which is about 10% of that of the full moon. Proxima Centauri is much cooler than the Sun and thus appears red to the eye, but it is actually even brighter in the NIR as may be seen on the left-hand image. Three exoplanets are currently known to be orbiting the star (not visible on these images), whose masses range from about a quarter to about seven times that of the Earth. (Reproduced by kind permission of ESO.)

---

[60] That is, speeds which are a significant fraction of the speed of light – 0.2 c (20% of the speed of light), for example, is 60,000 km/s.

[61] Proxima Centauri is about 4¼ ly from the Earth. A swarm of mini space probes travelling at 0.1 c would thus take 42½ years to travel to it. The radio signals broadcast back to us would travel at the speed of light (c) and thus get to us 4¼ years later. If the probes could reach a speed of 0.2 c, the results would be back in about 25½ years. Such intervals are probably feasible – the results currently being obtained by the New Horizons spacecraft, now exploring the outer parts of the Solar System, for instance, are arriving 16 years after its launch, while the Hubble Space Telescope (HST) is still producing cutting-edge results 33 years after its launch – and there are even older spacecraft than these still operating.

[62] Near Infrared e-m radiation (electro-magnetic radiation). Other, related, abbreviations are IR (Infrared), MIR (Mid Infrared) and FIR (Far Infrared).

Theme 1

Exoplanets

3 Exoplanets (#1)

INTRODUCTION

Exoplanets are just planets orbiting stars other than the Sun, or which are loners, free-floating by themselves in the spaces between the stars (Chapter 7) – and that is all.

Accordingly, therefore, our study of exoplanets initially needs to be based upon and informed by our knowledge of those nearer and dearer 'friends' which *do* populate the Solar System. In other words, we must preface our *beyond*-the-Solar-System-explorations by first undertaking some *within*-the-Solar-System-explorations.

THE SOLAR SYSTEM PLANETS

We know that the first normal exoplanet (51 Peg b – Chapter 6) was discovered in 1995 by Michel Mayor and Didier Queloz using the Elodie[63] spectrograph and the 1.93-m telescope at the Observatoire de Haute Provence in France.

But when, where and how were the first Solar System planets discovered? The answer to that is:

'we *do not* know – but we *can* make some semi-educated guesses.'

Nobody, now, has any clue as to when planets were first noticed by humankind and identified as moving in the sky,[64] in a manner that was different from that of the fixed stars.[65] After the Sun and Moon though, Venus and Jupiter are, at times, the brightest objects in the sky and over twice as bright as the brightest star, Sirius (Figure 3.1). Furthermore, the early humankind peoples were living without lights (save, perhaps, for a few fires) and had an unpolluted atmosphere, so all objects in the night sky would have been miraculously clear and bright compared with the pathetic views which are the experience of 99% of people living today.

On the other hand, many individuals amongst those ancient peoples must have had poor eyesight due to myopia and/or various diseases and/or parasites. Those with good healthy eyesight would then have developed reputations of being 'far sighted' or 'eagle-eyed' and perhaps it was such lucky individuals who first noticed the planets. Howsoever it may have happened, finding the naked eye planets seems likely to have been a discovery made independently many times over, in many different parts of the world and on many different historical occasions.

Following the stars' changing patterns in the sky must have been important to ancient peoples for multiple reasons, including:

- when to expect seasonal weather changes
- for planning crop planting and harvesting
- for navigating whilst travelling long distances

and

- from religious or spiritual motives.

---

[63] A name, not an acronym.

[64] In the Greco-Roman tradition, the moving 'stars' were called planets from the Greek words for 'errant' (πλανώμενος; pronounced planómenos) or 'stray' or 'wanderer' (περιπλανώμενος; pronounced periplanómenos).

[65] All stars and planets etc. are, of course, moving across the sky because of the Earth's counter rotation. Were we, somehow, to stop the Earth's rotation, then the stars would occupy fixed points in the observer's sky, except over time scales of centuries or more. The planets and all other Solar System objects would, however, still be seen to move around the sky. Hence, despite their diurnal motion, stars and all other objects beyond the Solar System are called 'Fixed'.

DOI: 10.1201/9781003246459-4

FIGURE 3.1 Changes in the relative positions of Venus and Jupiter arising from their motions with respect to the fixed stars, as seen from ESO's Paranal observatory. Venus is the brighter object in the centre and Jupiter is on the right, moving downwards with respect to Venus. The Moon can also be seen as the brightest object of all moving upwards from near the horizon in the first image to being above Venus and Jupiter in the third image. Some telescope domes of the VLT[66] are seen in silhouette in the foreground. (Reproduced by kind permission of ESO/Y. Beletsky.)

The movements of the naked eye planets, Venus and Jupiter (plus those of Mercury, Mars and Saturn[67]) with respect to the fixed stars, would thus surely have been noticed from an early stage in humankind's historical development.

Of these various reasons for observing stars' positions, that of long-distance navigation over both sea and land would have required the highest levels of accuracy, reliability and constancy in knowing the positions of the stars. Any 'star' which did not follow the diurnal rotation[68] across the sky of all the other stars would soon be noticed and discarded as not being navigationally useful (Box 3.1).

Perhaps, though, that oddly-behaved 'star' might then be seen as something of special interest and so observed independently for its own sake?

BOX 3.1 – RUNNING-DOWN-THE-LATITUDE NAVIGATION

One early navigation technique was 'running down the latitude'. That is, keeping the journey to following a path parallel to the Equator by maintaining a particular star at a fixed altitude above the horizon when it was highest in the sky. This technique was more used at sea than on land, where other clues, such as landmarks, could be used, but the principle is the same on land or sea.

Of course, those ancient navigators knew nothing about latitudes or the equator and almost certainly thought that the Earth was flat. But they did know (say) that after leaving city, C, they would arrive at town, T, fifty days later, provided that star, S, when at its maximum altitude above the horizon, was kept to a camel's height (or whatever) throughout their journey.

Suppose, then, some 2,000 to 3,000 years ago, a camel caravan set off from the westernmost point on the shore of the Wadi El Rayan Lake in Egypt for a 450 km journey to the Siwa oasis. It would need to keep heading out westward along the 29° 11′ N line of latitude for about 15 days. The navigators, though, knew just from previous experience that to get to their destination from that departure point, they just had to stick to a track which kept any selected

---

[66] Very Large Telescope. Comprising: four 8.2-m and four 1.8-m individual telescopes.

[67] Uranus, when closest to the Earth (at opposition), is bright enough to be seen directly by the unaided eye. This feat, however, requires the observer to have excellent eyesight and to be observing in ideal conditions from a really good observing site. There are no early records of Uranus ever having been recognised to be an object moving with respect to the background stars, so Sir William Herschel's telescopic discovery of the planet on 13 March 1781 was the first moment when the Sun became known to have, including the Earth, seven planets.

star at the same maximum height above the horizon (altitude) throughout their journey and keep going straight ahead for about 15 days.

Unfortunately, this time, they selected a nice bright reddish star in a convenient part of the sky for their navigation reference point. A 'star' which we now happen to call Mars. Because it is actually a planet, the position of this nice red star could easily move North or South by a degree or more in the sky over 15 days.

On this occasion, when they set out, Mars' altitude, when it was due South, was about 81° 40′.[68] After 15 days of travelling, Mars had moved to an altitude of about 82° 40′. The caravan, of course, directed its journey so as to keep Mars' altitude at 81° 40′ and so their path moved North of its correct line across the ground. After 15 days, they were thus some 1° of latitude North of the Siwa oasis – a distance on the ground of about 110 km. The caravan would also have passed through the Qattara depression and any still surviving travellers were now in the open desert[69] and some 150 km from the nearest part of the Mediterranean. They then probably proceeded to lynch their navigators before perishing themselves.

The motives, therefore, for travellers some millennia ago and especially for their navigators/guides, to identify correctly those stars which moved with respect to the vast majority of other (fixed) stars in order to *avoid* using them, was thus a very strong one. Hence, if not discovered earlier, it is highly likely that the planets were found once the stars had started to be used for long distance navigation.

Sea travel, until the last few centuries, was mostly undertaken close to the coast, so that landmarks would have been used as guides. However, navigating when out-of-sight of land, even if only because the boat had been driven miles off its intended course by storms, doubtless required the use of stars in some way.

The first recorded usage of stars for navigation dates back to ~3,000 to ~1,000 BCE[70] as revealed by Minoan frescos on the island of Crete. Whilst the ancient Polynesians were probably using stars as guides for voyages of hundreds of kilometres across the Pacific by ~1,500 BCE.

Even earlier examples though, of travelling out-of-sight of land, may possibly exist. Australia, for example, is estimated to have become inhabited by humankind some 40,000 to 60,000 years ago. Sea levels were lower at times between then and now, but it would seem that a journey to Australia would still have required sea crossings of hundreds of kilometres. So perhaps at least some early peoples were studying the stars and planets around 50,000 years ago.

The use of stars for navigation thus dates back at least to two or three millennia BCE and could date back 50 millennia or more. That, of course, does not mean that the planets have been known that length of time, but it would seem to be a very likely possibility. Furthermore, the planets' very peculiarity of changing positions with respect to all the fixed stars may well have brought them to the attention of early humankind many millennia even further into the past.

Nonetheless, whensoever, howsoever and by whomsoever the planets were first recognised, recognised they *were* – and have been of special interest ever since.

Of course, the differences between the physical natures of planets and stars were unknown until much more recently. In fact, it was not the possibility that the planets might resemble the Earth which was first postulated, but that the Earth might resemble the planets and so itself be moving

---

[68] These are actual figures from a recent close approach of Mars to the Earth in space, i.e., when it was near to its opposition.

[69] The climate could have been rather different from that to be experienced today, but the travellers were probably still pretty annoyed.

[70] Year dates prior to 1 CE will normally be indicated in this fashion. Year dates soon after 1 CE, where it seems likely that confusion could arise with dates before 1 CE, will be indicated by CE. More recent dates where such confusion does not seem likely will just have the year number, e.g., 2025, not 2025 CE.

through space; this was advocated by Philolaus around 400 BCE. Another hundred years later Epicurus was suggesting that there might be an infinite number of inhabited worlds.

So, What Else Can Solar System Planets Do to Help Us Understand Exoplanets?

Once (howsoever it may have happened) planets were recognised as being different from fixed stars, their study for non-navigational purposes soon overwhelmed and may also have preceded any navigational motivations which there might have been.

Those non-navigational purposes were, of course, in the main, astrological, celebratory, predictive, religious and so on in their natures – especially in respect of local VIP's fortunes and futures. Of course, if you believe in astrology, the stars and planets would similarly have been predicting the fates of lesser mortals, but only the richest could afford to pay to have their fates revealed to them by their Fortune-tellers or Witchdoctors. As Shakespeare has Calpurnia pithily express a somewhat similar idea in *Julius Caesar*:

> "When beggars die, there are no comets seen;
> The heavens themselves blaze forth the death of princes."

Predicting the movement of the planets soon became an important requirement for Astrology in order to try and probe into future happenings (especially for the aforementioned VIPs). This study more-or-less culminated in Claudius Ptolemy's *Almagest*[71] written around 150 CE.

Unfortunately, the *Almagest* was based upon a geocentric model of the Solar System and also assumed that the planets, Sun and Moon must move in circular orbits. In order to match the actual movements of the planets across the sky, therefore, the *Almagest*'s model of the Solar System became extremely complex, but its predictions did work to a certain extent.

For the study of exoplanets, little further relevant knowledge about the physical natures of the Solar System planets was obtained until the optical telescope was invented. This occurred (arguably[72]) in what is now the Netherlands during 1608 and was by the children of a spectacle-maker named Hans Lippershey whilst they were playing with some of his spare lenses.

Lippershey only saw his telescope as being useful in military contexts and sold several of the instruments for such usage to the local Stadholder, Prince Maurits of Orange. There is no record that he ever pointed it towards anything in the sky, not even at the Moon.

Thus, in astronomical contexts, the credit for the use of the telescope goes to Galileo Galilei (Figure 3.2). Galileo, from 1609 onwards and using instruments which he built himself, made a host of new discoveries about the Solar System, such as:

- The craters on the Moon
- The four large satellites of Jupiter
- The phases of Venus
- The spots on the Sun

and

- The innumerable stars which were not visible to the unaided-eye.

---

[71] From Arabic and Greek words meaning 'The' (al) and 'Greatest' (majesti). Also known earlier as the '*Syntaxis Mathematica*' in Latin and by various similar titles in other languages of the time.

[72] The reader may find claims being made in the literature for the earlier invention of the telescope; some placing the date back as far as 1266 CE (see Roger Bacon's *Opus Majus*). Any further discussion of such claims here, though, would be well outside the remit of this book. The interested reader may research these topics via books on the History of Astronomy and via searching the internet; the article at https://en.wikipedia.org/wiki/History_of_the_telescope – on the '*History of the Telescope*' provides a good starting point.

He also saw Saturn's rings but could not make out what they were and saw a new planet, now called Neptune, but although he noted that it moved relative to the 'real' stars, he did not infer that it was actually another planet.

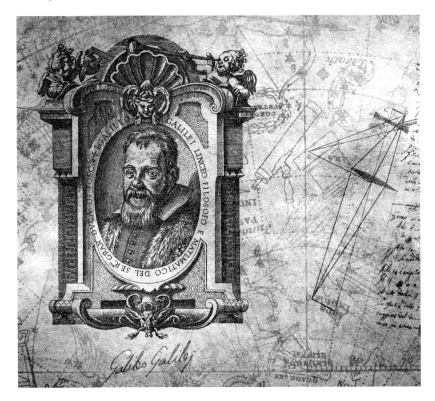

FIGURE 3.2 Galileo Galilei – His likeness, signature and the optical design of his telescopes. (Reproduced by kind permission of Pixabay and Dorothe.)

The seeds of much of the information which can be useful to us for understanding exoplanets and with which Solar System observations can provide for us already lie in Galileo's discoveries. Over the years and centuries following Galileo, those discoveries were confirmed, expanded and extended by a host of astronomers and astrophysicists, many, especially in the earlier years, being amateurs who frequently built their own instruments from scratch.

Extrapolating from these early Solar System observations together with our later knowledge, we may deduce that for exoplanets and exoplanetary systems we may expect to find:

- Numerous exoplanets around single host stars
- A hierarchy of sizes: from planets larger than Jupiter, through smaller planets like the Earth, to even smaller natural satellites and asteroids and on down to boulders, pebbles and eventually to dust grains – all orbiting the host star and/or a larger member of this hierarchy
- A major split in the physical natures of these objects; the smaller ones generally being mostly solid and formed from cold rocky materials and/or metals, the larger ones generally being mostly gases, especially hydrogen, though perhaps with small solid rocky cores.
- That all the objects are moving in orbits governed by Kepler's laws and Newton's law of gravity (Box 3.2). In a few extreme cases General Relativity may need to be invoked as well, to account for small discrepancies from Newtonian physics.

BOX 3.2 – NEWTON'S AND KEPLER'S LAWS

Newton's law of gravity: An attractive force operates between any two masses (material objects/particles of matter) and the magnitude of that force is proportional to each of the masses and inversely proportional to the square of the distance between their centres of gravity.

Kepler's law #1: The planets move in elliptical orbits with the Sun at one focus.[73]
Kepler's law #2: The planets' speeds around their orbits vary such that the line between the Sun and a planet sweeps out area at a constant rate.
Kepler's Law #3: A planet's orbital period, squared, is proportional to its mean distance from the Sun, cubed.

NB. Kepler's three laws are derivable from Newton's law of gravity and so are not independent of it.

Of course, for exoplanets and exoplanetary systems, we may also expect to find systems which are quite different from our solar system. Indeed, the observational methods which we currently use to discover exoplanets (see below) are biased towards picking up exoplanet systems which *are* quite different from the Solar System. Nonetheless, as the number of exoplanets in our catalogues continues to increase above the 5,580[74] mark, Solar System near-analogues form a noteworthy fraction of that total.

The detection of an exoplanet is one matter, studying its details, though, is quite another.

Modern investigations of Solar System objects, including visitations by landers (Figure 3.3) and close-up studies by orbiters and fly-by missions as well as today's incredibly powerful Earth-based instrumentation, mean that we can sometimes extrapolate from the fine details known for Solar System objects in order to aid our interpretation of the much vaguer information which we do have for some exoplanets. Computer modelling based on solar-system data can also be used to predict the possible properties of exoplanets even when those exoplanets are rather different from anything within the Solar System. Nonetheless, it remains the case that, for many exoplanets, their existence is about all we know about them.

---

[73] A mirror with the 3D shape of an ellipse has two points in space which are called its foci. A light-source placed at one of the foci will be imaged at the second focus and vice versa.

[74] This number is increasing rapidly. The figure given here is for March 2024 but will be well out-of-date by the time that you are reading this book. More current information can usually be found (at the time of writing), via following internet sites:
https://en.wikipedia.org/wiki/Lists_of_exoplanets
https://exoplanets.nasa.gov/faq/6/how-many-exoplanets-are-there/
http://exoplanet.eu/
http://var2.astro.cz/ETD/
https://exoplanetarchive.ipac.caltech.edu
http://cdsweb.u-strasbg.fr/
http://www.exoplanetkyoto.org/
or via an internet search for 'Exoplanets current total'.

FIGURE 3.3 An image of the surface of Mars obtained by NASA's Perseverance lander and showing fine detail of the surface and the Mars helicopter 'Ingenuity'. (Reproduced by kind permission of NASA/ JPL-Caltech.)

4 Exoplanets (#2)

FIRST CATCH YOUR EXOPLANET!

Just like catching a hare,[75] catching (discovering) an exoplanet is never easy, but once you know how to do it, it can be done.

Today's main observational methods for discovering and studying exoplanets are mostly mature, well-developed technologies and may be entitled:

- Astrometry
- Direct imaging
- Disc kinematics
- Eclipse timing variations
- Gravitational microlensing
- Orbital brightness variations
- Polarimetry
- Pulsar timing
- Pulsation timing variations
- Radial velocity (Doppler shifts)
- Transits
- Transit timing variations

and

- White dwarf atmospheres

From studies using these various approaches, some 5,580 individual exoplanets have been discovered at the time of writing. We will look into some of these discovery methods in a little more detail in Chapters 6 and 7. However, since the first exoplanets of any description were only detected in January 1992, what was happening in the years/decades/centuries before that? The next Section expands on a few of the most significant milestones.

BEFORE 22ND JANUARY 1992

Prior to 1992, a lot of effort had been put into trying to detect exoplanets, but without any success. Some scientists argued that this lack of success arose from a real lack of actual exoplanets. Many others[76] thought that there must be numerous exoplanets to be found, but that we just had not come up with the right way of detecting them yet.

---

[75] The title of this section is adapted from the beginning of a supposed recipe for Hare soup and serves to express its obvious, but usually unmentioned, initial requirement. That recipe and its command are generally attributed to the eighteenth-century cook and writer, Hannah Glasse. Unfortunately, her recipe actually begins *"Take and cut a large hare into pieces ..."*, so the true origin of the phrase remains obscure, although its implicit meaning is still in common use today.

[76] Including this author.

DOI: 10.1201/9781003246459-5

Despite what was written earlier about stars being 'fixed' in the sky, because they do not move around as do the planets, this is not quite true. The stars do move with respect to each other and with respect to our coordinate system. Their movements, though, are very small and slow compared with those of the planets and so they take much longer to become apparent. Thus, the slowest of the major planets, Neptune, typically moves at a rate of about 2° per year (7,200″/yr), whereas the 'fixed' star with the fastest known motion (Barnard's star) moves at just 10.39″/yr. Thus, the speeds at which stars move with respect to each other are very slow indeed when compared with planetary movements.

Nonetheless, stars *do* move, and even in the mid-nineteenth century, telescopes were of high enough quality to allow observers to measure a star's position sufficiently accurately to detect the 'faster' of those movements over a period of a few years. The measurement of accurate positions in the sky for stars and other objects forms the branch of Astronomy called Astrometry and this type of movement of a celestial object is usually called its Proper Motion.

Thus, by the mid-nineteenth century, the brightest star in the sky, Sirius (α CMa), had had its position measured many times. Friedrich Bessel[77] announced in 1844 that Sirius had the quite rapid proper motion of about 0.5″/yr, but that, unlike other stars, this proper motion was not in a straight line. The motion had a side-to-side wobble superimposed on top of its straight-line path (Figure 4.1).

Bessel correctly interpreted his observations to mean that Sirius was actually a binary star. The companion star was too faint to see, but both stars orbited their common centre of gravity[78] giving the visible star's proper motion its wobble as the invisible component pulled it from side to side. Eighteen years later, Alvan Clark, whilst testing a new telescope for the Dearborn Observatory, confirmed Bessel's prediction when he saw this companion as a faint star very close to the familiar super-bright, naked-eye primary (Figure 4.1).

After Bessel's successful prediction of and Clark's successful detection of the 'invisible' companion to Sirius A, many, perhaps most, astronomers expected that astrometry, using a similar approach, would provide the means whereby exoplanets would be found.

However, this turned out not to be the case and the reason for that lies in the relative masses of stars and planets. Sirius B is a white dwarf with a diameter of only 0.5% that of Sirius A, very small for a star. However, its mass is still half that of Sirius A and it is actually slightly more massive than our own Sun – and in this type of work, it is the mass, not the diameter, which counts.

The most massive planet in the Solar System is Jupiter, but it would take the masses of about 1,070 Jupiters (1,070 $M_{Jupiter}$ – Box 4.1) to equal that of Sirius B. If a single Jupiter mass were to replace Sirius B within the Sirius binary system, with all other parameters being unchanged, then the deviations of Sirius A from a straight-line proper motion would be reduced by a factor of 1,070, i.e., Sirius A's maximum sideways displacement, instead of being by about 2.5″ on each side would become about 0.002 5″ (2.5 milli arc-second). Put an Earth mass in place of Sirius B and the deviations of Sirius A from a straight-line path would then be just 0.000 000 8″ (0.8 micro arc-second) – and this would stretch even today's measuring capabilities,[79] never mind those prior to 1992.

---

[77] Bessel, in 1838, was also the first to detect the parallax movements of a star (61 Cyg), which arise from the changing positions in space of the Earth as it moves around its orbit.

[78] Usually known as the barycentre in this context.

[79] There are 1.296×10^{12} micro arc-seconds in a complete circle. At the time of writing, optical astrometry can achieve about 10 micro arc-second accuracy, whilst radio astrometry can get down to about 1 micro arc-second accuracy. An object 2 mm across would be about 1 micro arc-second in size when seen from the Earth at the distance of the Moon (384,000 km).

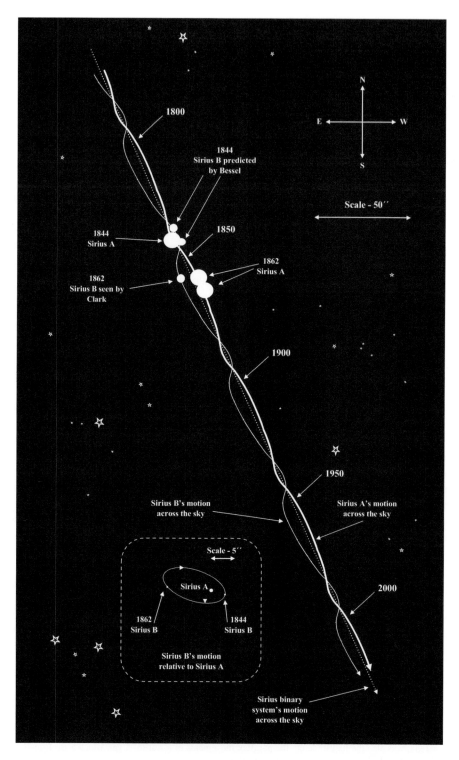

FIGURE 4.1 The Sirius binary star system. The main tracks show the meandering paths of Sirius A and B across the sky, while the dotted line shows how their centre of gravity moves in a straight line. The background stars are imaginary. Inset is a diagram of how Sirius B's orbit around Sirius A would appear to an observer at rest with respect to the binary system. The stars' positions at the prediction (1844) and detection (1862) dates are indicated. (© C. R. Kitchin 2022. Reproduced by permission.)

BOX 4.1 – SOME CONVENIENT UNITS FOR EXOPLANETARY WORK

The masses of the Sun, M_{Sun}, (1. 989 x 10^{30} kg), Jupiter, $M_{Jupiter}$, (1.898 x 10^{27} kg) and the Earth, $M_{Earth,}$ (5.972 x 10^{24} kg) are frequently used as measurement/comparison units when discussing exoplanets.

For most purposes, the following conversions and approximations will be found to be adequate:

$M_{Sun} \approx 2$ x 10^{30} kg
$M_{Jupiter} \approx 2$ x 10^{27} kg
$M_{Earth} \approx 6$ x 10^{24} kg

Giving:

1 $M_{Sun} \approx 1,000$ $M_{Jupiter} \approx 300,000$ M_{Earth}
0.001 $M_{Sun} \approx 1$ $M_{Jupiter} \approx 300$ M_{Earth}
$0.000\ 003$ $M_{Sun} \approx 0.003$ $M_{Jupiter} \approx 1$ M_{Earth}

The luminosity of the Sun (i.e., its total energy emission) is often similarly used as a unit and has the value: 3.86 x 10^{26} W ≈ 4 x 10^{26} W and the symbol L_{Sun}.

Other quantities of the Sun, Jupiter and Earth, such as radius, volume, etc., may be used in similar fashions

The luminosities of Jupiter and the Earth are due largely to the solar energy reflected from them and are not usually encountered as units in the same manner as their masses.

Despite the obvious practical difficulties involved, in 1855, seven years before Clark saw Sirius B, William Jacob, the then Director of the East India observatory, published a claim attributing the anomalies which he had observed in the orbital motion of the binary star 70 Oph to the effects of an exoplanet belonging to the system. Jacob was working with a 0.3 m telescope and making positional measurements by eye and was in poor health, but he still achieved accuracies of around 80 milli arc-second. This result was undoubtedly a remarkable achievement, but it was not an exoplanet discovery; Jacob's positional anomalies all turned out to be due to other causes.

Over the next century and a half, other claims were made for the discovery of exoplanets – including several more for 70 Oph.[80] All such claims were based upon astrometric (positional) measurements and almost all turned out to be false alarms. The exception was for the star γ Cep A for which, in 1988, Bruce Campbell, Gordon Walker and Stephenson Yang claimed to have found an exoplanet with a period of 2.7 years. That claim was withdrawn though in 1992. However, in 2003, with much more data, an exoplanet of about 1.7 $M_{Jupiter}$ and with an orbital period of 2.5 years *was* finally confirmed as being real. That, though, was too late; by then other teams had crossed the finish line and had found confirmed exoplanets elsewhere. Nonetheless, γ Cep A b remains the first exoplanet to be clearly discovered via astrometry (Figure 4.2).

---

[80] 70 Oph is only ~17 ly away from the Earth, so any orbital motions it might have will be angularly larger than for more distant stars – and so easier to observe.

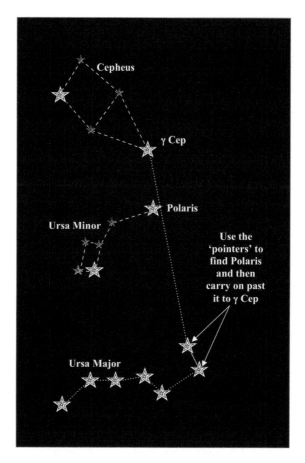

FIGURE 4.2 γ Cep b – Use the pointers for the pole star (Polaris) in Ursa Major to find γ Cep (and hence the position of γ Cep b). (© C. R. Kitchin 2023. Reproduced by permission.)

Even today there is a lot of confusion and some very optimistic claims associated with the announced discoveries of exoplanets via astrometry. This arises from trying to decide when an entity is an exoplanet and when it is a brown dwarf (Box 4.2). The difference is essentially that a brown dwarf is generating energy in its interior via nuclear fusion reactions, i.e., it is a STAR, whilst an exoplanet is *not* generating energy via nuclear reactions and so it is a PLANET.[81]

BOX 4.2 – WHAT'S WHAT IN PLANETS AND STARS?

There is a smooth variation in the masses of individual astronomical objects from about 250 M_{Sun} for the most massive stars to pico-gram, or less, dust particles. Although the mass variation of these bodies is smooth, for our own convenience we divide them into different classes.

---

[81] Although it may have significant alternative energy sources, such as gravitational energy (more frequently known in physics contexts as Potential Energy), being released as heavy/dense material sinks down towards core).

Amongst the macroscopic entities, the IAU's recent official/semi-official definitions, or the ones in common usage (or the ones being used by this author in this book anyway, – even if no one else agrees[82]), for the typical mass ranges of these classes are:

(i)   Star:
      ~80 $M_{Jupiter}$ (~0.076 M_{Sun}) to ~250 M_{Sun}.
      Generates energy via the fusion of *normal* hydrogen nuclei, or it is the remnant of such an object.
(ii) Brown dwarf:
      ~13 $M_{Jupiter}$ to ~80 $M_{Jupiter}$.
      The higher limit is where the body can no longer support the fusion of *normal* hydrogen.
      The lower limit is where the body can no longer generate energy through *deuterium* (heavy hydrogen) fusion.
(iii) Planet, Exoplanet, Dwarf planet, Dwarf exoplanet, Exomoon, Large planetary natural satellite, Large Exosatellite, Large asteroid, Large Exoasteroid[83]:
      ~10^{20} kg (~0.000 02 M_{Earth}) to ~3 x 10^{28} kg (~13 $M_{Jupiter}$, ~4,000 M_{Earth})
      Based upon the criteria that the entity is not massive enough to generate energy through fusion reactions and yet is massive enough to become spherical(ish) under gravity.
(iv) Smaller objects:
      ~10^9 to ~10^{20} kg.
      Irregularly shaped, but perhaps large enough for some ETs to live on?
      ~100 m to ~1,000 km in 'size'.
(v)   Boulders, pebbles, dust and lesser objects:
      <10^9 kg.

Similar entities, which are free-floating, are also to be included in these ranges, based upon their masses.
The third and fourth classes of this list form the primary concern of this book.

There are now a number of quite well-established criteria for separating these two types of objects (Box 4.2) and, in this book, their masses will mostly be used for the purpose. It needs to be noted though that for many exoplanets the inclination angle of their orbits to the plane of the sky, i, is unknown and this effects our estimate of their masses. If the orbital inclination is unknown, then what is determined is M_{Star} sin(i),[84] not the true value of M_{Star}, and the latter is almost certainly a larger number.

---

[82] Simple(ish) numerical formulae have been suggested several times for determining what is and what is not a planet, for example, Jean-Luc Margot's Π discriminant (Astron. J., 185, **150**, 2015). However, none has achieved widespread general acceptance yet. Wikipedia's article on 'Clearing the neighbourhood' would be useful for further reading for anyone interested in the topic.

[83] All these objects, which differ significantly from each other in some ways, are lumped together here because it is likely to be the local conditions which govern an object's viability for hosting ETs and/or ETIs, not whether it is orbiting a distant planet rather than a distant star, etc.

[84] You may not have encountered the trigonometrical function 'sine' (usually abbreviated to 'sin'), but worry not – all that you need to know is that whatsoever the angle involved, the sine of that angle always lies between –1 and +1. Thus, at most 'sin(i)' can take the value –1 or +1, but usually it lies between those two values. When the inclination is not known, therefore, the estimated mass of the exoplanet has to be smaller than its real mass or equal to it.

Even with the criteria in Box 4.2, confusion between exoplanets and brown dwarfs is still rife. Thus, one major catalogue of known exoplanets and their details lists 17 exoplanets which have been discovered via astrometry, whilst a second lists just two. A close examination, though, reveals that in the first catalogue, the masses given for 11 of the 'exoplanets' clearly make them brown dwarfs, with another three being on the borderline between exoplanets and brown dwarfs. Leaving just three which probably are exoplanets. Of the two candidates listed in the second catalogue, one is an exoplanet, but the other is a brown dwarf. Interestingly, neither catalogue includes γ Cep A b within its astrometrically detected list. Some brief details of γ Cep A b, the three more recent probable astrometrically discovered exoplanets and the three border-line cases are given in Table 4.1.

TABLE 4.1

The Currently Known Probable and Possible Exoplanets Discovered via Astrometry

Name	Orbit – Semi-major axis[85] (AU)	Orbital period (Days)	Mass ($M_{Jupiter}$)	Discovery date	Probable/ Possible
γ Cep A b	2.5	903	1.7	2003	Probable
HD 176051 b	1.76	1,016	1.5	2010	Probable
WISE J0458+6434 b	4	15,500	14	2011	Possible
2MASS J0249-0557 AB c	2,000	?	11.6	2018	Possible
TVLM 513-46546 b	0.3	220	0.38	2020	Probable
GJ 896 A b	0.64	?	2.26	2022	Probable
GJ 2030 c	16.8	15,500	12.9	2022	Possible

At the time of writing therefore, astrometry is still only to be credited with the discovery of seven exoplanets. The technique has, additionally though, been useful for studying other exoplanets once they have been discovered by other methods. A possibility for the future is the CHES[86] concept, which would place an astrometric spacecraft at the Earth's L2 point (Chapter 13) to observe the positions of stars within a few tens of light years to precisions of around a micro-arc-second.

---

[85] Half the length of the longest axis of the elliptical orbit. For a circular orbit it is the radius of the circle.
[86] Closeby (*sic*) Habitable Exoplanet Survey.

5 22nd January 1992 – The FIRST Exoplanet?

The answer to the question posed by the title for this Chapter requires the answer to another question beforehand. To wit:

What do you mean by an Exoplanet?

and it turns out that there are two possibilities:

- Exoplanets formed at the start of a star's life.

and

- Exoplanets formed at the end of a star's life.

Anticipating later discussions, we will call these two types of exoplanet:

- Normal Exoplanets

and

- Pulsar Exoplanets

respectively.

At the time of writing there are seven known examples of the Pulsar Exoplanets and 5,580 known examples of the Normal Exoplanets (hence the name which I have given to the latter[87]).

The first two Pulsar Exoplanets were found simultaneously in January 1992 by Aleksander Wolszczan (Figure 5.1) and Dale Frail and are now known as PSR1257 +12 b and PSR1257 +12 c. The first Normal Exoplanet was found in November 1995 by Michel Mayor and Didier Queloz and is known as 51 Peg b (Chapter 6).

Thus, there were two First Discoveries, which are of equal validity and which are of essentially different objects. The Normal Exoplanets are the ones with which we shall mainly be concerned in this book, as the possible hosts for ETs. However, we will first look a little further into the discovery and details of the Pulsar Exoplanets, before returning to concentrate on the Normal Exoplanets in subsequent Chapters.

---

[87] In fact, there are probably some 1,000,000,000 neutron stars currently within our galaxy. Thus, even if only 1% have exoplanets, that means there are potentially some 10,000,000 pulsar planetary systems 'out there' still waiting to be found.

DOI: 10.1201/9781003246459-6

FIGURE 5.1   Aleksander Wolszczan high on the structure of the 300-m Arecibo[88] radio telescope and with the enclosure for the radio dish's antennae and secondary reflector in the background. (Reproduced by kind permission of A. Wolszczan and A. Acevedo.)

PULSAR EXOPLANETS

The name 'Pulsar' is derived from 'Pulsating Radio source' and that latter name gives you the gist of how they are observed (see also Box 5.1). The first pulsar was discovered by Jocelyn Bell on the 28th of November 1967 and announced three months later. It comprised a series of brief, stronger than normal, radio signals (pulses) from one point in the sky, at intervals of 1.3 seconds of time. Since then, over 3,300 pulsars have now been identified.

BOX 5.1 – PULSARS

Pulsars are mostly the remnants of some type of supernova. They are extremely dense objects with masses of around 1 M_{Sun} compressed into spheres some 20 km in diameter. Their mean density (and it is higher at the centre) is thus around 10^{17} kg m^{-3}. A football made from average pulsar matter would weigh, on Earth, about 600,000,000,000 tonnes ® (400,000,000,000 tons for American/Gridiron footballs).

The core of a supernova collapses to form a neutron star (see below) during the supernova event. Not all neutron stars are pulsars, and there may be some pulsars which are not neutron stars, but the vast majority of pulsars are neutron stars.

Most normal matter (such as that making up you and me) is formed from atoms and molecules. Those in their turn are built up from the sub-atomic particles: electrons, protons and neutrons. The neutrons and protons form the atom's nucleus with the electrons making up a surrounding cloud.

Take a neutron out of an atom's nucleus and in about 14 to 15 minutes (on average) the isolated neutron splits into an electron and a proton. Conversely, take an electron and a proton and force them together hard enough and they will combine to become a neutron.

---

[88] A fixed 300-m diameter radio dish, sited in Puerto Rico and which operated from 1963 to 2020, when it collapsed.

The force required to combine electrons and protons into neutrons is enormous and if we have a plasma of many electrons and protons, that combining force is what we call the pressure of the plasma. In the centre of our Sun, the plasma pressure is some 2.5×10^{16} N m^{-2} (250,000,000,000 times the pressure of the Earth's atmosphere at sea level). But that does not even get us to the starting post when it comes to the pressures needed to form neutron stars. Forming a neutron star by pushing the electrons into the protons needs a pressure some 40,000,000,000 times higher than that at the centre of the Sun ®.

Few places within the entire Universe can achieve pressures of that magnitude – but the centre of a supernova is one of them and so that is where neutron stars and pulsars have their beginnings. A neutron star is then literally that; a star formed primarily from neutrons; the protons and electrons having been forced to combine.

Once formed inside a supernova, our neutron star/pulsar, through the conservation of angular momentum during the collapse, will be rotating very rapidly (think of an ice-skater rotating faster as he/she 'collapses' by bringing his/her arms in towards her/his body).

To make this object into a pulsar, all that is now required is to arrange for the radiation from the now very hot object, to be emitted in the form of a narrow beam (in practice, two opposed beams) and that occurs because the emitted radiation is channelled outwards along the magnetic axes of the pulsar.

We thus see pulses from the neutron star, not continuous emission, because the beamed radiation only crosses our line of sight once per revolution; this is like the flash seen at sea from the rotating continuous beam coming from a lighthouse (Figure 5.2). Because the neutron star is so massive, its momentum is huge and its rotational period is therefore extremely stable. The radio pulses which we observe are thus at very regular intervals and can act as a clock. It is this clock property which underlies the discovery of the pulsar exoplanets.

FIGURE 5.2　The beam from a lighthouse. We see a bright flash (pulse) when the beam points towards us and that flash repeats at the rotation rate of the lighthouse' mechanisms – just like the (radio) beam from a pulsar. (Reproduced by kind permission of Pixabay and Achim Scholty.)

Of course, many 'ordinary' neutron stars may also be pulsars, but we do not detect them as such because their emission beams never point towards the Earth. There may also be neutron stars which are not pulsars, these latter though may still have exoplanets and so references to 'pulsars' and 'neutron stars' are used more or less interchangeably in this section.

The radio pulses from pulsars arrive at very regular intervals (Box 5.1) – like the ticking of an extremely good clock.[89] If the pulsar is an isolated object, then the pulses will, indeed, be as regular as claimed. However, if the pulsar is a part of a binary system, whether it be with an exoplanet, another neutron star or any other type of star, then the two objects will both be orbiting their bary-centre. The distance from us to the pulsar therefore changes; being furthest from us when it is on the far side of the companion and nearest when on our side of the companion.

Now radio waves are low frequency versions of light waves and travel at the same enormous speed through space: 300,000 km/s. But, while fast, 300,000 km/s is not an infinite speed, so it still takes longer for radio waves to get to us from a distant object than from a nearer one. In particular, the radio emissions from a pulsar/exoplanet binary system take longer to travel to us when the pulsar is on the far side of the exoplanet as we see it than when it is on the near side. This change in distance may be observed as a change in the pulse intervals which we are receiving.

When the pulsar and exoplanet are in line as we see them, whether the pulsar is at the nearer or further point in its orbit from us than the exoplanet, then its orbital movement is across our line of sight to it and so its distance from us is not changing. We then receive the flashes at exactly the same intervals as the rotation period of the neutron star.

When the pulsar's orbital motion is taking it away from us, then each successive flash originates from a slightly greater distance away from us and so it takes just that little bit longer to get to us. The observed interval between two flashes is thus just slightly longer than the neutron star's rotation period. Conversely, when the pulsar is approaching us through its orbital motion, intervals between two consecutive flashes are slightly less than the neutron star's rotation period.[90]

Whether or not a pulsar is a component of a binary or multiple stellar/planetary system may thus, potentially, be revealed by timing its pulses and then searching for patterns of regular changes within them. This is why this approach to discovering exoplanets is called the Pulsar Timing method.

So, in February 1990 when Aleksander Wolszczan was searching the sky for new pulsars using the Arecibo radio telescope and found one 2,300 ly away in Virgo, which became named PSR 1257+12,[91] he was delighted. But it wasn't the end of the story. He and his postdoc colleague, Dail Frail, continued to monitor its pulses over the next several months. They soon found that PSR B1257+12's pulse timings had some complex changes buried within that data.

The pulsar's basic pulse period is 0.006 219 s (6.219 ms = 161 pulses per second) and it could be longer or shorter than this by some 0.000 000 000 01 s (10^{-11} s). Now such a small change cannot currently be measured. But there are 161 pulses every second and those changes accumulate. After 10 minutes observation, the last pulse will be about ±1 μs or so astray from when it should arrive – and that is measurable.

Wolszczan and Frail, in a few months, thus had enough data to suggest that PSR B1257+12 not only had one exoplanet but almost certainly two, and there might be a third. They named these Exoplanets PSR B1257+12 B and PSR B1257+12 C and announced their discovery on the 22nd of January 1992; these were undoubtedly the FIRST exoplanets, of any type, clearly to be discovered.

---

[89] Actually, pulsars' rotations do slow down very gradually and so, therefore, do their 'clocks'. Provided, though, that you can measure the slowing-down rate, the change can be allowed for and the pulsar then does become the very good clock stated.

[90] This change in the period observed for the pulses with changing speed is in fact, just an example of the Doppler effect and that effect will be encountered again in the discussion of the Radial Velocity method of discovering Exoplanets (below).

[91] The name of this pulsar and its exoplanet family is particularly confusing. Its basic name is as given in the text, with *PSR* standing for 'Pulsar', *1257* being its right ascension (12h 57m) and *+12* its declination (12° N). However, it is also named PSR B1257+12 and PSR J1300+1240 where the 'B' indicates that the coordinates are for its position in the sky in 1950, whilst 'J' indicates the higher precision and changed coordinates for 2000. The naming of the exoplanets, which adds further to the confusion, will be encountered shortly.

FIGURE 5.3 Artist's concept of how the PSR B1257+12 system might appear from a nearby visiting space-craft. The pulsar is the white star-like object surrounded by nebulous emissions in the top left of the image. The three planets are in their alphabetical order (b, c and d) moving outwards from the pulsar. (Reproduced by kind permission of NASA/JPL-Caltech.)

Two years later, the third exoplanet in the system was confirmed (Figure 5.3). The details of those three exoplanets[92] are given in Table 5.1. At the time of writing another four pulsars each with single exoplanet companions have been discovered. These latter have masses ranging from 2 M_{Earth} to 2.5 $M_{Jupiter}$ (Table 5.2).

TABLE 5.1

The Exoplanets of PSR B1257+12 – Measured Properties

Current name (Old name/ Proper name[93])	Orbit – Semi-major axis (AU/m)	Orbital period (days)	Mass (M_{Earth}/kg)	Discovery date
PSR B1257+12 b (PSR 1257+12 A/Draugr)	$0.19/2.8 \times 10^{10}$	25.3	$0.02/1.2 \times 10^{23}$	April 1994
PSR B1257+12 c (PSR 1257+12 B/Poltergeist)	$0.36/5.4 \times 10^{10}$	66.5	$4.3/2.6 \times 10^{25}$	January 1992
PSR B1257+12 d (PSR 1257+12 C/Phobetor)	$0.46/6.9 \times 10^{10}$	98.2	$3.9/2.3 \times 10^{25}$	January 1992

The reason why these pulsar exoplanets are generally regarded as being significantly different from all the normal exoplanets is that they have formed at the end of a star's existence, not, as is

---

[92] Two separate claims for the possibility of a fourth exoplanet, which had been suggested, have now been ruled out as being due to noise instead.

[93] The pulsar itself has also been given the proper name; 'Lich'.

thought to be the case for normal exoplanets, during the star's formation, i.e., at the beginning of its existence.

Anticipating, very briefly, discussions which follow, normal exoplanets are thought to form during a gravitational collapse, or condensation process from the material forming gaseous nebulae and/or the interstellar medium and which is also condensing to produce the protostar. Material left behind during the collapse gathers into a disc of material orbiting the protostar and within that disc, the planets form via further smaller scale condensations.

By contrast, pulsars are the result of some types of supernova explosions – and supernovae are some of the most violent events known to occur anywhere within the Universe. A supernova occurs as the core of a very massive star collapses, as a white dwarf collapses or as the two white dwarf stars in a close binary system collide with each other. Whichever way (and there are variants within these main types), the energy generated in a tenth of a second or so can reach a hundred times[94] the total energy that our Sun will produce throughout its 10,000,000,000-year existence ®. Temperatures within the supernova can peak at a thousand times the present temperature at the centre of the Sun and the bulk of the matter involved in the supernova is exploded outwards at speeds of up to 20,000 km/s ®. Anything much less like the quiet calm of interstellar nebulae collapsing over several million years, with temperatures close to absolute zero and speeds of 10 m/s (36 km/h) or so, is difficult to envisage.

Thus, for any pulsar or neutron star exoplanet, if it was in existence before the supernova, then it must have survived that incredible explosion and also *not* been expelled from its orbit during the process.

Alternatively,[95] it must have condensed after the supernova occurred, out of the debris or other material retained by the neutron star. Much of this 'debris' though is actually expanding outwards at up to 3% of the speed of light and even 10,000 years after the event is still four times hotter than the evaporating point of any known chemical element.[96]

Despite this apparent unlikelihood of such an event, pulsar exoplanets *do* exist. Therefore, there must be some means by which they can either survive the supernova or re-condense afterwards.

Pulsar (and neutron star) planets of the first type mentioned above, i.e., survivals from before the supernova, are called First Generation pulsar planets. Planets formed from debris from the supernova which has not been lost and now forms a fallback disc around the pulsar/neutron star are Second Generation, whilst Third Generation exoplanets are thought to come from material torn from a stellar companion to the supernova which survived the explosion and is now expanding as it ages and so loses matter to the neutron star.

At the time of writing, the (probably) most widely held view amongst Astronomers and Astrophysicists is that pulsar exoplanets are third generation objects. The reason behind this is that for four out of the five pulsars currently known to have exoplanets (see above and Table 3.5), the pulsar period is very short: 0.0035 to 0.011 ms. Such pulsars are called milli-second pulsars – and several thousand, over and above these four, are now known.

Milli-second pulsars are thought to have a complex history:

- A binary star system comprising a massive star (~10 M_{Sun}, or more) and a lower mass star form normally from interstellar gas clouds.

---

[94] The figures are: ~10^{46} J for the brightest supernovae and ~10^{44} J for the total solar emission throughout its existence.

[95] A third possibility would be for the neutron star, an independent exoplanet and a third object of some sort, to interact so that the independent exoplanet is captured by the neutron star and the third object removes the excess energy released during this process. This is very unlikely, although not impossible. However, for it to happen three times (PSR B1257+12's three exoplanets) does verge on the unbelievable, especially when the exoplanets are all in more-or-less coplanar and nearly circular orbits. This possibility is therefore not considered further here.

[96] Rhenium – Boiling (evaporating) point: 5869 K.

- Massive stars burn up their nuclear fuel proportionately much faster than do lower mass stars. The ~10 M_{Sun} star thus evolves,[97] perhaps in as little as 10 to 20 million years, off the main sequence, through the giant stage to a white dwarf, then explodes as a supernova and so becomes a neutron star.
- The binary system survives the supernova event.
- The neutron star and lower mass star continue to orbit each other with little happening except that the initial very rapid rotational period of the neutron star slows down, perhaps from tens of milli-seconds to tens of seconds.
- The lower mass star begins its own evolution off the main sequence. This could occur from 100 million to 10 billion years after the supernova.
- The lower mass star expands towards becoming a red giant. At some point though, its outer layers will start to be captured by the gravitational field of the neutron star.[98]
- The material captured by the neutron star from the evolving solar mass star falls to its surface, releasing huge quantities of gravitational energy and also increasing the neutron star's spin until is rotating every few milli-seconds; faster even than when it was first formed.

If this were a book about milli-second pulsars, this account would finish at this point. Because, though, we are concerned with possible exoplanets, there are a few more stages:

- Not all the material captured by the neutron star reaches its surface, some accumulates into a disc orbiting the neutron star.

and/or

- If the initial mass of the lower mass companion star in the original binary was some eight M_{Sun}, or more, it could explode as a supernova as well. Some of *that* supernova's material then being captured by the original neutron star. The system settles down to a single neutron star surrounded by a disc of condensing/condensed material if the second supernova disrupts the binary system, or, as a neutron star surrounded by a disc of condensing/condensed material together with being a member of a binary system and with a neutron star/white dwarf companion, if not disrupted.
- Either way, we have a neutron star surrounded by an orbiting disc of cooling material.
- The exoplanets then form from the disc material, much as they might 'normally'. The composition of the material though, is likely to be quite different from that which goes to form normal exoplanets (including the Solar System) with elements heavier than hydrogen and helium being much more plentiful than usual.

At the end of this long journey, we see that the processes thought to be forming milli-second pulsars could also produce pulsar exoplanets naturally along the way.

This scenario does not preclude other types of pulsars and neutron stars from having their own exoplanets. Indeed, the neutron stars left by supernovae originating from the collision and merger of a double white dwarf binary system might well also be left with an orbiting disc of material. On the whole, though, the milli-second pulsar theory does seem to fit the currently known observations fairly well.[99]

---

[97] Astronomers use the term 'evolve' to mean aging. In this context therefore, it does not imply the changes in a large population of living beings as mutations and survival-of-the-fittest cause new generations to differ from the older ones, but the change in a single entity as time passes. My apologies for this to all Biologists and to most other people, but the practice does not now seem likely to change amongst Astronomers.

[98] See the Author's book; *Stars, Nebulae and the Interstellar Medium*, Adam Hilger, 1987, or undertake an internet search for 'Roche Lobes' if you would like more details of this process.

[99] A recent review of pulsar exoplanets, though, lists another 11 possible ways in which pulsar exoplanets might be formed.

So, after this long discussion we seem to have arrived at the conclusions that the pulsar exoplanets were formed by the same process as normal exoplanets, i.e., by collapse and condensation from a disc of gas and dust orbiting the central object. Why, then, make a distinction between the two types of exoplanet?

The answer to that question is that normal exoplanets are made up of materials containing (initially) about 91% hydrogen and 8.9% helium by number. Only ~0.1% of an interstellar gas cloud is in the form of any of the heavier elements.[100] The disc around a milli-second pulsar contains material that has been processed inside a star and one or two supernovae and so will contain significant quantities of the heavy elements; in fact, all the way up to uranium and beyond. So, the composition (and perhaps the internal structure) of pulsar exoplanets may be expected to differ considerably from those of normal exoplanets.[101]

For PSR B1257+12, it is now some billion years (or more) since its supernova exploded. The neutron star may initially have had a surface temperature in the region of ≥ 1,000,000 K and each square metre of its surface would have been emitting around 1,000,000,000 times more energy than each square metre of the Sun's surface does today[102] ®.

The pulsar planets at that time were probably still in the process of condensing within the accretion disc around the pulsar. But if they had been formed, then their temperatures must have been sky high as well surely? – Right?

Wrong!

The innermost planet might have had a temperature[103] at its surface of ~ 80 °C (~ 350 K), which is on the warmish side for us. The two outer planets, though, would have had temperatures of ~ –18 °C (~255 K – about the temperature inside your freezer) and ~ –50 °C (~220 K) ®.

The reason that the exoplanet temperatures seem so unexpectedly low is that the neutron star is *very* small. The Sun's radius is some 696,000 km. The radii of pretty well all neutron stars and pulsars are close to 10 km. Thus, although the neutron star's surface emits prodigious quantities of energy over every square metre, it actually does not have very many square metres of surface from which to emit. Just after the supernova, therefore, the newly formed neutron star has only about one-fifth of the Sun's present luminosity ®.

SCENARIO #1: PSR B1257+12'S EXOPLANETS – WHAT THEY MIGHT BE LIKE TODAY

By the present day, perhaps a billion years since PSR B1257+12's supernova, the exoplanets are in existence and the neutron star has cooled down a lot. A great deal of the neutron star's initial energy was lost via neutrinos, but by now it is mostly just electromagnetic radiation (radio, infrared,

---

[100]Rather confusingly, Astronomers often refer to all the chemical elements, other than hydrogen and helium, as 'metals' and their relative abundance to hydrogen and helium within an object as its 'metallicity'. The normal exoplanets would thus be described as forming within a disc of normal, or galactic, metallicity. The pulsars exoplanets would be forming within a disc of high metallicity and/or of unusual metallicity. Generally, within this book I try to use the term 'Heavy Elements' as a synonym for this usage of 'metals'.

[101]Even if normal exoplanets born before the supernova survive and now form the pulsar exoplanets, their outer layers will surely have been lost during the supernova and what we now see will be the cores of those normal planets. Those cores are probably from Jupiter-like gas giant planets and the heavy elements will have accumulated there through gravity, so these will also quite different in their compositions from a 'normal' normal exoplanet.

[102]Although I have used the ® symbol here, these figures *are* very approximate. They are not too far adrift, though, but nonetheless, do not go betting the farm on them.

[103]There are many ways of estimating planetary temperatures. This estimate is the one usually called the Equilibrium Temperature and it is based just upon the energy that the planet receives from its central star/neutron star balanced against the energy radiated away by the planet. It does not take into account any effects of an atmosphere, such as the Greenhouse effect, nor any heat leaking outwards from the hot central parts of the planet, etc. The Earth's equilibrium temperature is thus ~ –18°C (~ 255 K), while its average surface temperature, due to the Greenhouse effect, is ~ 15 °C (~ 288 K). N.B.: Conservationists please note; the Greenhouse effect ain't *all* bad.

visible light, UV[104], x-rays, γ rays) that does the job. Some particle emissions also may help, though, together with magnetic interactions. Estimating the actual temperature which PSR B1257+12's surface has now reached however, is a tough task – and convincing cases can be made for temperatures ranging from less than 10,000 K to more than 100,000 K. To give some idea of what the exoplanets might now be like, their equilibrium temperatures for pulsar temperatures of 10,000 K and 100,000 K are:

10,000 K (pulsar temperature):

- PSR B1257+12 b: ~3.5 K
- PSR B1257+12 c: ~2.6 K
- PSR B1257+12 d: ~2.3 K

100,000 K (pulsar temperature) :

- PSR B1257+12 b: ~35 K
- PSR B1257+12 c: ~26 K
- PSR B1257+12 d: ~23 K

These are very low temperatures indeed for the exoplanets, although prospective colonists will be relieved to know that they are on top of the 2.7 K microwave background temperature of space (bringing PSR B1257+12 b at the hotter end of the pulsar temperature range to a balmy ~40 K, – 235 °C).

If these temperatures are anywhere near to being the actual temperatures of the exoplanets (but see Scenario #2 below, though), then we may picture them as being:

- Shape: spherical or very nearly so (even if formed from solid iron, the material's strength would be quite overwhelmed by their gravitational fields).
  - Atmospheres: none, or only extremely tenuous, atmospheres (all gases save hydrogen and helium will have been frozen out and hydrogen would also freeze at the lower ends of the ranges).
  - Solid surfaces, probably with liquid interiors for the two outer exoplanets, at least (the Earth is ~4.5 billion years old and still has a liquid core).
- Composed, as discussed above, mostly of the heavier and denser elements and compounds, probably with a thin surface layer of frozen gases.
- The central regions will be denser than the outer layers through compression by gravity for the two outer exoplanets. Additionally, the very densest materials are likely to have sunk towards the centre forming an even denser core.
- Radii of: [105]
  - PSR B1257+12 b: ~1,600 km (= ~0.9 R_{Moon})
  - PSR B1257+12 c: ~8500 km (= ~2.6 R_{Earth})
  - PSR B1257+12 d: ~8,000 km (= ~2.5 R_{Earth})
- Surface gravities of:[105]
  - PSR B1257+12 b: ~3.1 m s^{-2} (~0.32 g[106])

---

[104] Ultra-Violet.

[105] The diameters and surface gravities are calculated using the average density of the Moon (~3,300 kg m^{-3}) and the Earth (~5,500 kg m^{-3}) as guides. The Moon is the analogue for PSR B1257+12 b and the Earth for PSR B1257+12 c and d. After some allowance is made for the greater internal compressions, denser materials and higher masses of the exoplanets than their Solar System analogues, we may guess at mean densities for these exoplanets of around twice the Moon and Earth values; i.e., ~ 7,000 kg m^{-3} and ~11,000 kg m^{-3}.

[106] Where 'g' is the surface gravity of the Earth; 9.81 m s^{-2}.

- ○ PSR B1257+12 c: ~ 25.4 m s^{-2} (~2.6 g)
  - ○ PSR B1257+12 d: ~ 24.6 m s^{-2} (~2.5 g)
- Pulsar as seen from the exoplanets: Star-like (no disc) – Apparent magnitude estimates range from:
  - ○ about six times brighter than Venus as seen from the Earth (m[107] ~ −6.6, for PSR B1257+12 d and a pulsar temperature of 10,000 K)

  to

  - ○ about 370,000 times brighter than Venus as seen from the Earth (m ~ −18.5, for PSR B1257+12 b and a pulsar temperature of 100,000 K).

  For PSR B1257+12 c and PSR B1257+12 d, the pulsar will be about a quarter and a sixth of these brightnesses, respectively.
- Surface illuminations of the exoplanets in the visible from the pulsar are thus estimated as:
  - ○ If the pulsar is six times brighter than Venus:

    On the Earth, Venus at its brightest can cast visible shadows. So, for PSR B1257+12 b, the illumination would be about six times brighter than that from Venus on the Earth (see previous bullet point) and it would be like a clear dark night on the Earth,[108] without any Moon in the sky and with no significant light pollution. It would be very difficult for us to move around the surface without an artificial light of some sort.

  to

  - ○ If the pulsar is 370,000 times brighter than Venus:

    (~250 times brighter than the Full Moon), being on PSR B1257+12 b would be similar to being under bright twilight[109] soon after sunset or just before sunrise on the Earth. It would probably be just about light enough still to be driving without lights.

FIGURE 5.4 Earth light illuminating the dark side of the Moon, with the much brighter, thin crescent old Moon still showing on the left-hand edge. PSR B1257+12 d would be about twice as bright as the dark side (but without the bright crescent) at the maximum brightness estimated to be possible for it. (© C. R. Kitchin 2022. Reproduced by permission.)

---

[107] The symbol, in this context, is for apparent magnitude.
[108] About 0.002 lux.
[109] About 80 lux.

For PSR B1257+12 c and PSR B1257+12 d, the illuminations will be about a quarter and a sixth of these figures, respectively.

- What they would look like from a nearby spacecraft:
  - The Full Earth as seen from the Moon is about 30 times brighter than the Full Moon seen from the Earth. Thus, the illumination at the pulsar's maximum estimated brightness level would be about ten times higher than what we observe as the Earth-light on the Moon for PSR B1257+12 b and it would be about twice as bright for PSR B1257+12 d (Figure 5.4).
  - For the pulsar's lowest estimated brightness, all three planets would just be small circular regions of (black) space which lacked any stars.
- The nature of the surface structure is still largely guess-work and could range from an almost featureless, smooth and polished 'ball bearing' through mountains and valleys produced by 'geological' activity, to extremely densely cratered surfaces like those of Callisto and Pallas.
- Although the exoplanets' masses and orbital semi-major axes differ quite a lot, they are all probably largely formed from heavy elements and compounds (see earlier discussion). They have also all probably been completely molten at some time in the past and their outer layers have solidified producing stresses in their crusts. There may therefore be surface features such as mountain ranges and rift valleys. Without significant atmospheres, even small dust particles will produce impact craters, but larger craters will need larger impacting bodies. It is possible that those larger bodies were 'moped-up' whilst the exoplanets were still liquid and would thus leave no evidence to be seen now. Similarly, a possibly very thin crust and the strong gravity on the two outer planets might mean that any craters, etc., which are produced are then quickly engulfed into the interior. Of course, almost certainly, there are other possible and as yet unknown conformations for planetary surfaces, so that conceivably none of the above suggestions is correct.

Thus, were we to be explorers visiting the PSR B1257+12 system in our FTL (Faster-Than-Light ✿✿✿✿✿) spacecraft, we would see the three exoplanets as rather dark objects orbiting an intensely bright central star (Figure 5.3). The pulsar would appear as a steady light because, with its 161 rotations per second, our eyes would not respond quickly enough to any variations to notice them; just as movie frames, etc., flashing 24 times a second, give us the impression of smooth, continuous motions. The exoplanets would be incredibly cold and directly exposed to any x-rays and γ rays still being emitted by the pulsar. The radio beam, which we detect as pulses, would only pass across them twice per orbit, since their orbits are inclined at about 45° to our line of sight. Their surfaces could take almost any structure.

Will we ever find out which of these projections is correct, or even if something completely different has occurred? Well, a surface feature 100 km in size at the exoplanets' distance of 2,300 ly would be just 0.001 micro-arc-seconds across as seen from the Solar System. Since our current astronomical tools can reach resolutions of near to 1 micro-arc-second, it is perhaps thinkable that future developments, such as space-based optical interferometers[110] ✿✿ and/or using the Sun as a 'lens' for gravity waves[111] ✿✿✿, might do the job. The answer could thus be 'Yes' – we might yet see features on these exoplanets ✿✿.

---

[110] An interferometer is an observing or transmitting instrument which combines two or more individual telescopes in order to obtain a better angular resolution/angular beamwidth than is possible using one of the individual telescopes. Theoretically an interferometer can operate at any wavelength, for astronomical purposes, only the spectral regions from the radio through to the visible have been used so far. See also Chapters 7, 15 and 18, the author's book *Astrophysical Techniques* 7th Edn., CRC Press, 2020, most books on observational astrophysics or search the internet for 'optical/radio interferometers'.

[111] A recent suggestion for solar gravitational lens imaging suggests a surface resolution of 10 km for an exoplanet 100 ly away might be achievable. The spacecraft, though, would need to be placed ~16 times the distance of Pluto from the Sun, requiring the project take around a century and a half at the speeds achieved by the New Horizons mission to Pluto.

Suppose we now don our spacesuits and land on one of the exoplanets 🪙🪙🪙🪙 (such possibilities are also discussed in more detail in Chapter 18), then it would be like being on the Earth during a dark night or early to late twilight. The pull of the exoplanet's gravity would be much weaker than Earth's on PSR B1257+12 b and much stronger on PSR B1257+12 c and PSR B1257+12 d. There would also almost certainly be no atmosphere and so no wind and the outside temperature would be vastly colder than anywhere on Earth save inside a cryogenic laboratory's specialist apparatus. This latter point, though, is not as significant as it might seem – if your spacesuit keeps you warm and snug in space, where the temperature is around 2.7 K when not in direct sunlight/starlight, then it will also do so on the exoplanet's slightly warmer surface. You might though, need to put extra insulation on your feet to prevent frostbite from conduction through the bottom of your feet into the surface.

Whether ETs live on PSR B1257+12's exoplanets is another question and one whose possibilities are addressed in much more detail in Chapter 13. Here we will just note that for us and other forms of terrestrial life, given that we could get there, it would not be much more difficult to land on, explore, even live on for a while, but perhaps not to colonise, one of these exoplanets, than it would be to do the same on our Moon ®. It would just need more energy because of the lower temperatures and *much* better protection from high energy particles, x-rays and γ rays. So, it is quite possible that ETIs could do the same.

Whether life could originate on the exoplanets seems much less likely, although several SF authors[112] have 'invented' ETs based upon liquid helium 🪙🪙🪙🪙🪙. Such ETs, though, might find these exoplanets too hot ® (helium's boiling temperature is 4.2 K).

SCENARIO #2: PSR B1257+12's EXOPLANETS – WHAT THEY MIGHT BE LIKE TODAY

Scenario #1 of what PSR B1257+12' exoplanets might be like, paints a fairly forbidding picture by our human standards. But is it correct? The definitive answer to that question is likely to take some time. There are people, though, who argue that Scenario #1 is quite mistaken and that PSR B1257+12's exoplanets may have been in the past and may still be, more like one of our Paradises than one of our icier versions of Hell.

Thus, it has been postulated[113] that the two outer exoplanets, PSR B1257+12 c and d, with masses around four times that of the Earth (Super-Earths) would have been able to retain atmospheres and so have had much higher surface temperatures than those we encountered in Scenario #1:

> "We have shown that the harsh environment around neutron stars can still accommodate planets with warm atmospheres, provided they are Super-Earths. … if part of the pulsar power is injected in the atmosphere, all three of the PSR B1257+12 planets may lie in the habitable zone.[114] … the two Super-Earths may have retained their atmosphere for at least a hundred million years … with the atmosphere possibly still being present to these days. … if a moderately strong planetary magnetosphere is present, the atmospheres can survive strong pulsar winds and reach survival timescales of several billion years."

Most of the exoplanets' details for this scenario remain the same as those in Scenario #1. The main difference between the two models is the contention that atmospheres around two (or three) of the exoplanets may be retained. The temperature of the neutron star is assumed to have fallen to ~6,000 K; lower even than the temperatures assumed in Scenario #1 and making the exoplanet's heating from the pulsar similar to that which the Earth would receive from the Sun if the Earth's orbit were 100 AU in radius (i.e., > three times the distance of Neptune from the Sun).

---

[112] For example, Larry Niven's *Flatlander* short story in *Neutron Star*, Futura Publications, 1968, although this was actually anti-matter helium and Niven did not explain the chemistry involved. Arthur C. Clarke's *Crusade* short story in *The Wind from the Sun*, Corgi, 1974. Clarke did invoke electric currents flowing 'forever' in superconductors as a life-force.

[113] Patruno A., Kama M. *Neutron Star Planets: Atmospheric Processes and Irradiation*. Astronomy and Astrophysics, A147, **608**, 2017.

[114] Chapter 13.

Additional sources of heating for the exoplanets may be the pulsar wind (very high-speed particle emissions and the γ rays which those particles produce in the upper atmospheres) and x-rays and γ rays from the neutron star. The concept, in particular, takes account of the heating of the atmosphere by those x- and γ rays which penetrate deeply enough into the atmosphere for their energies to be retained, rather being lost through evaporating the outer layers of the atmosphere into space.

Once the pulsar wind comes to an end; in about a billion years for milli-second pulsars, the exoplanets will cool down rapidly and move out of the habitable zone. Under some circumstances, it is suggested that other energy sources; heat leaking from the interior of the exoplanet, energy from accretion of interstellar medium material by the neutron star or tidal effects might continue to heat the exoplanets atmospheres and surfaces. At the pessimistic end, the authors give lifetimes for the atmospheres of a million years (Earth-mass exoplanets, 1 AU from their neutron star), at their most optimistic (Super-Earths with very thick initial atmospheres and continuing energy sources), they suggest that lifetimes of 1 trillion years might be possible.

CONCLUSION

So, what are PSR B1257+12's exoplanets *really* like? – icy cold hells, or nice warm home-from-homes?, … or … ?

For PSR B1257+12 c and d (the ~4 M_{Earth} Super-Earths), assuming that scenario #2 has some validity, they may have spent some time within the neutron star's habitable zone. They might still be there now, but it seems more likely that they are now too cold for terrestrial-type life. Whether life of any type (Chapter 13) ever originated on them and if it still survives in some form is quite unanswerable at the moment, as is whether some space-travelling ETs from elsewhere have landed on them and explored or colonised them. The author's 'gut feeling' is that all these possibilities are of very low probability – but that they are not of zero probability.

If Scenario #1 is nearer the actuality, then the probabilities of life-forms seem likely to be several orders of magnitude closer to zero (but still *not* zero).

For PSR B1257+12 b, although the original suggestion for scenario #2 holds out the remote possibility that it could have had an atmosphere for an appreciable time, its mass (0.02 M_{Earth}) is so low and the resulting gravitational field is so weak that it seems very implausible. Most likely PSR B1257+12 b never acquired a significant atmosphere or lost it very quickly. The cold, airless, blasted cinder described in Scenario #1 would thus seem to be its most credible fate.

In the end, 'you pays your money and you makes your choice', or, like me, you sit back with a glass of your favourite tipple and await developments.

ALL THE REST

As mentioned earlier, to date, only another four pulsar exoplanets have been found (Table 5.2).

TABLE 5.2
The Other Pulsar Exoplanets – Measured Properties

Name	Orbit – Semi-major axis (AU/m)	Orbital period (Days)	Mass (M_{Earth}/kg)	Pulsar period (s)	Discovery date
PSR B1620-26 b	23/3.4 × 10¹²	~100 years[115]	795/4.8 × 10²⁷	0.011 0	2003
PSR J1719-1438 b	0.0044/6.6 × 10⁸	0.09	>383/>2.3 × 10²⁷	0.005 8	2011
PSR B0329+54 b	10.3/1.5 × 10¹²	10140	2.0[116]/1.2 × 10²⁵	0.7145	2017
PSR J2322-2650 b	0.01/1.5 × 10⁹	0.32	252.6/1.5 × 10²⁷	0.0035	2017

---

[115] The orbit is around a pulsar-white dwarf binary system. The pulsar is also very old (~12.7 billion years), so this pulsar exoplanet may be rather different from the others.
[116] M sin i.

Apart from PSR B0329+54 b, these exoplanets are all similar to Jupiter in their masses. It would seem that such large planets could not have survived from prior to the supernova explosion unless they themselves were previously small stars or brown dwarfs, so their existence somewhat supports the re-condensing models for their formation. PSR B0329+54 b probably resembles the PSR B1257+12 exoplanets, except for its much larger distance from its pulsar. The diameters of the other three exoplanets would likely be similar to that of Jupiter (140,000 km) if they had similar compositions to Jupiter, but if their compositions are the heavier elements and compounds, as suggested for the smaller exoplanets, then gravitational compression could have made them a good deal smaller.

One suggestion, therefore, for PSR J1719-1438 b, is that it is mainly composed of carbon and oxygen and has a density 23 times that of water. This density would make its diameter ~60,000 km; just a little larger than Uranus and Neptune and it would have a surface gravity around 17 times that of the Earth. Humans would have weights on such a planet of over a tonne and so we would probably not wish to colonise it, even if we could. However, ETs built like rhinoceroses on steroids might like it, though they could get dizzy from whirling around their pulsar every 2 hours.

PSR J2322-2650 b is classed as a gas giant (like Jupiter) in some sources but, being so close to its pulsar, this seems somewhat unlikely. Beyond that, the natures of these massive exoplanets are unknown.

As already mentioned, the pulsar exoplanets are to be considered as a separate class from the normal exoplanets and the latter will be our main concern. They are also expected to be <0.5% of all exoplanets which we have discovered and will discover. Apart from occasional inter-comparisons, we shall therefore concentrate on the latter henceforth.

6 1st November 1995 – The FIRST NORMAL Exoplanet

'Normal' is probably a misnomer in this context because Exoplanets are prodigiously varied in their natures. Since I intend to be master[117] in this book, however, I shall define it to mean *all exoplanets which are not pulsar exoplanets* – and it is these *normal* exoplanets which we shall, in due course, consider for visitation, occupation, colonisation by ourselves and/or consider whether ETs and/or ETIs might inhabit them.

As mentioned earlier, the first normal exoplanet (51 Peg b) was discovered in 1995 by Michel Mayor and Didier Queloz using the Elodie[63] spectroscope[118] (Box 6.1) and the 1.93-m Telescope at the Observatoire de Haute Provence in France. The discovery approach is called the Radial Velocity method and it looks for variations in the velocities of stars away from or towards the Earth.[119]

BOX 6.1 – SPECTRA, SPECTROSCOPY AND SPECTROSCOPES

Most people will have seen a rainbow at some time. If not in the sky after a rain storm has passed over (Figure 6.1), then, perhaps, in the spray generated by waterfalls or even just in the spray from a garden hose or decorative fountain in a pond or lake. These beautiful natural phenomena are examples of spectra, situations wherein the white light, by which we normally see, has been revealed to be made up of the seven colours:

Red, orange, yellow, green, blue, indigo and violet

If you want to remember these colours, then there are many mnemonics:

"Richard of York Gave Battle in Vain"

or

"Respect Others, You Grow By Including Variety".

and so on.

In practice, many people (including the author) have never managed to see the 'indigo' bit of the rainbow, but perhaps did think there was a different colour between the green and the blue, a shade which is now given the name 'Cyan'. The colours of the rainbow are thus more correctly given as:

Red, orange, yellow, green, cyan, blue, and violet

---

[117] "When I use a word, Humpty Dumpty said in rather a scornful tone, 'it means just what I choose it to mean — neither more nor less. … The question is,' said Humpty Dumpty, 'which is to be master — that's all' ". *Alice through the Looking Glass*, Lewis Carroll, Macmillan, 1871.

[118] a.k.a. spectrographs.

[119] More usually called the radial velocity or line of sight velocity. By convention it is negative for approaching stars and positive for receding ones. Elodie was able to measure the stars' velocities to a precision of ± 15 m s^{-1}. Currently, state-of-the-art spectrographs like the KPF (Keck Planet Finder) now measure to ± 0.3 m s^{-1} (~ ⅔ miles per hour, a very slow walking pace for a human).

DOI: 10.1201/9781003246459-7

FIGURE 6.1   A natural rainbow after a rain storm. (© C. R. Kitchin 2010. Reproduced by permission.)

If you want to stick with the older definition, though, that is fine.

A rainbow, howsoever beautiful it may be, is a very poor-quality spectrum. Even 350 years ago, Newton was producing much better spectra with a simple glass prism and used the Latin word 'spectrum' to describe what he saw. Better optics and new devices such as diffraction gratings[120] soon improved on Newton's very simple spectroscope immeasurably.

The second and third crucial discoveries about the spectroscopic phenomenon were by made Thomas Melville in 1752 and William Wollaston in 1802. Melville discovered that various salts, when sprinkled into a flame, emitted yellow light, and this, when seen through a spectroscope, was a bright yellow emission line[121] against a dark background. He was thus the first to see the spectrum characteristic of the chemical element sodium (present as sodium chloride in his salt mixtures) – and, indeed, he was the first to see the spectrum characteristic of any chemical element.

Wollaston's discovery was the inverse of Melville's, that of dark lines within the bright solar spectrum (Figure 6.2). Despite an initial suggestion that these dark lines were the divisions between the basic colours, it soon became apparent the dark lines in a bright spectrum were the *exact* inverses of Melville's bright lines against a dark background. That is to say, if an element, when heated, emitted a bright line at a particular point in the spectrum,[122] then it

---

[120] Going much further into the optics of spectroscopes would be a diversion too far for this book. However, an internet search for 'diffraction gratings' will rapidly produce far more information than the reader probably desires. The interest reader could also consult two of the author's other books: *Astrophysical Techniques* 7th Edn., CRC Press, 2020, and *Optical Astronomical Spectroscopy*, Institute of Physics, 1995.

[121] Although the word 'line' is invariably used in this context, the word 'feature' would be much more appropriate. The spectrum features only appear as lines because the entrance apertures to most spectroscopes are narrow slits (i.e., of a linear or line shape). If the shape of the entrance aperture slit was bent into (say) an 'S' profile, the spectrum features would also have 'S' shapes.

[122] The ' … particular point in the spectrum' is measured by the wavelength and/or frequency of the light at that point. For light waves, its usual unit is nanometres (nm) and their wavelength values are shown in Figure 6.2 at the top of each spectrum. Frequency is also used as the measure, though normally at the much longer microwave and radio wavelengths. As an example, though, the sodium D1 Line (Figure 6.2 (ii)), whose wavelength is 589.6 nm, could equally well be identified by its frequency of 5.085×10^{14} Hz.

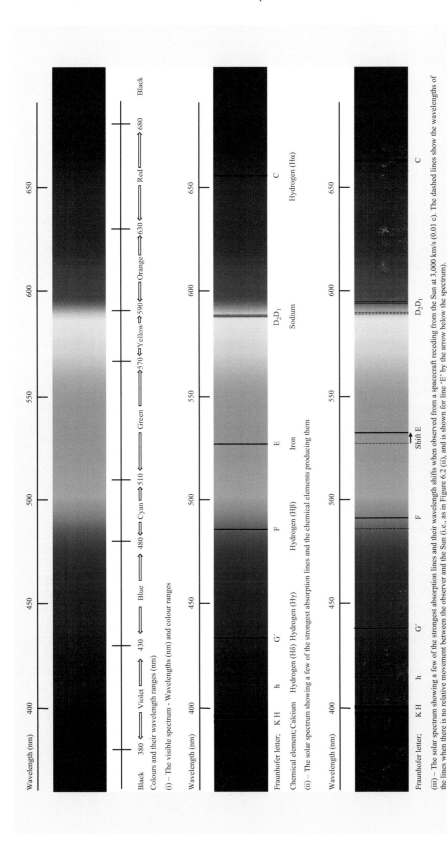

FIGURE 6.2 Spectra. (i) A colour drawing of a continuous spectrum over the visible region showing the natural colours and their wavelength ranges. (ii) A schematic of the prominent absorption lines in the visible Solar spectrum showing their Fraunhofer[123] and other identities and the chemical elements producing them.

---

[123] The labelling system started by Joseph von Fraunhofer in 1814 by his labelling of seven prominent solar absorption lines with the letters A to G.

FIGURE 6.2 (CONTINUED) N.B., the visible solar spectrum actually contains many tens of thousands of such absorption lines. This only shows the very strongest ones. (iii) As discussed below and elsewhere (see Doppler effect), when a source of waves, such as electromagnetic radiation, including light, moves away from an observer (or vice versa), the wavelengths of the observed emissions are increased. When the source moves towards the observer, the observed wavelengths are decreased. In this illustration, an observer, with a spectroscope, is imagined to be moving away from the Sun at 3,000 km/s (0.01 c). The lines shown in (ii) are here shown at their shifted positions corresponding to this speed. The unshifted positions are indicated by the dotted lines. Measuring the wavelength difference between the unshifted and shifted position of a line or lines enables the relative velocity between the source and observer to be calculated. (© C. R. Kitchin 2022. Reproduced by permission.)

would absorb light at the same point when the element was illuminated by a bright continuous spectrum (Figure 6.2 (i)) and a dark line would then be seen across the spectrum at that point (Figure 6.2 (ii)).

It is these spectrum lines (absorption and emission) which make spectroscopy one of the most powerful tools in the scientific toolbox. It may already be apparent that if particular spectrum lines appear in the emission spectrum of a substance being heated, then the element producing those lines must be present within that substance and thus spectroscopy is of major importance to almost all forms of chemical analysis. The dark lines in the solar (and stellar, planetary, interstellar medium, galactic, etc., etc.) spectra (Figure 6.2 (ii)) permit the apparently magical ability of Astronomers to measure the composition of stars and galaxies billions of light years away from the Earth. Since light is a wave, the Doppler shift (Chapter 4, below and Figure 6.2 (iii)) which causes the lines to change their wavelengths when the emitting object moves towards or away from the Earth from their normal values and so gives Astronomers the equally apparently magical ability to determine the speeds[119] with which objects are moving through the outer reaches of the Universe.

However, just as we're getting interested again, this time in spectroscopy, I have to refer the reader to other sources[120] for further information, since otherwise we would stray too far from this book's remit.

In 1994, the Elodie instrument was state-of-the-art and could obtain spectra and measure them for radial velocities to a precision of about ±10 metres per second. Mayor and Queloz started observing a star some 51 ly away from us which is just about bright enough to be seen with the unaided eye (if you've got good eyesight and it is a clear and dark night). The star is part of the Pegasus constellation and is called 51 Peg (Figure 6.3 (i)). It is very similar to the Sun, save that it is some 25% larger and so is classed as a sub-giant star against the Sun's main sequence or dwarf classification. It is currently approaching us at about 33.33 km/s.

Mayor and Queloz used Elodie to make measurements of 51 Peg's radial velocity from about September 1994 onwards and within about six months they had sufficient data to be sure that 33.33 km/s average approach velocity increased regularly to 33.40 km/s then 2.1 days slowed to 33.26 km/s before returning to 33.40 km/s another 2.1 days later (Figures 6.3 (ii) and 6.3 (iii)).

Analysis of the velocity changes showed that the star, 51 Peg, was being orbited (and so pulled towards and away from us every 4.2 days) by an object whose mass was about 46% that of Jupiter. An object with such a mass could not be a star or brown dwarf (Box 4.2) – so it *must* be a planet – now called 51 Peg b.[124] Now Jupiter itself is 5.2 AU out from the Sun and takes 11.86 years to

---

[124] Also given the proper name 'Dimidium'; from the Latin for 'half', since it is almost half Jupiter's mass.

complete one orbit. The radial velocity changes for 51 Peg b repeat every 4.2 days and so that must be *its* orbital period. Astonishingly, therefore, 51 Peg b is just 0.053 AU out from its star, a seventh of Mercury's distance from the Sun. Instead of the icy cold exoplanets found around PSR B1257+12 (Chapter 5) Mayor and Queloz had found an exoplanet with a surface temperature close to 1,280 K (1,010 °C) ®; more than hot enough to evaporate sodium and zinc into gases (Figure 6.3 (iv)). Furthermore 51 Peg b is nearly 75 times more massive than PSR B1257 c – the most massive of those three pulsar exoplanets (Chapter 5).

51 Peg b could not be much less like the PSR B1257+12 exoplanets discussed in Chapter 5, if it tried. In fact, it is probably much less inhabitable by humankind life-forms than the PSR B1257+12 exoplanets would be. At least with adequately heated space suits, we could survive and explore those pulsar exoplanets. Though it seems doubtful (Chapter 13), there may be ET life-forms capable of living at 1,280 K. For ourselves and others like us, sometime in the future, we *might* be able to design machines which *might* be able to survive the temperatures (and pressures) on 51 Peg b ✿✿. Spacesuits capable of doing the same, if not quite totally impossible are probably pointless (what would some future Neil Armstrong *do* after taking his/her next great leap for mankind into a pool of boiling zinc?).

Despite these inhospitable and unusual properties, Mayor and Queloz announced their discovery at a conference in Florence in October 1995 and in just a few days; by mid-October, Geoff Marcy and Paul Butler using the 3-m telescope at Lick observatory in California had confirmed it.

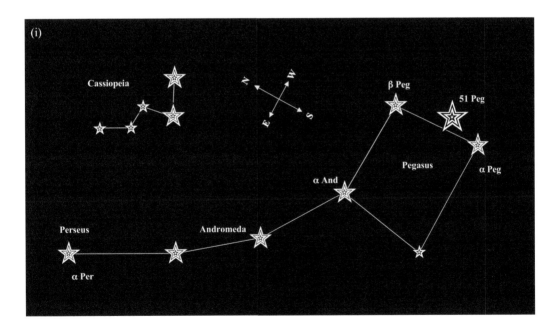

FIGURE 6.3 (i)   51 Peg and 51 Peg b. A star map showing the location in the sky of 51 Peg; about halfway between α and β Peg.

N.B.: The star symbol for 51 Peg is shown as larger/brighter than those for the other major stars in the region, but this is just to highlight its whereabouts. 51 Peg is actually only about 4% as bright as α Peg.

(© C. R. Kitchin 2022. Reproduced by permission.)

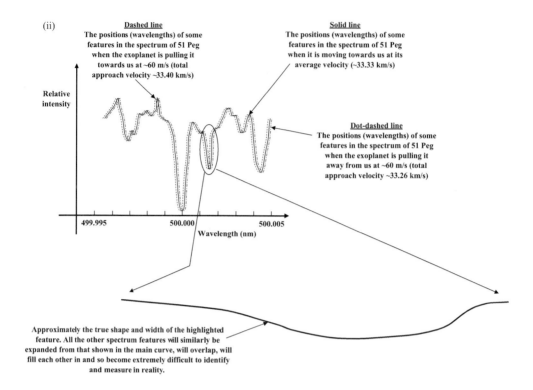

FIGURE 6.3 (ii)   A schematic plot of the intensity within the spectrum of 51 Peg near the 500 nm part of the spectrum (cyan). The star and its exoplanet are in mutual orbits around their barycentre. So, at times, the star (51 Peg), in addition to its normal ~33.33 km/s approach velocity (solid line) towards the Earth, has that velocity increased by its ~60 m/s orbital motion (dashed line) and so the wavelengths decrease slightly (Box 6.1). Half an orbit later (2.1 days) and the exoplanet pulls the star away from us at ~60 m/s (dot-dashed line), increasing the wavelengths.

N.B.: For clarity as to what is happening, the spectrum features have been drawn as being very much sharper (narrower) than actuality. The wavelength changes are shown accurately to scale. In actuality, though, the spectral features in this part of the star's spectrum are some ~30 times wider than shown, making the problem of measuring the Doppler shifts hugely more difficult than it might appear from this illustration (see expanded detail at the bottom of this diagram).

(© C. R. Kitchin 2022. Reproduced by permission.).

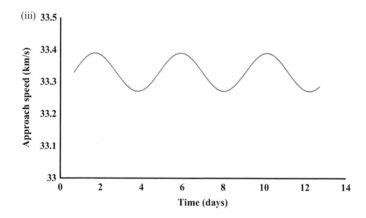

FIGURE 6.3 (iii)   A schematic plot of the velocity variations of 51 Peg due to the varying directions of the gravitational pull from its exoplanet, 51 Peg b.

(C. R. Kitchin 2022. Reproduced by permission.)

FIGURE 6.3 (iv) 51 Peg and its exoplanet, 51 Peg b – an artist's impression of their possible close-up appearances. Given, though, the enormous differences between the temperatures of 51 Peg b and Jupiter (1,280 K and 88 K), it seems unlikely that 51 Peg b will actually look quite so much like Jupiter. Far more and far more violent meteorological activity would seem more credible; say, Jupiter with a couple of dozen red spots. (Reproduced by kind permission of ESO/M. Kornmesser/Nick Risinger (skysurvey.org).).

24 years later[125] Mayor and Queloz were sitting down in Stockholm to a well-deserved banquet of:

Duck stuffed with black chanterelles and lemon thyme, potatoes with caramelised garlic, spiced pickled golden beetroot and a roast duck jus. Served with Savoy cabbage, smoked shiitake, baked onion and spruce oil.
Washed down with
Taittinger Folies de la Marquetterie and
Lunadoro Vino Nobile di Montepulciano DOCG Pagliareto 2015[126]

in celebration of their quarter-shares of the 2019 Nobel Prize for Physics arising from their 51 Peg b discovery.

---

[125] Actually, quicker than the average for the award of the Physics Nobel Prize.
[126] *Nobel Prize menu* site at https://www.nobelprize.org/ceremonies/nobel-banquet-menu-2019/.

7 Après Moi, le Déluge[127]

Moi, in this case, was the discovery of 51 Peg b and a *Déluge* did indeed follow it; with some 5,580 additional, normal exoplanets discovered since 1995.

Chapter 4 listed the main methods used to date for discovering/detecting exoplanets. Thus far we have looked into how Astrometry, Pulsar timing and Radial velocity measurements do that job. Of the 13 discovery methods listed, though, most have 'only' discovered a very small number of exoplanets (Table 7.1). Prominent amongst the discovery methods is the Radial velocity approach, claiming nearly 20% of the discoveries, but, way out in front, is the Transit discovery method. The latter has made the initial discovery of over three-quarters of all currently known normal exoplanets (Table 7.1). In this book, we are concerned more with the natures of the discovered exoplanets rather than the method of their discovery. So, we will not look into further details of those methods producing very few discoveries.[128]

TABLE 7.1
Proportions of exoplanets discovered by various methods

Discovery method	Proportion of all discoveries (%)
Transit	75.33
Radial velocity	19.52
Gravitational microlensing	2.70
Direct imaging[129]	1.17
Transit timing variations	0.44
Eclipse timing variations	0.35
Orbital brightness variations	0.17
Astrometry	0.13
Pulsar timing variations[130]	0.13
Pulsation timing	0.04
Disc kinematics	0.02
BEER	0.02[131]
Polarimetry	0.00[132]

---

[127] After Me, the Flood. (Also 'Après Nous, le Déluge' – 'After Us the Flood'.)

Attributed to the French king, Louis XV (1715 – 1774) and/or Madam de Pompadour (1721–1764). Nowadays the phrase has, *inter alia*, the meaning: '... now that I've set this thing in motion, there will be a torrent of follow-on imitations' and that is the intended implication here.

Louis, though, perhaps used it in the mournful expectation of further disasters (he had just ruinously lost the Battle of Rossbach to the Prussians) which would come after the passage of Halley's comet and which Halley had predicted would occur in 1758, the following year.

[128] Readers interested in finding out further details of some or all of these small-hitters will find plenty of material via an internet search or in some of the author's other books; *Astrophysical Techniques*, 7th Edn., CRC Press, 2020, *Exoplanets; Finding, Exploring and Understanding Alien Worlds*, Springer, 2011.

[129] A concept study at the time of writing, is envisaging the HWO (Habitable Worlds Observatory); hoped to be capable of '... directly imaging an Earth-twin around a sun-like star ...', so Direct Imaging could be making a bid for the lead in this table in 50(?) years from now.

[130] As discussed in Chapters 5 and 6, these are not regarded as normal exoplanets. They are included in this table for completeness.

[131] Beaming, Ellipsoidal and Reflection; detection of exoplanets from brightness variations outside transits due to relativistic and other small effects.

[132] Polarised light from exoplanets has been detected, but none have yet been discovered via such observations.

DOI: 10.1201/9781003246459-8

TABLE 7.1 (CONTINUED)

Proportions of exoplanets discovered by various methods

Discovery method	Proportion of all discoveries (%)
White dwarf atmospheres	0.00[133]
Interferometry	0.00[134]

Since we have looked into how the radial velocity discovery method works in Chapter 6, that leaves only the transit discovery method, therefore, now to be looked at in more detail. Of course, once an exoplanet *has* been discovered and by whatever method, some of the other observing methods will then join in to determine its properties in as much detail as possible.

THE TRANSIT DISCOVERY METHOD FOR EXOPLANETS

If you have ever seen a partial solar eclipse (Box 7.1), or even better, a transit across the Sun of Mercury or Venus (Figure 7.1), then you already know some 99% of all there is to know about the Transit Discovery Method for Exoplanets.

BOX 7.1 – APPULSES, CONJUNCTIONS, ECLIPSES, OCCULTATIONS, OPPOSITIONS, SYZYGIES AND TRANSITS

All these terms are used to denote alignments, or near-alignments, of three astronomical bodies in space and/or in the sky. Syzygy is the most general of the terms, meaning just any alignment of any three objects. The remaining terms are special cases of alignments for particular astronomical objects and usually the Earth is one of the three objects involved.

- Appulse – When any two objects in the sky, at least one of which is moving with respect to the fixed stars, are fairly close to each other in the sky. The two objects, plus the Earth are then near to being in a straight line in space.
- Conjunction – When any two objects in the sky, at least one of which is moving with respect to the fixed stars, have the same right ascension or the same celestial longitude. In principle, the two objects' declinations or celestial latitudes could be very different, but the term is usually only applied when the declinations or celestial latitudes have similar values. Any pair of objects can be in conjunction (with the Earth forming the third body), but the term is particularly used when one of the other two objects is the Sun. In the latter case, the order Earth-Object-Sun is termed Inferior Conjunction, whilst the order Earth-Sun-Object is Superior Conjunction.
- Eclipse – When two objects of similar angular sizes in the sky are almost exactly aligned. The more distant of the objects is then completely, or nearly completely, obscured. A slight miss-alignment from a straight-line results in a partial eclipse. A lunar eclipse seen from the Earth is not strictly an eclipse, although it is a syzygy. Were an astronaut on the surface

---

[133] This is not a discovery method for currently existing exoplanets. Instead, the previous existence of an exoplanet orbiting the white dwarf is revealed by the presence in the white dwarf's atmosphere of unusually high levels of elements, such as silicon, which are likely to have comprised the exoplanet's mantle, etc. These elements may have formed part of an exoplanet(s) and/or exoasteroid(s) which has broken up and crashed into the white dwarf some time ago.

[134] The Darwin concept was a spacecraft-based NIR interferometer system designed to observe exoplanets directly by combining the outputs from several telescopes in such a manner as to cancel the star light, whilst leaving the exoplanets visible (see 'Nulling Interferometers' if you wish to research this topic further). The distances between the several spacecraft would have needed to be controlled to within a nanometre, though, so it is probably not feasible at present.

of the Moon to witness the event, though, then it could, quite correctly, be called a solar eclipse seen from the Moon, with the Earth being the obscuring intermediate object.

- Occultation – When an object with a large angular size in the sky passes in front of an object with a much smaller angular size, obscuring the latter completely.
- Opposition – When any two objects in the sky, at least one of which is moving with respect to the fixed stars, have right ascensions or celestial longitudes which differ from each other by close to 180°, i.e., they are on opposite sides of the sky as seen from the Earth. The term is usually used for planets and asteroids and the like so that the syzygy is Sun-Earth-Object and then it indicates that the planet (etc.) is near to its closest approach in space to the Earth.
- Transit – The inverse situation of an occultation. The angularly smaller object passes in front of the angularly larger object and so may be seen in silhouette (Figure 7.1).

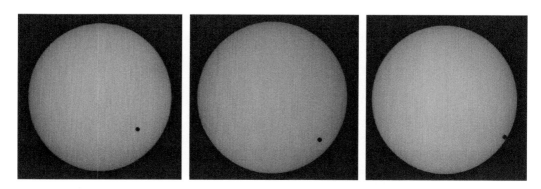

FIGURE 7.1   The transit of Venus across the Sun on 8 June 2004. Three images showing the last hour of the transit as Venus moves across the solar disc. Venus is the small, round, black silhouette towards the bottom right of the solar disc. The full duration of the transit was 6 h and 12 m. The last transit of Venus was on the 5th/6th of June 2012, and the next will be on 10th/11th of December 2117. (© C. R. Kitchin 2004. Reproduced by permission).

Figure 7.1 shows just what a transit looks like, whether it be of Venus across the Sun's disc or an exoplanet many light years away passing across the disc of its host star. Of course, we cannot (yet 🌀🌀🌀) resolve the details of the event for the exoplanet; it just appears star-like.

There is though, an aspect of the transit which we can detect and this can be deduced from the images in Figure 7.1. Where we see the black silhouette of Venus, it is obscuring a part of the solar surface and so we no longer receive energy from that part of the Sun. If we were to measure the total energy which we receive from the Sun, it would therefore reduce slightly[135] whilst the transit was in progress. So, whilst we cannot see the exoplanet silhouetted against the disc of its star, we can measure the energy which we receive from the star and exoplanet system and find that it, too, reduces slightly during the transit.

In fact, this is cart-before-the-horse stuff; we do not see the transit and so deduce that the energy which we receive has reduced. What we actually do is to see that the energy which we receive from

---

[135] By about 0.027% – if you try to check this, remember that when Venus is in transit, its distance is about 0.28 AU from the Earth. The black dot which we see is therefore covering an area of the Sun nearly 13 times larger than the area of the actual cross-sectional area of Venus. A similar geometrical situation means that the Moon, with a physical diameter of 3,475 km, has an angular diameter sufficient to obscure the whole solar disc (physical diameter 1,393,000 km) during a solar total eclipse. For exoplanets, the relative distance between the planet and its host star is so small compared with the distance from the Earth to the exoplanet system, that the exoplanet obscures an area of the star's surface which equals the cross-sectional area of the planet.

the host star has reduced and so deduce that a transit by an exoplanet is in progress – and that is really *all* there is to detecting exoplanets via the transit detection method.

The principle of the Transit Detection Method for exoplanets is thus almost trivially simple; just monitor the brightness of a star for a year or two and see if it fades somewhat at times. The practice, though, requires very considerably more effort on the part of the observer – as we shall see next.

IT IS ALL VERY CONFUSING

Events other than an exoplanet's transit can also reduce the energy which we receive from a star. Thus, the star may have large star spots (c.f. sunspots), it may be intrinsically variable or it may be a member of a binary system wherein the companion is not a planet (i.e., the companion could be a faint brown dwarf or a condensation in a part of a disc of material surrounding the star). Generally, though, the light curve for an exoplanet transit (Figure 7.2 (i)) has a characteristic boxy shape which distinguishes it from the light curves for these other phenomena (Figure 7.2 (ii)).

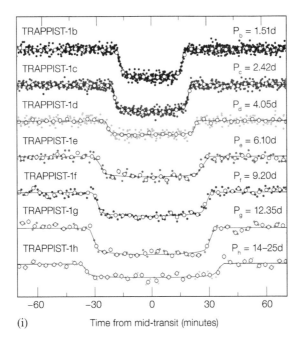

(i) Time from mid-transit (minutes)

FIGURE 7.2 (i) Transit light curves for the seven exoplanets of the host star TRAPPIST-1[136] (c.f., the first light curve in part (ii) of this Figure). The observations for these light curves were made in the infrared (3.6 μm) using the Spitzer space telescope[137]. (Reproduced by kind permission of ESO/M. Gillon *et al.*).

---

[136] Transiting Planets and Planetesimals Small Telescope. Actually, TRAPPIST comprises two 0.6-m robotic telescopes. One of these is located in Morocco at the Oukaïmeden Observatory in the Atlas Mountains, and the other at ESO's La Silla Observatory in Chile. Both telescopes are remotely controlled and operated from Liege (Belgium). The star TRAPPIST-1 was first detected in 1999 during the Two Micron All-Sky Survey (2MASS) project and is a very cool (2,570 K) red dwarf with a mass about 9% that of the Sun. It is about 41 ly from the Earth in Aquarius.

[137] Launched 25th August 2003. Deactivated 30th January 2020. It used a 0.85-m Ritchey-Chrétien IR telescope which was cooled to 5.5 K for the first five years and eight months of its lifetime (after which its liquid helium coolant supply ran-out).

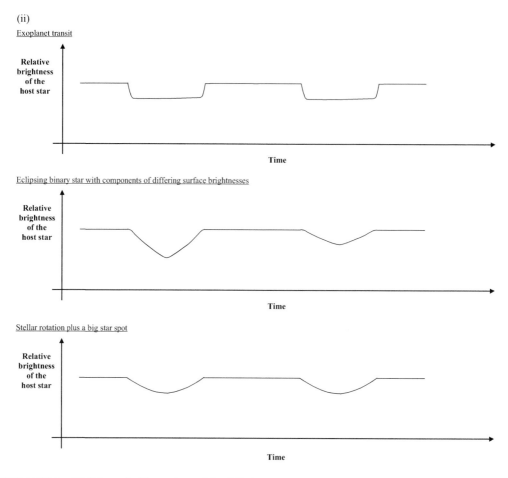

FIGURE 7.2   (ii) Schematic illustrations of the differing shapes of light curves for assorted periodic variable stars;

- o An exoplanet transit – the boxy, flat-bottomed shape (see also part (i) of this figure) arises because the exoplanet obscures a constant and usually small area of the disc of the host star except, briefly, at the start and finish of the event. (© C. R. Kitchin 2022. Reproduced by permission.)
- o An eclipsing binary star with components of differing surface brightnesses. The constantly changing brightness arises because a large and constantly changing area of the eclipsed star is obscured throughout each eclipse. The differing depths of the eclipses occur because, although the areas obscured are equal each time, the surface brightnesses of the stars differ. The depth is thus greater when the star with the higher surface brightness is partly obscured. N.B., if the eclipses were total and annular (i.e., the stars of different sizes), the light curve would resemble that of a transiting exoplanet, with flat bottoms to the curves, however the differing depths for the eclipsing binary events would still reveal the true nature of what was happening. (© C. R. Kitchin 2022. Reproduced by permission.)
- o Stellar rotation plus a big starspot. The continuous variation in intensity throughout the event arises because the visible area of the starspot increases and then decreases as the spot is rotated across the stellar disc; being seen edge-on at the start and finish and face-on around the mid-point of the event[138]. (© C. R. Kitchin 2022. Reproduced by permission.)
- O Cepheids and related variable stars. These and other variable stars change their brightnesses through physical changes in the natures of the stars themselves; Cepheid variables expand and contract in size, for example. The light curves come in many shapes, but will usually reveal their true natures by not maintaining the rigid constancy of repetition of the events shown by transiting exoplanets, etc. wherein the orbital period or stellar rotation period governs the timing of the events. (© C. R. Kitchin 2022. Reproduced by permission.).

---

[138] For the mathematically minded reader, the shape of the curve is part of a sine wave.

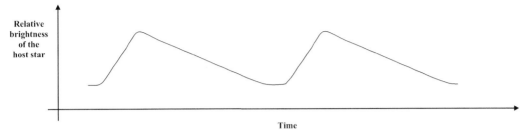

Cepheids and related variable stars

FIGURE 7.2 (CONTINUED)

It takes a long time

From the caption to Figure 7.1, we may see that there will be no transits of Venus for the 105 years between 2012 and 2117. That interval can even reach 121 years, but it can also be as short as 8 years. This variation in the interval between transits arises because we observe Venus from the Earth and the Earth (as well as Venus) is orbiting the Sun.

Were we able to observe from a spot that is more-or-less stationary with respect to the Sun and from where we could see Venus transit the Sun, such as from close to a distant star, then the transits would occur regularly at intervals of 0.616 years (Venus' orbital period around the Sun). Still, even 0.616 years is a long time to watch on the off-chance of a transit happening and were it to be Neptune instead of Venus, the wait could be 165 years. Furthermore, until you have seen a transit, you do not know if you are in the right place to see a transit, nor do you even know if there is a planet there to do the transiting.

The solution to the 'long wait' problem is to monitor many stars at a time, not just one (see next section). The waits, though, can still be quite lengthy. Thus, TESS (next section) monitors about 200,000 stars ® and currently averages about one confirmed exoplanet transit discovery every five days. The use of machine learning/artificial intelligence algorithms to make at least the initial identifications of possible transits within the vast amounts of data produced by both terrestrial and space-based transit searches is becoming increasingly common and necessary at the time of writing.

When we look out into the Universe therefore and monitor a star's brightness to see if it has an exoplanet which will obliging transit the star for us, we may well have a long wait – perhaps even an infinite wait. There is consequently a bias in the observations towards discovering short orbital period exoplanets since their transits are much more likely to occur during any given observational period than are the longer ones. It is also usually necessary to observe several transits before a detection can be confirmed – and this strengthens the bias a great deal more.

Accordingly, of the transiting exoplanets known at the time of writing (Figure 7.3), half had orbital periods of eight days or less. For a solar-type host star and an Earth-mass exoplanet, an 8-day orbit would mean that the host star-exoplanet separation would be just 11,700,000 km (0.078 AU). For a 0.1 M_{Sun} red dwarf as the host star, the orbital radius would be around 5,400,000 km (0.036 AU).

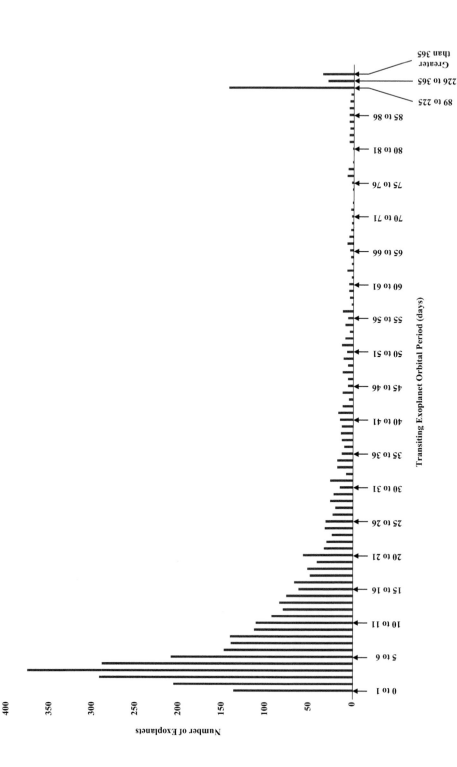

FIGURE 7.3 A histogram of the orbital periods of transiting exoplanets in 2022. Except for the last three vertical bars, each bar gives the number of exoplanets with orbital periods in a particular 24-hour time slot. Thus, for example, 111 exoplanets have orbital periods between 10 and 11 days and this is indicated as '10 to 11' on the horizontal axis.

For the three right-most bars, the numbers of exoplanets from several 24-hour time slots have been binned to give the numbers whose orbital periods lie between those of Mercury, Venus and the Earth. These are labelled: 89 to 225 (Mercury to Venus), 226 to 365 (Venus to the Earth) and greater than 365 (periods longer than 1 year). (Based upon data from *The Extrasolar Planets Encyclopaedia* (http://exoplanet.eu/.) (© C. R. Kitchin 2022. Reproduced by permission)

The current record holder for the shortest period for a transiting exoplanet[139] is ZTFJ0220+21 b, the orbital period of which is just 53 minutes. The mass of ZTFJ0220+21 b is estimated to be around 14 $M_{Jupiter}$ and that of its white dwarf host star is estimated at 0.8 M_{Sun}, resulting in their separation being just 300,000 km, quite a bit less than the distance between the Earth and the Moon. The host star has a surface temperature of around 14,200 K (13,900 °C) so any ET managing to live on ZTFJ0220+21 b would see it as covering an area of the sky some 20 times that covered by the Sun as seen from the Earth and having an apparent luminosity around 800 times that of the Sun.

The Solar System planet with the shortest orbital period, at 87.97 days, is Mercury, so 94% of the currently known transiting exoplanets have orbital periods shorter than that of Mercury (Figure 7.3). About 4% have periods between those of Mercury and Venus (224.70 days), 0.8% between those of Venus and the Earth (365.26 days) and just 0.9% take longer than one year to orbit their host stars.

An additional contributory factor to the observational bias will be that short period planets will be closer to their host stars and therefore their transits will be visible over a wider viewing angle than those of the more distant planets. Thus, a transit of the Sun by Mercury will be visible to all observers within a band around the sky 1.4° wide, whilst a transit of Neptune will be visible from a band around the sky just 0.002° wide, i.e., Mercury-type transits are 700 times more likely to be seen than Neptune-type transits because of this factor alone.

It perhaps needs emphasizing at this point, though, that this skewed distribution of transit periods, though largely due to observational bias, probably does have some reality as well. Current ideas on the formation of the Solar System and therefore also probably of exoplanetary systems (Box 7.2) suggest that small rocky planets will be produced in orbits close to the star. Whilst much larger (more massive) largely gaseous planets with much larger separations will form in the more distant orbits.

BOX 7.2 – FORMATION OF THE SOLAR SYSTEM AND OF EXOPLANETARY SYSTEMS

In brief, planetary systems are thought to originate alongside their host stars from an interstellar gas cloud collapsing under its own gravity. The material with lowest angular momentum (rotational energy) collapses all the way down to form the star. The material with higher angular momentum forms a thin flat disc orbiting the proto-star.

From the energy released during the collapse and/or once the star begins to radiate significant amounts of heat, the disc becomes warm close to the star and is colder further out. The material forming the disc, though, will be quite uniform in its composition; mostly hydrogen, some helium and then small quantities (<0.1% each) of all the remaining chemical elements. The radial temperature gradient within the disc, however, means that only the elements and compounds with the higher evaporating temperatures can condense to form solids and liquids in the inner part of the disc. These are the elements and compounds which go to form rocks, metals and the like – and they are present only in small quantities. The planets formed close to the proto-star are therefore small and rocky (e.g., Mercury, Venus, Earth and Mars within the Solar System).

Further from the proto-star, temperatures are lower and hydrogen-rich compounds, such as methane, can freeze out of the gas. Since hydrogen is very abundant, so are its compounds and much larger amounts of the material can liquefy and solidify than was the case for the inner part of the disc. This material quickly forms much larger (more massive) proto-planets than the small rocky ones nearer to the star. Those massive proto-planets eventually accumulate

---

[139] Several even shorter transit intervals have been detected but the companion's mass in these systems makes it almost certain to be a brown dwarf.

material to the point where their gravity can capture the still gaseous hydrogen and helium. Once this happens, those proto-planets grow very rapidly, like a snowball rolling down a snow-covered hill. The planets formed further out from the proto-star are therefore large and largely gaseous (e.g., Jupiter, Saturn, Uranus and Neptune within the Solar System).

Most theories of the origin of the Earth and Moon currently suggest that the proto-Earth underwent a collision with another protoplanet in a late(ish) stage of the formation of the proto-sun. That proto-planet has been named 'Theia' and some of the debris from the collision is suggested to have coalesced into the proto-Moon and some to have been absorbed into the proto-Earth. This may make the Earth unusual and therefore imply that life originating on it was in some way 'special'. However, it is often suggested that Theia was about the present size of Mars and the four inner planets would need some 20 or so Mars-sized proto-planets for their formation. Thus, the inner Solar System would have been crowded 4,500 million years ago[140] and collisions would have been quite likely. Collisions may even have been quite probable in the outer reaches of the Solar System; Uranus' very high tilt of its rotation axis is often attributed to a collision with an Earth-sized proto-planet.

Since these large outer planets will use up much more material than the small inner ones (we could have 317 Earths for 1 Jupiter), then, with a limited supply of material to form planets, it is likely that there will generally be many small planets close to the star and only a few large planets further out – thus also leading to more short period transits than long period ones. This latter supposition, however, is not a hard and fast rule – see the case of ZTFJ0220+21 b, above.

The Transit Method Needs **BIG** Telescopes, Yes?

Not quite true.

From their invention in 1608, optical telescopes (and telescopes for other parts of the spectrum) have continually grown in diameter; from a centimetre or two to ~10 metres. In about half a decade, that size will quadruple with the completion of the 39-m ELT.[141] The reason for this growth is twofold; bigger telescopes can collect more light and so fainter objects can be detected and larger diameter optics have better (finer/smaller) angular resolutions[142] and so are able to see smaller features.

This monotonic size-growth in state-of-the-art astronomical equipment is so ingrained in most astronomers' mind-sets that it can blind observers to the abilities of smaller-than-the-maximum instruments. Yet the invention of electronic imaging devices[143] which are hundreds-of-times-over improvements on the photographic imagers used almost to the end of the 20th century, means that an amateur astronomer's 0.2-m telescope with an electronic imager could nowadays compete on equal terms with a 2-m telescope as used only three decades ago – and there are still plenty of research topics that could be done today with a photographically-equipped 2-m telescope.

Thus, although light grasp and angular resolution matter and large telescopes *are* used to observe transiting exoplanets, a very great deal of the work on exoplanet transits is undertaken with small,

---

[140] Minerals within a meteorite named Erg Chech 002 and found in Algeria some 1,500 km south of Algiers in 2020 have been dated reasonably reliably (from radioactive aluminium-26 decay products) to being 4,565.6 million years old. These minerals, though probably formed in an early proto-planetesimal, before the Earth itself was 'properly' in existence and floated around the Solar System for billions of years until colliding with the Earth perhaps a century or more ago.

[141] European Large Telescope. Currently under construction at the top of Cerro Armazones in Chile

[142] This is only true if the telescope's optical quality is adequate and the atmospheric distortions do not degrade the image quality. Almost all large optical telescopes, however, are now equipped with adaptive optics which, in a limited way, correct for atmospheric distortions.

[143] Principally CCDs (Charge Coupled Devices), although many other types of electronic detector are also used. See the author's book *Astrophysical Techniques*, 7th Edn., CRC Press, 2020, other practical astronomy books or the internet for further information.

even tiny, instruments. If you are a keen photographer, some of your spare telephoto lenses could probably do the job – indeed, commercial, off-the-shelf telephoto lenses are *exactly* what some transit telescopes[144] *have* used for the job. For example, the two, now decommissioned, SuperWASP (Super Wide-Angle Search for Planets) instruments each used eight Cannon 200 mm f8 off-the-shelf telephoto lenses with apertures of 110 mm (Figure 7.4) and discovered 192 exoplanets. Most of SuperWASP's optics, though, have now been re-used in a new instrument monitoring spacecraft in geostationary orbits around the Earth.

FIGURE 7.4   SuperWASP. (Reproduced by kind permission of Prof. D. Pollacco.)

The currently operating NGTS[145] (Figure 7.5) uses 12 commercial (ASA[146]) 200 mm diameter, f2.8 robotic telescopes[147] in an array based at the Paranal observatory.[148] NGTS operates at the red end of the spectrum and just into the infrared in order to observe optimally the smaller and cooler stars. The combined fields of view of the component telescopes cover 96 square degrees (~490 times the area of the Full Moon) and they can measure stars' brightnesses to an accuracy of 0.015%. At the time of writing it had discovered 25 exoplanets, however, its observations are now mostly devoted to following up possible, but not yet confirmed, transits identified by TESS (see below). It is currently the most accurate ground-based, wide-angle, photometric instrument available and if several of its telescopes concentrate on one object then it equals and may even surpass TESS' abilities. The 400 mm ASTEP[149] telescope in Antarctica is being upgraded at the time of writing. It observes in the infrared where the contrast is likely to be higher between the (hot) host star and (cooler) exoplanet.

---

[144] There are two types of transit telescopes. The ones currently under discussion are used for observing the transits of exoplanets across the discs of their host stars. A much older, but still current, usage of the term is for the telescopes used to observe stars transiting the line (meridian) due south of the observer (due north in the southern hemisphere) for timing purposes and in the past, for measuring stars' positions in the sky.

[145] Next-Generation Transit Survey.

[146] Astro Systems Austria.

[147] Many amateur astronomers own telescopes of this size and larger.

[148] ESO's observatory on Cerro Paranal (Chile) where many major telescopes are sited.

[149] Antarctic-Search for Transiting ExoPlanets.

FIGURE 7.5 The NGTS array of telescopes. (Reproduced by kind permission of Simon Walker.)

Of course, it is slightly misleading to stress the smallness of the telescopes/cameras used in some transit detectors because a great deal more equipment and expenditure is needed in addition to the basic optics. We have already seen in the previous section that finding exoplanets transits is likely to need long periods of constant observation of the same stars. Few observers are likely to have the patience or stamina to make such observations personally. The first 'extra' therefore is to make the telescope work automatically (at least), or to be robotically controlled (at best).

The second 'extra' is a sophisticated computer system and purpose-written software to download the images and process them to search for the desired brightness changes. The third 'extra is a very good observing site and this may require the instruments to operate remotely from the observer. That in turn will carry with it some very expensive health and safety requirements. However, these comments apply mostly if you want to *discover* new exoplanet transits. Observing already known and predicted transits is much easier and it can be and is carried out by both amateur and professional astronomers with little-modified over-the-counter cameras and telescopes.

At the other end of the financial scale, some spacecraft may be used to observe exoplanet transits, although the telescopes still remain of modest sizes. Such observations obtained using the Spitzer telescope[137] are shown in Figure 7.2 (i). The HST has been used to observe many already-known exoplanets and this includes NIR observations of the transits of the star HD 189733 by its Jupiter-sized exoplanet. Similarly, the JWST[150] has already been used to undertake spectroscopy of the exoplanet WASP-39 b during several of its transits.

However, three spacecraft have been purpose-built for and used to discover transiting exoplanets: CoRoT,[151] Kepler[152] and TESS.[153] Whilst PLATO (Planetary Transits and Oscillations of stars, possible launch in 2026) and ARIEL (Atmospheric Remote-sensing Infrared Exoplanet Large-survey, possible launch in 2029) are planned for the future.

CoRoT used a 270 mm off-axis mirror for its telescope and had the aim of discovering small rocky exoplanets. It was in a polar orbit about 900 km above the Earth's surface, it could observe its target stars for periods of up to 180 days and it discovered 32 exoplanets.

---

[150] James Webb Space Telescope.

[151] Convection, Rotation et Transits planétaires. Launched December 2006, Deactivated June 2014.

[152] Launched March 2009, Deactivated November 2018. Discovered 2,662 exoplanets.

[153] Transiting Exoplanet Survey Satellite. Launched April 2018. Still operating at the time of writing.

Kepler used a 0.95-m telescope and was in a heliocentric orbit which trailed the Earth by a distance which increased by ~26,000,000 km each year. It observed about 100,000 stars continuously throughout its mission and these were located in Cygnus and Lyra. Some 2,778 exoplanets were discovered, although there may be a few more yet to come since the processing of Kepler's data is still continuing.[154]

TESS (Figure 7.6) is still operating in its extended mission at the time of writing. Its aim is to look for exoplanets around small stars and it covers most of the sky, examining some 200,000 stars. Its observations are made using four identical lens-based f1.4 cameras each with an aperture of 105 mm[155] and giving a field of view 24° by 24°. At the end of its primary mission (July 2020) it had discovered 66 confirmed exoplanets with another ~2,100 possibilities (candidates) which await further observations before being confirmed as exoplanets, or otherwise. At the time of writing the total of TESS' confirmed discoveries has risen to 415 and the number of candidates to ~19,000.

FIGURE 7.6 The TESS spacecraft. An artist's impression of the spacecraft in space. (Reproduced by kind permission of NASA's Goddard Space Flight Center.)

Transits' Triumph!

As discussed earlier, the first normal exoplanets were discovered in 1995 via the radial velocity method. The transit method did not make its first discovery until 2002, though it has since greatly surpassed[156] the radial velocity method in its achievements. That first success now has the name OGLE-TR-56 b.[157] It is slightly larger and more massive than Jupiter. Its orbital period, though, is

---

[154] Scientific support for Kepler (and now TESS), especially in respect of enabling young scientists to attend conferences and for paying publication charges, etc., has been provided for a number of years by a Crowd-Funding programme. The programme is now called *Nonprofit Adopt a Star* (adoptastar.org). Previous named 'Adopt a Star' and 'Pale Blue Dot', to date it has raised well over US$500,000.

[155] Actually, *smaller* than the cameras used in SuperWASP - although TESS's cameras are purpose-built.

[156] The volume of data from Kepler, TESS, etc. is so great that AI (artificial intelligence/machine learning) is routinely used for the initial classification of observations with, currently, false positives (errors) down to levels of 1% or 2%.

[157] OGLE – Optical Gravitational Lensing Experiment. OGLE was designed to detect exoplanets via the Gravitational Microlensing method (below). Since gravitational microlensing changes the brightnesses of host stars slightly, OGLE is essentially a very sensitive astronomical photometer. As such, it is also suited to detecting transiting exoplanets and this is how it detected OGLE-TR-56 b.

just 1.2 days and this places it only about 2,500,000 km above its host star's surface. Since the host star is similar to, but slightly more massive than the Sun, the exoplanet's outer temperature is close to 2,000 K.

Just what this would mean for life-forms living on this planet was expressed very vividly at the time of its discovery and I cannot better that depiction; the life-forms of OGLE-TR-56 b, like us, might well get caught out in rain storms, but *their* rain drops would be composed of *molten iron*.

THE LONESOME COWBOYS[158]

So far, we have been concerned with exoplanets which 'belong' to a host star or stars. However, as noted briefly in Chapter 3, there are some exoplanets which do not 'belong' to host stars. These are usually called Free-Floating[159] Exoplanets, and although they are gravitationally bound to stay within the Milky Way galaxy, or other galaxies, they are otherwise free to wander around the Galaxy, just like stars, but without having any internal fusion nuclear energy sources.

We have already noted (Box 4.2) that there is:

"… a smooth variation in the masses of individual astronomical objects from about 250 M_{Sun} for the most massive stars to micro-gram dust particles."

and, while in that Box we were concerned with exoplanets hosted by stars, there is no reason to think that it will not apply to free-floating exoplanets as well.

Now, within the Milky Way Galaxy there are some 100,000,000,000 to 400,000 000,000 (10^{11} to 4×10^{11}) stars. The proportion of these which are massive (>5 M_{Sun}) is around 0.5% to 2% while the proportion which a low mass (< 0.5 M_{Sun}) is perhaps 50% to 75%.

Thus, as you might well expect on general principles, the low mass stars are much more numerous than the high mass stars. Taking median values for the above figures, the number of low mass stars is thus some ~60 times higher than the number of high mass stars, i.e., within the Milky Way galaxy there may be ~ 2,000,000,000 high mass stars and ~120,000,000,000 low mass stars.

If we extrapolate this progression in a simple linear fashion to the planetary region (i.e., to, say, 0.1 $M_{Jupiter}$ (or about 0.0001 M_{Sun}; a factor of ~5,000), then we arrive at a figure of ~10^{15} to 10^{16} exoplanets for the Milky Way Galaxy. Now, even assuming that each and every star in the Milky Way Galaxy hosts 100 exoplanets (a very extravagant assumption), that gives us only around 20,000,000,000,000 (2×10^{13}) hosted exoplanets, so that free-floating exoplanets would out-number stellar-hosted exoplanets by a factor of at least x50. Recently a claim for the discovery of three free-floating brown dwarfs with masses below eight $M_{Jupiter}$ (one is just three $M_{Jupiter}$) based on JWST observations has been made. Their surface temperatures, though range from ~1,800 K (~1,500 °C) down to ~1,100 K (~800 °C), so they seem much more likely to be free-floating exoplanets still in the process of cooling after their formations.

Of, course the estimate just made for the relative numbers of hosted and free-floating exoplanets is WRONG[160] – and perhaps by a large factor. But I do not know by how much it is wrong, nor

---

[158] Title of a 1968 Western film produced by Paul Morrissey and Andy Warhol (on a budget of $3,000).

[159] a.k.a., Orphan, Rogue, Starless, Unbound Exoplanets, IPMOs (Isolated Planetary Mass Objects) and a number of other alternatives.

[160] One independent and indubitably more rigorous estimate than mine, at the time of writing, places the figure at ~50,000,000,000 (5×10^{10}) free-floating exoplanets ®. However, that is still probably at least comparable with and possibly larger than the (credible) number of stellar-hosted exoplanets.

Another independent estimate, suggests that the number of free-floating Jupiter-mass exoplanets is about twice the number of main sequence stars in the galaxy (i.e., several times 10^{11}) (T. Sumi *et al*, Nature, 349, **473**, 2011). A third, on the basis of the microlensing of Quasars, suggests 2,000 exoplanets for every main sequence star (i.e., more than 10^{14} within a typical galaxy ®) (X Dai, E Guerras, ApJ Letters, L27, **853**, 2018).

even in which way it is wrong. Nonetheless, it indicates that we should not necessarily assume that it is usual for exoplanets to be hosted by stars; it may be that the free-floating exoplanets are in the majority, perhaps overwhelmingly so.

The methods which we have so far encountered for discovering exoplanets rely upon observing the effects of the exoplanet upon the host star (changes in positions in the sky, pulse timings, radial velocity changes, brightness changes) so how do we go about finding exoplanets which do not have a host star to be observed? The answer to this problem lies with two of the methods of exoplanet detection which have been mentioned earlier, but not examined in detail: Gravitational Microlensing and Direct Imaging.

GRAVITATIONAL MICROLENSING

We all know that the gravity of an object pulls other objects towards itself (Newton's falling apple, etc.). But fewer people realise that this attractive force also affects light beams as well as material objects. Nonetheless, that is the case – and it means that massive objects can act like lenses and bend passing light beams towards a 'focus'. Unfortunately, such gravitational lenses are of very poor quality compared with even the cheapest normal optical lens and do not produce a true image (Figure 7.7). They do though concentrate the light to some extent. Thus, if from our Earthly point of view, a free-floating exoplanet passes in front of a distant star, then the exoplanet, acting as a gravitational lens will concentrate the light from the star somewhat and we will see it brightening briefly (Figure 7.8 Left).

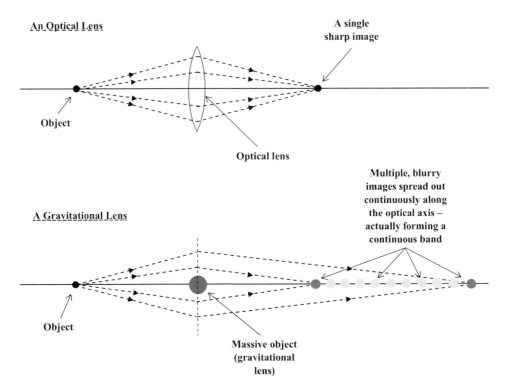

FIGURE 7.7 Optical and gravitational lenses. Top: The lens refracts light more strongly (through a larger angle) the further away the light beam is from the optical axis and so produces a sharp image. Bottom: The gravitational field weakens with increasing distance from the massive gravitating object. This gravitational 'lens' therefore refracts light less strongly (through a smaller angle) the further away the light beam is from the optical axis and so does not produce a single sharp image but a band of overlapping blurry images strung out along the optical axis.

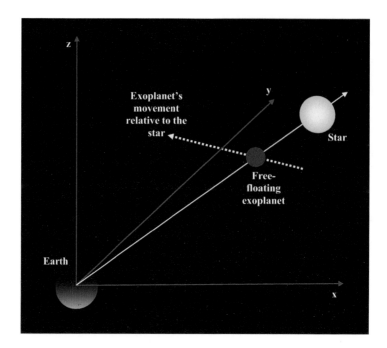

Sketch of the arrangement of objects in space for a
gravitational lensing event to occur

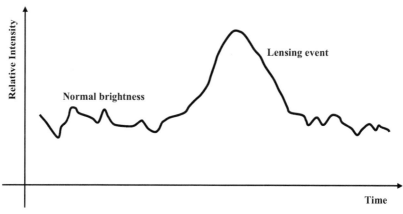

Schematic graph of the star's light curve
during the gravitational lensing event

FIGURE 7.8A Free-floating exoplanets. The brightening of a distant star due to a free-floating exoplanet crossing in front of it. Top: A colour sketch of the arrangement in space of the Earth, free-floating exoplanet and star for the gravitational lensing event to occur. N.B.: NOT to scale. Bottom: Schematic graph of the star's light curve during a gravitational lensing event. N.B.: The duration of a gravitational microlensing event varies but is typically expected to be a few hours for Earth-mass free-floating exoplanets up to a few days for exoplanets of a few times Jupiter's mass. The shape of the curve though, is the same for both types of exoplanet. (© C. R. Kitchin 2022. Reproduced by permission.)

FIGURE 7.8B Free-floating exoplanets. An artist's impression of a freefloating exoplanet as might be seen from a spacecraft near one of its natural satellites. (© C. R. Kitchin 2022. Reproduced by permission, the M4 image is reproduced by kind permission of ESO.)

Finding free-floating exoplanets via gravitational microlensing events is obviously a matter of chance (when the alignment of the Earth, exoplanet and distant star happens to occur and if we happen to be looking in that direction at the time) and so very few such events have been observed to date. However, the ideal method for catching them would be to follow the brightnesses of as many stars as possible for as long as possible – and this is exactly what the Kepler spacecraft[152] did. Kepler monitored the brightnesses of stars in a 115 square-degree[161] area of the sky in Cygnus and Lyra from May 2009 to November 2018. Part of its campaign from April to July 2016 was specifically aimed at detecting free-floating exoplanets through their gravitational microlensing and via observing a part of the galactic bulge at 30-minute intervals. After a great deal of data processing, five possible exoplanet lensing events were identified.

One of the Kepler microlensing events showed a double peak and is likely to be an exoplanet with a stellar host. The other four events are single-peaked and whilst it is possible that they have very distant stellar hosts, the paper concludes that they are likely to be free-floating in space;

"Whilst we cannot yet unambiguously confirm FFPs[162] from the … dataset, we have identified several short-time-scale events that are promising candidates."[163]

Other, ground-based surveys, such as MOA[164] and OGLE (above) have observed a number of microlensing events which *might* be due to free-floating exoplanets, but only one which has a reasonable certainty. The OGLE-2016-BLG-1928 event lasted just 41.5 minutes and is estimated to have been produced by a free-floating 0.29 M_{Earth} exoplanet. The possibility of a stellar host within 8 AU of the exoplanet is ruled out, but there could still be a more distant host star.

Gravitational microlensing has detected far more exoplanets with stellar hosts than free-floating exoplanets. The broad light curve of the microlensing event due to the host star then has a very sharp event superimposed upon it due to its orbiting exoplanet. Some 225 such discoveries are listed at the time of writing.

---

[161] About 0.25% of the whole sky.
[162] Free-floating Planets.
[163] I. McDonald *et al.* Nature, 5584, **505**, 2021.
[164] Microlensing Observations in Astrophysics collaboration. It uses a 1.8-m telescope at the Mount John Observatory (New Zealand).

The currently-under-construction Roman space telescope[165] is expected to be launched in 2027. It has a 2.4-m primary mirror and one of its two primary scientific objectives will be to discover and study exoplanets via gravitational microlensing; including free-floating exoplanets with masses down to less than 1 M_{Earth}. Thus, a quantum leap in our information about such exoplanets may, perhaps, soon be expected.

DIRECT IMAGING

A free-floating exoplanet does not (by definition) have a nearby star to illuminate its surface[166] (Figure 7.8 Right). They are all, therefore, going to be very faint objects and, in the absence of significant internal energy sources, they will only be able to be seen by us via the reflected radiation received from stars many light years distant from them. In practice, that will mean that we will not currently be able to detect any of them.

However, free-floating exoplanets which have only recently been formed by condensation from interstellar gas clouds will still be warm, even hot, from the gravitational energy released during their collapses, through the decay of radioactive elements and through stored thermal energy inside the planets which is still leaking out towards their surfaces.

Thus, as an illustration, the total energy emitted from Jupiter presently is ~14.1 W m^{-2} and this comprises ~6.6 W m^{-2} originating from the currently incoming absorbed solar energy and ~7.5 W m^{-2} internal heat[167] dating from billions of years ago. So that even for Jupiter, which is some ~4,500,000,000 years away from its formation, the internal energy source is still larger than the energy received from its host star (i.e., the Sun).

If Jupiter were magically whisked off to become a free-floating planet, without any other changes, then it would still be emitting more than half the energy which it does now as a planet with a host star. However, the internal energy contribution will decrease slowly with time. So, going back in time, the internal energy source must have been more and more dominant, the further back we go towards Jupiter's formation. Perhaps, therefore, Jupiter was once warm enough to be emitting significantly at infrared wavelengths, i.e., having a temperature of hundreds of kelvins, even a thousand or two kelvins.[168] With our current infrared telescopes and spacecraft, we would now be able to detect it, perhaps even out to distances away from us of a few hundred light years.

In fact, quite a number of warmish objects, which could be free-floating exoplanets, or brown dwarfs or still condensing objects (which could become either of those possibilities, or even become 'proper' stars) *have* been seen directly on infrared images.

Recently somewhere between 70 and 170 Jupiter-sized such objects were identified in the Upper Scorpius OB stellar association. That star association is about 420 ly distant from us and stars' ages are put at somewhere around 3 million and 10 million years. The 80,000 images supporting this claim had been gathered over two decades using several large telescopes in collaboration. Almost certainly some of these objects are not free-floating exoplanets, but the advantage of direct image detections over gravitational microlensing detections is that we can go and look at the objects, time and again, until it *is* clear exactly what they are.

---

[165] Named for Nancy Roman, NASA's first Chief of Astronomy.

[166] For exoplanets which do have host stars and which are therefore well illuminated, it is still difficult to obtain direct images. In this case it is because the light from the much brighter host star swamps that from the exoplanet. Sometimes an almost normal mage can be obtained when the contrast is naturally small. Thus, 2M1207 is a brown dwarf star with a large exoplanet some 41 AU away from it. The exoplanet, 2M1207 b, was imaged directly in 2004 using the VLT. More normally, the contrast between the host star and exoplanet must be reduced in some other way (Chapter 8[189]).

[167] Even the Earth is still radiating some ~5 x 10^{13} W from its stored internal primordial heat. This, though, works out at about 0.1 W from each square metre of the Earth's surface. So, it would take about 10 days to heat up a mug of water until it was hot enough to make your coffee (assuming no heat was lost from the mug of water whilst it is being heated). On Jupiter, the same mug of coffee could be ready in just 3 hours 20 minutes. Either way, you are better off with an electric kettle.

[168] The wavelength for the peak emission of a hot(ish) object is around 6 μm for a 500 K (~200 °C) temperature and 1.5 μm for 2,000 K (~1,700 °C); the MIR (Mid-Infrared) and NIR, respectively.

Observations in 2013 using the Pan-STARRS 1[169] telescope discovered a ~6 $M_{Jupiter}$ free-floating exoplanet; PSO[170] J318.5-22 at a distance of about 80 ly from us. It is to be found in Capricornus and may have formed about 12 million years ago. It may have a surface temperature of ~1,100 K (~800 ℃) below a layer of clouds comprising hot dust and molten iron (though there is still discussion over this aspect of the object). Nonetheless it is clearly a definite free-floating exoplanet and not a brown dwarf.

A striking and very recent discovery using the JWST is of *Binary* free-floating planets, with each component having roughly the mass of Jupiter. They have thus been named 'JuMBOs' (Jupiter Mass Binary Objects). One such system would be a curiosity, but the JWST (at the time of writing) has detected around 40 examples, all clustered within the Orion Nebula (M42), and *that* observation is decidedly difficult to explain.

What would it be like to visit, or live-on, a free-floating exoplanet? If, as seems likely, free-floating exoplanets start off at temperatures in the 1,000s K and that they are also likely to end up eventually at temperatures close to 0 K (c.f., PSR B1257+12 b, c and d; Chapter 5), then somewhere between that beginning and that end, the free-floating exoplanet must pass through a temperature regime which suits almost any form of life. However, if you require light, like terrestrial photosynthetic plants and the terrestrial animals which use such plants as the beginning of their food chains, then, even if the temperature is OK for a few billion years, survival-wise, you will be out of luck.

Of course, on Earth there are life-forms which do not depend on light (Chapter 12); chemoautotrophs of various types survive by using chemical reactions to provide their energy and synthesise their biological requirements from carbon dioxide (Chapter 12). Thus, should we ever need to colonise a free-floating exoplanet which has a temperature around 373 K (100 ℃), plus having water, then bacteria from hydrothermal vents along Earth's mid-oceanic ridges might form the basis of the food chain and tube worms, limpets and shrimps could be the staple food for our humankind colonists.[171]

Apart from their perpetual and near-total darkness, free-floating exoplanets should come in all the varieties of hosted exoplanets (Table 8.1) and perhaps a few other types not to be found near stars, as well. For humankind visitors and explorers, and ETs and ETIs of all types (Chapter 13), there should therefore be opportunities enough to go round amongst the free-floaters to satisfy all needs and tastes. Whether life could evolve from scratch on a free-floating exoplanet is more of a problematic question, but if life can evolve from scratch on hosted exoplanets (Chapter 10), then my bet would be that it will be able to succeed on the free-floaters as well.

DIY – FINDING YOUR OWN EXOPLANET

Actually, as you may infer from some of the above discussions, discovering your own, new exoplanet is probably beyond your financial means, unless you should happen to be one of the 3,000 or so $ billionaires currently on the Earth.

With far fewer financial resources,[172] you may, however, observe already known exoplanets relatively easily.

As we have seen, astonishingly small telescopes can observe exoplanet transits and for those exoplanets already discovered, the periods and timings of their transits are widely published on the internet. You thus need a lot of patience, a 0.2-m or so class telescope and an instrument for measuring a star's brightness (a.k.a., a photometer). Photometers can be bought over-the-counter from some commercial telescope suppliers. However, a CCD imager designed for a telescope is probably more

---

[169] A 1.8-m telescope on Haleakala on Hawaii. Now joined by the Pan-Starrs 2 instrument.

[170] Pan-Starrs Object.

[171] 100 ℃ is a bit hot for humankind, but then you can't have everything. Also, you may think, that a staple diet of tube worms is a price too high for a colony (but, Indigenous Australians thrived on bush tucker, including Witchetty grubs, for millennia, so you could be wrong).

[172] Though still not altogether cheaply, think – second-hand car prices.

widely available from the same sources, it may even be cheaper and will do just as good a job.[173] Having selected your target, therefore, all you need to do is align your telescope onto it and, starting a little time before the transit is due to occur, measure your target's brightnesses at intervals of a few seconds or a few minutes. With even a small amount of luck you will detect the star dimming as the exoplanet crosses its disc.

There are also citizen-science internet sites which you may join to help process data obtained on professional instruments, or just to watch the observations as they happen (e.g., NASA: https:// exoplanets.nasa.gov/citizen-science/; NASA: https://exoplanets.nasa.gov/exoplanet-watch/, UNITE; https://science.unistellaroptics.com).

---

[173] It is what the professional astronomers use – the photometers of three decades ago, which had a single detector observing a single star, are all now in museums or on junk heaps.

8 Exoplanets: The Devil Is in the Details

Usually, one of the first tasks in any new scientific activity is to organise and classify the objects of the study so that the 'Wood can be seen for the Trees'. Some such groupings of types of exoplanets may have become apparent earlier, and Box 4.2 discusses some classifications extending beyond exoplanets.

Here, though, we start to look at the attributes of exoplanets in more detail and this is easiest done by looking at the common and defining properties of exoplanets which place them into their various groupings.

SOLAR SYSTEM

We start (again) with what happens within the Solar System because this is the place which we know best. For centuries, therefore, the planets within the Solar System have been divided into two main groups, called the Terrestrial Planets and the Jovian Planets (Boxes 4.2 and 7.2).

For the purposes of *this* book, these two groups will comprise:

- The Terrestrial Planets: Earth, Mars, Mercury and Venus (planets)

  Ceres, Eris, Gongong, Haumea, Makemake, Orcus, Pluto, Quaoar and Sedna (dwarf planets[174]).

  Plus, another 15 satellites with masses greater than or equal(ish) to that of Ceres (~10^{21} kg; the left-overs[175])

and

- The Jovian Planets: Jupiter, Saturn, Uranus and Neptune.

Their defining characteristics are:

- Terrestrial Planets, etc.
  - Small masses (for a planet). The upper mass limit is still imprecisely defined but probably lies between 5 and 10 M_{Earth}. At the upper mass end, their classification will also depend upon their composition.
  - Solid surfaces
  - Thin or no atmospheres

---

[174] As noted in Chapter 4, whilst there are subtle and not-so-subtle differences between dwarf planets, asteroids and planetary satellites, the main distinguishing criterion used in this book is their masses. Orcus is currently the lowest mass dwarf planet with a mass of 5.5×10^{20}, so in this book, unless otherwise stated, the limiting mass for a terrestrial-type planetary object will be taken to be $\geq$ ~10^{21} kg. N.B.; this is an arbitrary definition so you are quite at liberty to decide your own preferences.

[175] These 15 objects are;
   Satellite of the Earth – Moon
   Satellites of Jupiter – Callisto, Europa, Ganymede, Io
   Satellites of Saturn – Dione, Iapetus, Rhea, Titan
   Satellites of Uranus – Ariel, Oberon, Titania, Umbriel
   Satellite of Neptune – Triton
   Satellite of Pluto - Charon

DOI: 10.1201/9781003246459-9

- ○ More-or-less spherical
- ○ Rocky composition, perhaps with metallic and/or liquid cores
- Jovian Planets[176]:
  - ○ Large Masses – From the upper limit for terrestrial planets to around the lowest mass for a brown dwarf of ~13 $M_{Jupiter}$ (Box 4.2).
  - ○ Thick deep atmospheres. There may be no identifiable surface as such, just a continuing increase in density to beyond that typical for materials normally solid under terrestrial conditions.
  - ○ Composition largely hydrogen and helium. Heavy elements and compounds are likely to have become concentrated towards their centres.
  - ○ In many cases there may be a significant amount of internal energy still being released from the continuing gravitational collapse and/or the heavy-element differentiation.

Looking outside the Solar System at the ~5,580 now-known exoplanets, perhaps surprisingly, the same two classes still prove to be useful. However, they do need some refining, mostly on the bases of their surface temperatures and their masses.

Attempts to classify exoplanets have been made since well before any exoplanets were known. *Star Trek*,[177] for example, recognised nine classes of exoplanets. Thus, its class M planets could sustain human-type life, while its class K planets were habitable, but not by human-type life. While such an inhabitability-based classification may be useful once numerous examples of extra-terrestrial type life-forms have been found, it is not very relevant for us at the moment.

Our division of the Solar System planets into terrestrial and Jovian types (above) is essentially based on mass. In addition to inhabitability and mass, other classification schemes have been proposed based on composition, density, heat source(s), orbits and sizes. After some detailed considerations of the various ways of classifying exoplanets, though, Alan Stern and Harold Levison concluded[178] in 2002:

> "... we find that no *single* criterion can adequately define the conditions for planethood. Instead, we argue, a series of criteria must be used in concert to form a good test."

and they proposed (yet another) classification system based on size and composition.

With no generally agreed way, then, of classifying exoplanets, what should we do here? Well, the answer is obvious:

> We will devise our own system.[179]

based upon what is needed for the purposes of this book.

We have already mentioned one criterion for defining planets, brown dwarfs and stars (Box 4.2) and that is that the object should *not* be generating energy via nuclear fusion reactions in order for it to be a planet or exoplanet. We will now extend that just slightly by requiring that the object should not be generating energy by nuclear fusion *now*, but also that it has not generated energy in that fashion in the *past* and that, as far as we can tell, it will not generate it that in way in the *future*.

---

[176] a.k.a. Gas Giants

[177] A popular cult science-fiction series started on television in the 1960s by Gene Rodenberry, now merchandised in many other media.

[178] *Regarding the Criteria for Planethood and Proposed Planetary Classification Schemes*. Highlights of Astronomy. 205, **12**, 2002.

[179] Actually, its results are not very different from those of most of the other usages and definitions to be encountered in respect of exoplanets.

The 'no-nuclear-fusion' requirement is what makes the division between large exoplanets and small brown dwarfs at a mass of ~13 $M_{Jupiter}$. So, our second classification criterion will be based on the objects' masses. The masses though are often also associated with bulk compositions, so we will glance at those as well.

Finally, for the existence of life in any form, the ambient temperature is likely to be important (Chapter 10). So, we will also include temperature as a criterion.[180]

Thus, in summary, our classification hereinafter will be based on:

- No nuclear fusion reactions
- Mass (with composition helping a bit)
- Temperature (in some form).

and we end up with the following classes (Table 8.1):

---

[180] Given our discussion of the various temperatures of PSR B1257+12's exoplanets (Chapter 5), you may well ask 'Which / What temperature?'. The basic answer to that is that it is; whatsoever temperature we can observe directly and / or infer from the exoplanet's proximity to its host star. There will be many cases where this is not relevant to whether the exoplanet is habitable in some other, but unobserved, parts of its atmosphere or interior.

Find this unsatisfactory? - well, it is the best we can do at the moment.

TABLE 8.1
Exoplanet Classification System as Used in This Book

Main class	Sub Class	Mass - M_{Earth} / $M_{Jupiter}$ [181] (see also above)	Mass - kg	Temperature[182]	Diameter D_{Earth}/ $D_{Jupiter}$ (very approximate)	Diameter – km (very approximate)	Notes	Example
Small bodies (Figure 8.1)	---	< ~ 0.0002 M_{Earth}	< ~1.0 × 10^{21}	Cold/ Temperate/ Hot	< ~ 0.1 D_{Earth}	< ~ 1,000	Rocky (silicate)/metallic compositions. No Atmospheres. May be irregularly shaped.	Phobos – (Mars satellite) Mass: 1.8 × 10^9 M_{Earth} (1.1 × 10^{16} kg) Temperature: ~ 230 K (~ −40 ° C) Size: ~ 0.0018 D_{Earth} (18 × 22 × 27 km)
Terrestrial	Mini-Earths	~ 0.0002 → ~ 0.1 M_{Earth}	~ 1.0 × 10^{21} → ~ 6 × 10^{23}	Cold Temperate Hot	~ 0.1 → ~ 0.5 D_{Earth}	~ 1,000 → ~ 5,000	Rocky (silicate)/metallic compositions. Some with atmospheres	Charon – (Pluto satellite) Mass: 2.7 × 10^{-4} M_{Earth} (1.6 × 10^{21} kg) Temperature: ~ 50 K (~ −220 °C) Diameter: 0.1 D_{Earth} (1,210 km) Moon – (Earth satellite) Mass: 0.0123 M_{Earth} (7.34 × 10^{22} kg) Temperature: ~ 100 → 390 K (~ −170 → 120 °C) Diameter: 0.273 D_{Earth} (3,476 km) Kepler[181]-1520 b – (Exoplanet – 2,000 ly away, in Cygnus – N.B., this planet is so hot that it is evaporating)) Mass: ~ 0.02 M_{Earth} (~ 10^{23} kg) Temperature: ~ 2,200 K (~ 2,000 °C) Diameter: ~ 0.2 D_{Earth} (~ 3,000 km) – NB very uncertain
	Earth-Type (Figure 8.2)	~ 0.1 → ~ 5 M_{Earth}	~ 6 × 10^{23} → ~ 3 × 10^{25}	Cold Temperate	~ 0.5 → ~ 1.5 D_{Earth}	~ 5,000 → ~ 15,000	Rocky (silicate/metallic compositions. Mostly with atmospheres	No current examples Earth Mass: 1 M_{Earth} (6 × 10^{24} kg) Temperature: ~ 290 K (~ 20 °C) Diameter: 1 D_{Earth} (12,700 km)

[181] Reminder; M_{Earth} = 5.97 10^{24} kg, $M_{Jupiter}$ = 1.90 × 10^{27} kg. D_{Earth} = 1.27 × 10^4 km, $D_{Jupiter}$ = 1.40 × 10^5 km.

[182] Anthropometric values/terminology. Based upon a range for 'temperature' which is just about survivable for humankind using only relatively minor artificial aids, i.e.:
Cold: < ~ 240 K (~ −30 °C),
Temperate: ~ 240 K → ~ 325 K (~ −30 °C → ~ 50 °C),
Hot: > ~ 325 K (~ 50 °C).
Planets which are not rotating or rotating very slowly with respect to their host star are likely to have very different temperatures on their light and dark sides (c.f., the Moon in Table 8.1).

TABLE 8.1 (CONTINUED)
Exoplanet Classification System as Used in This Book

Main class	Sub Class	Mass - M_{Earth} / $M_{Jupiter}$[181] (see also above)	Mass - kg	Temperature[182]	Diameter D_{Earth}/ $D_{Jupiter}$ (very approximate)	Diameter – km (very approximate)	Notes	Example
				Hot				K2-138 g[184] – (Exoplanet, ~660 ly away, in Aquarius. N.B., some mass estimates are higher than given here and it may be better classed as Neptunian or Super Earth) Mass: 4.3 M_{Earth} (2.5 × 10^25 kg) Temperature: ~520 K (~250 °C) Diameter: 3.0 D_{Earth} (38,000 km)
	Super-Earths (Figures 8.3 & 8.4)	~5 → ~10 M_{Earth}	~3 × 10^25 → ~6 × 10^25	Cold	~1.5 → ~3.0 D_{Earth}	15,000 → 30,000	Rocky (silicate/metallic compositions. Mostly (all?) with atmospheres	No current examples (although you could look at PSR B1257+12 c and d in Chapter 5, bearing in mind their differing natures from normal exoplanets)
				Temperate				Kepler-20 d – (Exoplanet, ~930 ly away, in Lyra. Borderline with the Neptunian class) Mass: ~10 M_{Earth} (~6 × 10^25 kg) Temperature: ~370 K (~100 °C) Diameter: 2.75 D_{Earth} (35,000 km)
				Hot				WASP-47 e – (Exoplanet, ~870 ly away, in Aquarius) Mass: 6.8 M_{Earth} (4.1 × 10^25 kg) Temperature: ≥ 1,000 K (≥ 700 °C) Diameter: 1.8 D_{Earth} (23,000 km) Rocky materials start to melt at ~1,000 K (~700 °C). WASP-47 e may therefore be completely or partially covered in lava. Such an object is often called a Lava World.
Gas Giants	Neptunian (a.k.a. Ice Giants)	~10 → ~50 M_{Earth} (~0.03 → ~0.15 $M_{Jupiter}$)	~6 × 10^25 → ~3 × 10^26	Cold	~3.0 → ~5.0 D_{Earth} (~0.2 → 0.4 $D_{Jupiter}$)	~30,000 → 60,000	Thick atmospheres of ammonia, helium, hydrogen, methane and water. Probably have solid cores.	Uranus Mass: 14.5 M_{Earth} / 0.046 $M_{Jupiter}$ (8.68 × 10^25 kg) Temperature: ~50 K (~-220 °C) Diameter: 4.0 D_{Earth} / 0.36 $D_{Jupiter}$ (51,000 km)

[184] Discovered by the Kepler spacecraft during the second half of its mission.

TABLE 8.1 (CONTINUED)
Exoplanet Classification System as Used in This Book

Main class	Sub Class	Mass - M_{Earth} / $M_{Jupiter}$ [181] (see also above)	Mass - kg	Temperature [182]	Diameter D_{Earth} / $D_{Jupiter}$ (very approximate)	Diameter – km (very approximate)	Notes	Example
	Jovian (Figure 8.5)	~0.15 → ~3 $M_{Jupiter}$	~3 × 10²⁶ → ~6 × 10²⁷	Temperate	~0.4 → ~2.0 $D_{Jupiter}$ [186]	60,000 → 280,000	Thick atmospheres mostly of helium and hydrogen. Probably have solid cores.	Kepler-22 b – (Exoplanet, ~600 ly away. in Cygnus) Mass: 36 M_{Earth} / 0.11 $M_{Jupiter}$ (2.1 × 10²⁶ kg) Temperature: ~ 260 K (~ -10 °C) Diameter: 2.4 D_{Earth} / 0.22 $D_{Jupiter}$ (30,000 km)
				Hot				GJ 436 b [185] – (Exoplanet, ~32 ly away. in Leo) Mass: 22 M_{Earth} / 0.07 $M_{Jupiter}$ (1.3 × 10²⁶ kg) Temperature: ~ 710 K (~ 440 °C) Diameter: 4.3 D_{Earth} / 0.39 $D_{Jupiter}$ (55,000 km)
				Cold				Kepler-1654 b – (Exoplanet, ~3,700 ly away. in Cygnus, orbiting a binary star) Mass: 0.5 $M_{Jupiter}$ (9.5 × 10²⁶ kg) Temperature: ~ 200 K (~ -75 °C) Diameter: 0.82 $D_{Jupiter}$ (115,000 km)
	Super Jovians (Figure 8.6)	~3 → ~13 $M_{Jupiter}$	~6 × 10²⁷ → ~2.5 × 10²⁸	Temperate			Thick atmospheres of ammonia, helium, hydrogen, methane and water. Probably have solid cores.	CoRoT-9 b [151] – (Exoplanet, ~1,500 ly away. in Serpens) Mass: 0.84 $M_{Jupiter}$ (1.6 × 10²⁷ kg) Temperature: ~ 250 → 430 K (~ -25 → 160 °C) Diameter: 1.05 $D_{Jupiter}$ (150,000 km)
				Hot				Kepler-54 c – (Exoplanet, ~1,600 ly away. in Cygnus) Mass: 0.37 $M_{Jupiter}$ (7.0 × 10²⁶ kg) Temperature: ~ 420 K (~ 150 °C) Diameter: 0.11 $D_{Jupiter}$ (15,000 km)
				Cold				14 Her c – (Exoplanet, ~ 60 ly away in Hercules). Mass: ~ 6.9 $M_{Jupiter}$ (~1.4 × 10²⁸ kg) Temperature: $\leq$ ~70 K ($\leq$ ~-200 °C) Diameter: ~1.3 $D_{Jupiter}$ (180,000 km)

[185] From the *Gliese Jahreiß Catalogue of Nearby Stars*.

[186] Most planets over about 1 $M_{Jupiter}$ in mass have sizes which are fairly close to that of Jupiter. The reason for this is that if the mass increases, the planet's core becomes more compressed and so the outer diameter stays much the same.

TABLE 8.1 (CONTINUED)

Exoplanet Classification System as Used in This Book

Main class	Sub Class	Mass - M_{Earth} / $M_{Jupiter}$[181] (see also above)	Mass - kg	Temperature[182]	Diameter D_{Earth}/ $D_{Jupiter}$ (very approximate)	Diameter – km (very approximate)	Notes	Example
				Temperate				Kepler-1625 b – (Exoplanet, ~ 8,000 ly away in Aries). Mass: ~ 11.6 $M_{Jupiter}$ (~2.3 × 10^{28} kg) Temperature: ≤ ~ 350 K (≤ ~ 80 °C) Diameter: ~1 $D_{Jupiter}$ (140,000 km)
				Hot				ZTF[187] J0220+21 b – (Probably exoplanet rather than brown dwarf, ~ 1,200 ly away in Aries. Host star is a white dwarf). Mass: ~ 14 $M_{Jupiter}$ (~2.8 × 10^{28} kg) Temperature: ~ 1,300 K (~1,000 °C) Diameter: 0.54 $D_{Jupiter}$ (76,000 km)
Free-Floating (Figure 8.7)		≤ ~ 13 $M_{Jupiter}$	≤ 2.5 × 10^{28}	Cold/ Temperate/ Hot	≤ ~ 2? $D_{Jupiter}$	≤ ~ 280,000	Mostly have atmospheres – of hydrogen and helium for the most massive and coldest examples. Atmosphere-less (perhaps) for the lowest masses. Overall composition mirrors that of the interstellar medium, except where gases, especially hydrogen and helium, have escaped into space. Dense / heavy elements and compounds likely to form the cores of the more massive exoplanets. Less massive exoplanets likely to be fairly uniform throughout. External illumination very low; almost totally dark.	CFBDSIR 2149-0403[188] (Exoplanet almost certainly, 75 to 150 ly away in Aquarius. No plausible star found that could be host to this exoplanet. Has an atmosphere containing methane and water). Mass: 4 → 7 $M_{Jupiter}$ (8 × 10^{27} → 1.4 × 10^{28} kg) Temperature: ~700 K (~ 430 °C) (from internal stored energy) Diameter: ~0.9 $D_{Jupiter}$ (130,000 km)

[187] ZTF – Zwicky Transient Facility.

[188] This cumbersome name derives from Canada-France Brown Dwarfs Survey Infrared plus its position in the sky. RA 21h 49 m, Dec -04° 03'.

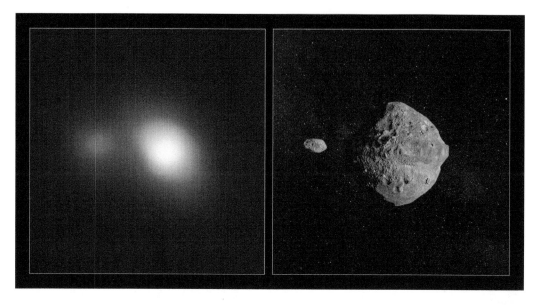

FIGURE 8.1   The binary Asteroid 66391 Moshup, an example of a cold small object (Table 8.1). The larger body is about 1.3 km across. Left: Direct image using the SPHERE[189] instrument on the VLT. Right: Artist's impression of the two asteroids. (Reproduced by kind permission of ESO.)

---

[189] Spectro-Polarimetric High-contrast Exoplanet Research. SPHERE uses a coronagraph to reduce the light from the host star so that the exoplanet can be seen. Many exoplanets are observed in this way using various types of Coronagraph. Coronagraphs were first invented for observing the very faint solar corona outside solar total eclipses by shielding off the much brighter solar photosphere; hence its name (see the Author's book *Solar Observing Techniques*, 2002, Springer, or the internet, for further information on solar coronagraphs). A coronagraph basically works by casting a shadow of the bright source onto the detector, so that the fainter source may then be seen more easily to one side of the shadow.

Future concepts, such as HabEx (Habitable Exoplanet Observatory), envisage space-based coronagraphs utilising two spacecraft flying in formation thousands of km apart. The leading one would be the occulting disc (star shade), the trailing spacecraft would carry the telescope(s) and camera(s). Such instruments might enable the observation of exoplanets 10,000 million times fainter than their host stars thus allowing terrestrial-type exoplanets to be observed close-in to red dwarf stars – and these pairings are probably the commonest potentially habitable exoplanet systems.

FIGURE 8.2 An artist's impression of the Hot Earth (Table 8.1), L 89-59 b. The exoplanet, on the left, has a mass of ~0.4 M_{Earth} (~3 × 10^{24} kg) and a temperature of ~ 600 K (~ 330 °C). (Reproduced by kind permission of ESO/M. Kommesser.)

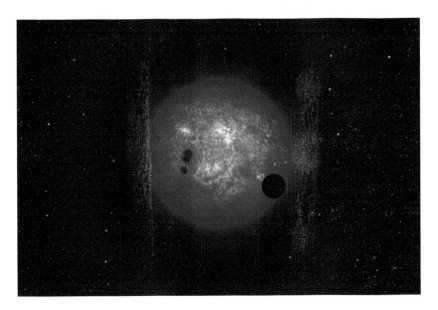

FIGURE 8.3 An artist's impression of the Hot Super Earth (Table 8.1), GJ-1214 b in transit across the disc of its host star, GJ-1214. The exoplanet has a mass of ~8.2 M_{Earth} (~ 5 × 10^{25} kg) and a temperature of ~ 480 K (~ 200 °C). (Reproduced by kind permission of ESO/L. Calçada.)

FIGURE 8.4 An artist's impression of the Cold Super Earth (Table 8.1), OGLE-2005-BLG-390L b. The exoplanet has a mass of ~5.5 M_{Earth} (~3 × 10^{25} kg) and a temperature of ~ 50 K (~ −220 ºC). (Reproduced by kind permission of ESO.)

FIGURE 8.5 An artist's impression of the Temperate Jovian Gas Giant (Table 8.1), CoRoT-9 b during a transit. The exoplanet has a mass of ~0.84 $M_{Jupiter}$ (~1.7 × 10^{27} kg) and a temperature of ~ 250 to 430 K (~ −20 to 160 ºC). (Reproduced by kind permission of ESO/L. Calçada.)

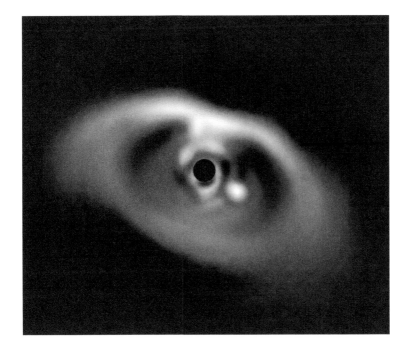

FIGURE 8.6   Direct image using the SPHERE instrument on the VLT of the Hot Super Jovian Gas Giant (Table 8.1), PDS 70 b. The host star is obscured (black central circle) and the exoplanet is the bright spot to its right, surrounded by the remnants of the circumstellar gas and dust disc. The exoplanet has a mass of ~7 $M_{Jupiter}$ (~ 1.4 × 10^{28} kg) and a temperature of ~ 1,200 K (~1,000 °C). (Reproduced by kind permission of ESO/A. Müller *et al*).)

FIGURE 8.7   Direct image using the SOFI instrument on ESO's NTT[190] of the Hot Free-Floating Exoplanet (Table 8.1), CFBDSIR 2149-0403 (bluish, arrowed). Mass: 4 to 7 $M_{Jupiter}$ (8 × 10^{27} to 1.4 × 10^{28} kg), temperature: ~700 K (~ 430 °C). (Reproduced by kind permission of ESO/P Delorme).

---

[190] NTT – New Technology Telescope, a 3.58-m telescope sited at ESO's La Silla observatory. SOFI – Son OF ISAAC – ISAAC was SOFI's predecessor and its name derives from Infrared Spectrometer And Array Camera.

AND FINALLY – THE JUICY BITS[191]

There should be something for everyone here. Indeed, I hope that there will be quite a few some-things for quite a few readers. At least you'll have something to talk about at your next cocktail party as a change from United's prospects in the cup (although your listener will probably look blank and switch the conversation to how United are getting on, anyway).

MOST SUITABLE FOR HUMANKIND

The Earth.

As Rhysling, the Blind Poet of the Spaceways, expresses the thought:

> "I pray for one last landing
> On the globe that gave me birth;
> Let me rest my eyes on fleecy skies
> And the cool, green hills of Earth."[192]

MOST SUITABLE FOR TERRESTRIAL-TYPE LIFE

Apart from the Earth, TRAPPIST-1 e is probably the current top favourite[193] (Cover Image). It is a Terrestrial, Earth-type, Temperate exoplanet (Table 8.1) some 40 ly away in Aquarius. This exo-planet is slightly less massive and smaller than the Earth. Its host star is a red dwarf with a mass of ~0.09 M_{Sun} – which is only just above the upper limit for a brown dwarf (Box 4.2) – with a surface temperature of 2,570 K (2,300 °C) and a luminosity of ~ 0.055% that of the Sun. The exoplanet, however, orbits every ~6 days, just 0.029 AU out from this red dwarf and so its surface equivalent temperature is around 246 K[194] (–27 °C). No hydrogen-based atmosphere has been detected on TRAPPIST-1 e, but it may still have a thin atmosphere of some sort. If so, even a minimal green-house effect might raise its temperature to more comfortable levels for us humans. Additionally, the exoplanet is tidally locked (Appendix D) onto its host star, so the sub-stellar point is likely to be sig-nificantly warmer than that of the anti-stellar point (at the centre of the dark side). Almost certainly, therefore, water, if present in any quantity, could in some places be in a liquid form. The exoplanet's density of 5,650 kg m⁻³ is similar to that of the Earth (5,510 kg m⁻³), suggesting, perhaps, that it has a similar rock/metal composition.

The star, TRAPPIST-1, is also host to another six exoplanets, all of which, including TRAPPIST-1 e, have had their discoveries confirmed by the Spitzer space telescope via their transits (Figure 7.2). The exoplanets, TRAPPIST-1 f and g, also lie within the star's habitable zone. Its current age is ~7.6 billion years, but its main sequence lifetime, during which it will undergo few changes, could be up to 10,000 billion years or so, i.e., ~ 1,000 times the Sun's main sequence lifetime. Thus, if TRAPPIST-1 e and its companions have yet to evolve life of some sort, to be seeded with life from elsewhere or to be colonised (Chapter 12), then they potentially still have a very long time left in

---

[191] American idiom referring to sensational gossip; i.e., the best, most interesting, spiciest, most fascinating, most shocking and / or most unbelievable topics within the subject.

[192] *The Green Hills of Earth*, Heinlein, R. A., 1947, The Saturday Evening Post. The story and the poem also appear in several Heinlein anthologies, it has some slight variations and several other Heinlein SF publications incorporate refer-ences to the poem.

[193] Taken from a catalogue of some 17,000 relatively nearby stars which might be able host habitable planets ('HabCat') was produced in 2002 mainly from Hipparcos observations (https://arxiv.org/pdf/astro-ph/0210675.pdf), however this does not seem to be available openly.

[194] The Russian town of Oymyakon lays claim to being the coldest permanently inhabited town on Earth. It is almost due North of Tokyo, but about 3,140 km closer to the North Pole. The average winter temperature there is ~223 K (~ -50°C) and a temperature of 202 K (-71 °C) has been recorded. The Oymyakon-ians might therefore find TRAPPIST-1 e to be quite a genial spot to colonise.

which to do so. The TRAPPIST-1 system is expected to be an early and frequent target for the JWST and JWST spectra of TRAPPIST-1 b have already been obtained, at the time of writing, in a search for an atmosphere on that exoplanet (Chapter 13).

Most Suitable for ETs and/or ETIs

BOX 8.1 – PIPE DREAM INTERLUDE

Now, for the keen-to-move-planet Rigel-c-based readers of this book, the author takes great pride in bringing them an advance copy[195] of the UNIVERSE-MOVERS Inc., Intergalactic Estate Agents' brochure for stardate, 33166.2 (Figure 8.8).

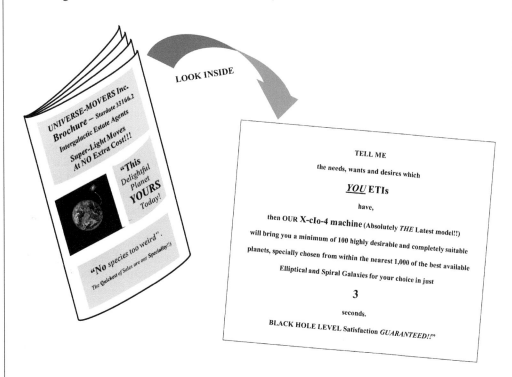

FIGURE 8.8 The UNIVERSE-MOVERS – Intergalactic Estate Agent's brochure for stardate, 33166.2 – an imaginary notion. (© C. R. Kitchin 2023. Reproduced by permission.)

Of course, we are currently only on stardate, 17724.5, so it will be some time before your burning desire for an extra-large super-Jovian planet with a chlorine/phosgene atmosphere and a minimum temperature of 800 K can so easily be fulfilled. However, with your species' 400,000-year lifetimes, that is not long to wait.

Away from such pipedreams (or maybe they *will* be real at some time – who knows?), these spoof ETIs and their spoof Estate Agents are making an important point for this book:

we cannot tell if an exoplanet is suitable for life until we know
what form that life takes and what it requires in order to live.

---

[195] My time travel secret is for sale, at a cost of *exactly* π U.S. dollars (no other figure acceptable).

Further progress on this Section (and whether or not the X-cIo-4 machine will find the clients' their hearts' desires (always assuming that they have hearts) within the required 3 seconds) will thus have to await later discussions (Chapter 18).

NEAREST

After the Sun, the nearest star to the Earth and Solar System is Proxima Centauri[196] (Figure 2.5), a small red dwarf flare star just 4.25 ly away. In theory, it can be seen if you have good eyes, a clear moonless night, a good observing site and at least a 100 mm (4-inch) telescope.[197] But you will also have to be further south than about latitude 24° N[198] because Prox Cen is only about 27° from the South Celestial Pole. If you *are* south of latitude 24 ° N, then Prox Cen, in the sky, lies about 2° 11′ south-west of the very bright star, α Cen A (Rigil Kentaurus).

It seems appropriate, somehow, that the nearest exoplanets(s) to the Earth should be orbiting the nearest star. Furthermore, one of the Prox Cen's three exoplanets is Earth-sized and within the host star's habitable zone.[199] It also seems likely that one or more of the ways of sending spacecraft probes to the stars (Chapter 18) will be tried out soon[200] in order to examine these exoplanets more closely.

Prox Cen's three planets are named Prox Cen b, c and d in the usual fashion. Prox Cen d is closest to the star at 0.029 AU and has an orbital period of just 5.1 days. It is also the least massive of the three exoplanets at about a quarter of the Earth's mass. Little else is currently known about Prox Cen d, though it could have a surface temperature in the region of 360 K (~90 °C).

Prox Cen b is the exoplanet that might, just, be habitable. It is 0.049 AU from the star, so it could have a surface temperature near 235 K (~ −40 °C). Its mass is estimated to be fairly similar to that of the Earth but probably a little bit higher. It may well therefore have an atmosphere, and greenhouse effects might then bring its temperature up to tolerable levels for humankind, or at least for polar bears. But, before you book your ticket to go there, there are some downsides:

- The exoplanet's orbital period is 11.2 days and it is probably tidally locked (Appendix D) to its host star. It will thus have a permanently hot and a permanently cold side with, if it has an atmosphere, unusual and possibly very violent, circulatory patterns (i.e., storms). If the orbit is quite elliptical, then, like Mercury, it might be tidally locked for only a part of its orbit; making for even more complex meteorology (i.e., *even more* storms).
- Prox Cen, the host star, is a flare star,[201] so it can change its brightness by 10,000% in just a few hours; briefly bringing the exoplanet's temperature up to, perhaps, ~ 750K (~500 °C – hot enough to melt zinc).

---

[196] 'Nearest' in Latin is 'Proxima'. Prox Cen is also known as α Cen C.

[197] 150 mm or 200 mm (6-inch or 8-inch) telescopes will be a *lot* better.

[198] Canton (China: Latitude 23° 7′ N) and Havana (Cuba: Latitude 23° 8′ N) are the northern-most large cities from whence Prox Cen might be visible, light pollution permitting.

[199] It is not clear yet whether this proximity and the (potential / possible / perhaps) habitability is a quite remarkable coincidence or whether exoplanets are just so common that it is what would be expected – the author inclines somewhat towards the latter view, but has been wrong many times before.

[200] Decades rather than centuries.

[201] Larry Niven's short story 'Flare time' (in the anthology 'Limits', 114, 1985, Futura) is an entertaining account what life might be like on a tidally-locked exoplanet with, in this case, two flare stars as its hosts. The exoplanet, named 'Medea', was envisioned by a number of SF authors led by Harlan Ellison and those authors then went on to write their own stories based upon the Medea concept.

Niven's contribution is complete with human colonists, flourishing ETs and ETIs, the latter in the form of six-legged centaur-types. He, probably correctly, envisages some life-forms which flourish only in the intense light from the flares and so their lives go from start to finish within the duration of a flare; an hour or so. Other life-forms bury themselves during flares. However, his vision of a nice, calm, stable and comfortable-for-humankind climate is probably highly optimistic.

- Its orbit may be quite elliptical; an eccentricity (ellipticity) some six times that of the Earth's orbit is often quoted, but all that is reasonably certain is that Prox Cen b's orbit is less than 20 times more eccentric than the Earth's orbit. Even at six times the Earth's orbital eccentricity, Prox Cen b would be 6,500,000 km from the host star at periastron and 8,200,000 km from it at apoastron causing its basic temperature to cycle from ~250 K through ~235 K to ~220 K and back again every 11 days – and, when you consider how we are battling on Earth with a possible change of ~1.5 degrees occurring over many years (i.e., climate change), the mind boggles at what might be happening to the climate on Prox Cen b.

worst of all:

- You will have humans as your nearest neighbours.[202]

The existence of the third exoplanet, Prox Cen c is unconfirmed at the time of writing, but it may be a super Earth of about 7 M_{Earth} and in a much larger orbit than the two exoplanets just discussed. If the planet and its orbital size (~1.5 AU from the host star) are confirmed, then Prox Cen c, when at the section of its orbit closest to the Earth, will, for brief intervals, be able to claim the title of the Nearest EVER Exoplanet to the Earth, whilst in-between those intervals, Prox Cen b or Prox Cen d will be closest, though they will then be further away than Prox Cen c is at *its* closest moments.

Furthest

Most exoplanets discovered to date lie within the Milky Way Galaxy, however, if they can exist within our galaxy, then there is every reason to think that they can exist in other galaxies as well. Furthermore, there is some slight observational evidence that such extra-galactic exoplanets are indeed out there.

Almost all possible detections of extra-galactic exoplanets have been via gravitational microlensing (Chapter 7) and these events are usually one-off, so that confirming observations are impossible. However, at the time of writing, three possible detections have been announced:

- A ~3 M_{Earth} exoplanet about half way (~ 4,000 million ly) between the Earth and the Double Quasar[203] galaxy (QSO 0957+561).
- A ~6 $M_{Jupiter}$ exoplanet within the Andromeda galaxy (M31) at a distance of ~ 2.2 million ly.
- Several free-floating exoplanets lying within an elliptical galaxy which also act as a gravitational lens on the quasar RX J1131-1231. The masses may range from that of our Moon to several $M_{Jupiter}$ and the lensing galaxy is ~ 3,800 million ly away from us.

Other than their possible masses, no further details of these candidate exoplanets are known.

A possible extra-galactic exoplanet detected via an x-ray transit found in the Chandra[204] spacecraft's data for 2012 was announced in 2021.[205] The event occurred within Whirlpool galaxy (M51) whose distance is ~30 million ly. The exoplanet might be of about Saturn's size and in a ~19 AU orbit around its 20 M_{Sun} host star. Unfortunately, the orbital period would then be ~70 years, so confirmation will have to await the year 2082. Nonetheless, this observation is probably the most certain of those mentioned in this section. Other Chandra observations may indicate microlens-

---

[202] Unless, of course, α Cen A, only 0.21 ly away from Prox Cen, should have exoplanet(s) with lifeforms on it / them.
[203] Quasi-Stellar Object; the nucleus of extremely active radio galaxy, a.k.a. QSO. It is thought that a black hole with a million solar masses, or even a billion solar masses, lies at the Quasar's centre.
[204] Launched July 1999 and still operating at the time of writing.
[205] Di Stefano R., *et al*. Nature Astronomy, 1297, **5**, 2021

ing, by planetary-mass objects, of two other quasars by intervening galaxies whose distances of ~4,000 million ly and ~9,000 million ly, respectively, but these are not resolvable into individual exoplanets.

Most Massive

It's impossible to identify the most massive exoplanet at the moment and that situation is likely to continue forever. The reason for this uncertainty is that if you take a massive exoplanet and gradually add more mass to it, then at some point it stops being a large exoplanet and becomes a small brown dwarf star. The transition point (Box 4.2) is near 13 $M_{Jupiter}$, but no more precise figure is known and anyway, when you get close to that turn-over point, the actual metamorphosis probably depends on the age, temperature, composition, etc., of the object as well as the mass and so it will vary over a range of masses.

A single example of a large exoplanet is detailed below, but even finding that is not simple. Of the 5,580 currently known exoplanets only 30% have had their masses determined as being 15 $M_{Jupiter}$ or less and sometimes that is the actual mass (M) and sometimes it is M sin i and so is only a lower limit.[84] Additionally, almost all the mass determinations have large uncertainties in their values.

However, with the above caveats and cautions, the exoplanet HATS-70 b[206] is chosen as an example for this section.

HATS-70 b was discovered in 2018 from its transits. Its mass is 12.9 ± 1.7 $M_{Jupiter}$ and so it is probably an exoplanet, though it could still be a very small brown dwarf. It is in a 1.9-day orbit some 0.036 AU[207] out from its large (1.8 M_{Sun}) and hot (7,900 K, 7,600 °C) host star,[208] moving at about 200 km/s. The exoplanet's calculated surface temperature is ~ 2,700 K (2,400 °C) which is more than the lowest temperature (1,000 K, 700 °C) possible for brown dwarfs, but this is almost certainly due to its proximity to the host star and not to internal nuclear fusion reactions. The exoplanet's radius is ~1.4 $R_{Jupiter}$ (~200,000 km), it lies in Canis Major about 20° SSE of Sirius (α CMa) and is at a distance away from us of ~4,800 ly. By this book's classification system (Table 8.1), it is a hot, super-Jovian gas giant.

The most massive rocky exoplanet (a Mega-Earth), known at the time of writing, is BD +20 594 b which has a mass of ~16 M_{Earth} (± 6 M_{Earth}). It orbits a red dwarf host star in a period of 41.7 days and lies about 600 ly away from us in Taurus. Little more is known about the exoplanet, but, if its estimated radius (2.2 R_{Earth}) is about right, then its surface gravity will be around 3.2 g, i.e., when on BD +20 594 b's surface an average(ish) adult human being would weigh about the same as an average(ish) donkey does on the Earth.

Least Massive

For the least massive exoplanet, there is really no lower limit to the possible mass. For the purposes of this book a lower limit of $\geq 10^9$ kg (Box 4.2 – boulders, pebbles, dust and lesser objects; size ~ 100 m or less) has been chosen; but, although convenient, this is quite arbitrary.

Amongst the currently known normal exoplanets, the least massive is WD 1145+017 b (see also the Section 'Hottest') at 0.00067 M_{Earth} (4×10^{21} kg – about the same as Haumea). WD 1145+017 b orbits a white dwarf star in Virgo every 4.5 hours. Its orbital radius is ~0.005 AU (750,000 km) so its surface temperature is around 4,000 K (~ 3,700 °C); hotter than many normal stars and so even rocky material is being vaporised and lost to the exoplanet. The exoplanet is thus likely to continue surviving for only another 100,000,000 years or so.

---

[206] HATS – Hungarian Automated Telescope network – South. The system comprises six basic units. Each unit, in turn, comprises four 0.18-m telescopes giving an 8.2° × 8.2° field of view. The units are located in Australia, Chile and Namibia.

[207] The host star's habitable zone extends from ~ 2.5 AU to ~ 8 AU.

[208] Spectral type; A7, probably main sequence.

However, there seems to be no reason why the two interstellar objects – 1I/'Oumuamua and 2I/Borisov, which recently passed through our Solar System on hyperbolic[209] orbits, should not be included here since they are, after all, just small examples of free-floating exoplanets.

1I/'Oumuamua (Figure 8.9) was discovered in 2017 having come into the Solar System from roughly the present position in the sky of Vega (α Lyr). BUT, it has NOT come from Vega. When, 300,000 years ago, 1I/'Oumuamua was as far away from us as Vega is now, the star was nowhere near that point in space. 1I/'Oumuamua is now heading roughly towards Pegasus.

FIGURE 8.9   An artist's impression of 1I/'Oumuamua. (© C. R. Kitchin 2023. Reproduced by permission.)

1I/'Oumuamua was/is difficult to observe since it was only discovered when it was on its way out from the Solar System. However, it seems to have been unusually elongated (cigar-shaped) and this shape gave rise to speculation that it was an ETI artefact. Spectroscopy, though, suggests a rocky composition. Estimates of its physical dimensions vary by large amounts, but taking it to be in the region of $200 \times 50 \times 50$ m and having a density of 1,500 kg m^{-3} would place its mass at around 10^9 kg, just massive enough to scrape into this book's remit as a 'smaller object' (Box 4.2).

2I/Borisov was first detected in 2019 and probably has a cometary nature, being seen as surrounded by a gaseous coma and having a tail – like Solar System comets. It may originally have been some 400 m in size, but it fragmented into smaller pieces during its Solar System passage. Taking a density of 1,000 kg m^{-3}, though, would place its original mass at ~ 10^{11} kg.

Largest

As already remarked (Chapter 7), Jovian and super-Jovian exoplanets are all fairly similar in their sizes, because, as their masses increase, more material is compressed into their cores.[186] However, the hot super-Jovian exoplanets do have their outer layers inflated by their high temperatures and so may be somewhat larger.

---

[209] Most objects 'belonging' to the Solar System are in orbits shaped like ellipses. Some ellipses are very close to being circles (e.g., Venus' orbit) and some are very close to being parabolic (e.g., Halley's comet's orbit), but they are all a part of the Solar System. An object in a genuine parabolic orbit has just enough energy (but only just enough) to leave the Solar System. An object in a hyperbolic orbit has more than enough energy to leave the Solar System and so is not a member of the Solar System.

The largest exoplanet at the time of writing is thus ROXs 42B (AB) b.[210] This exoplanet's mass is ~ 9 $M_{Jupiter}$ and its diameter is ~ 2.5 $D_{Jupiter}$ (~ 350,000 km[211]). Although a long way out from its host star (~140 AU) it has a measured surface temperature of ~ 2,000 K (~ 1,700 °C). Its mass seems too small for it to be a brown dwarf, so this high temperature must be due to relict energy from its formation.

SMALLEST

Similar considerations apply to 'smallest' as to 'least massive' in respect of exoplanets. So 1I/'Oumuamua and 2I/Borisov are probably the smallest exoplanets known as well as being the least massive. Amongst the currently known 'regular' exoplanets, SDSS J1228+1040 b[212] at ~ 130 km diameter claims the title. It has a calculated temperature of ~1,800 K (~1,500 °C), an orbital period of 2 hours and lies about 500,000 km out from its white dwarf host star.

HOTTEST

At the time of writing, KELT-9 b[213] holds the high-temperature record at 4,600 K (~ 4,300 °C[214]). KELT-9 b lies within the hot Jovian gas giant class, though its temperature is much higher than is normal for that class. Its host star is of spectral type[215] B 9.5 with a surface temperature of ~10,200 K (~9,900 °C) and the exoplanet is only 0.034 AU (~5,000,000 km) above that surface. The exoplanet is also tidally locked to the star and so it is the sub-stellar point which reaches the aforementioned temperature of 4,600 K.

The temperature inflates the KELT-9 b's size to almost 2 $D_{Jupiter}$ (280,000 km) and so it loses mass at a rate of around 1 to 5 lunar masses every million years.

WD 1145+017 b (see also the Section 'Least Massive') follows KELT-9 b with a temperature of ~ 4,000 K (~ 3,700 °C).

The hottest host star of an exoplanet found to date is b Cen which is actually a naked-eye-visible binary star system about 330 ly distant from us. The primary star has a surface temperature of about 18,000 K. Its exoplanet (b Cen (AB) b), though, is ~ 560 AU out from the binary pair, so its equilibrium temperature is ~ 55 K (−220 °C). Unless the exoplanet has significant amounts of internal energy (which may well be the case), then with an estimated mass of ~11 $M_{Jupiter}$, it will be a cold super-Jupiter.

COLDEST

Free-floating exoplanets (Chapter 7) without their own internal energy sources and similar hosted exoplanets which are very distant from their host stars will eventually cool down to close to the temperature of the cosmic microwave background radiation (CMB; left over from the first few moments of the Big Bang). The current temperature of the CMB is 2.73 K (−270.42 °C). In theory, to cool down until their temperatures exactly equal that of the CMB will take an infinite time – and the CMB temperature is decreasing slowly anyway. However, we may expect many exoplanets, perhaps even most exoplanets, if the free-floaters are in the majority (see also PSR B1257+12's planets

---

[210] Rho (ρ) Oph X-ray source.
[211] It would just about fit between the Earth and the Moon, when the Moon is at perigee.
[212] Sloan Digital Sky Survey
[213] KELT – Kilodegree Extremely Little Telescope. a.k.a. HD 195689 b
[214] 1,500 °C hotter than the boiling point of iron.
[215] A star's spectral type is a classification system largely based upon the star's surface temperature. The basic classes, which range in surface temperature from ~3,500 K (3200 °C) to ~50,000 K (50,000 °C) are given the labels; M, K, G (Sun), F, A, B and O. Each of these is then subdivided into ten sub-classes, so that the Sun's spectral type is G2. Further details may be found via the internet or in the author's books; *Stars, Nebulae and the Interstellar Medium*, Adam Hilger, 1987 and *Optical Astronomical Spectroscopy*, Institute of Physics, 1995.

– Chapter 5), to be down to one or two tens of kelvin above the CMB temperature, if not by now, then at some future time.

MOST NONCHALANT

For a real laid-back customer, you can hardly better WD1054-226's Moon-sized exoplanet, which orbits just 2.5 million km out from a white dwarf star. Furthermore, since the star has a surface temperature of about 8,000 K it also is real cool (for a white dwarf) – and its exoplanet is right in the centre of its habitable zone.

In all honesty, though, the above sentences should have a few 'possibly', 'perhaps' and 'may-be' caveats incorporated into them since the interpretation of the data is not *that* certain.

What does seem to be (fairly) certain is that photometric observations of the white dwarf show that it fades regularly with several periodicities, including ones of 23.1 minutes and 25.02 hours. The latter might correspond to transits of an object with a diameter around 3,000 km (~0.9 D_{Moon}) orbiting the white dwarf at a distance ~2.5 million km. If so, the object's equilibrium temperature would be ~320 K (~50 °C). The shorter periods in the light curve could then be due to concentrations in an orbiting ring of debris caused by resonances with the larger object.

So, after all and *if* it exists, WD 1054-226's exomoon could be a trifle too hot to be truly non-chalant. But,

- there is an awful lot of uncertainty in the above figures, so it could be laid back after all
- if life is present on the object, evolution will have ensured that it is suited to the conditions
- if life is not present on the object, then the white dwarf is cooling down, although it will take a long time to do so. As a result, though, the orbiting object will be spending millions, perhaps billions, of years at cooler and cooler temperatures in the future. These changing conditions may then be suited to the origin and evolution of life – and possibly suited to the origin and evolution of different types of life at different stages in the cooling process.

So, could some suitably insouciant ETs be living there now or in the future? Well, that would, indeed, be way 'out-of-this-world'.

MOST MIRROR-LIKE

This is a little misleading – no exoplanet known reflects like a mirror (i.e., so that you see a reverse image of yourself when you look into it) and it seems unlikely that such an exoplanet will ever be found, although 'ever' is a *very* long time.

However, exoplanets and all other objects in space,[216] except black holes, do reflect light. The Earth, for example, reflects about 30% of the solar light and other solar electromagnetic radiations which fall onto it, straight back into space, without them transferring any of their energy to the Earth. This reflectivity is called the Earth's Albedo, and although there are several detailed definitions for 'albedo', it will suffice here to say that if an object reflects 30% of the radiation incident onto it, then its albedo will be ~0.3 (and ~0.4 if the reflections are 40%, ~0.5 for 50% and so on). Our Moon is a poor reflector, with an albedo of around 0.12 (12%), whilst Venus and some of the outer planet's natural satellites have albedos close to 0.8 (80%). Thus, within the Solar System, these latter examples win the 'Most-Mirror-Like' award.

For exoplanets, the 'Most-Mirror-Like' award currently goes to a hot Neptune, named LTT 9779 b, which orbits a star slightly cooler than the Sun, lying some 260 ly away in Sculptor. Parts of the

---

[216] Even stars reflect light when it falls onto them from some other source; it's just that the reflected light is swamped by the star's own emissions so that we do not notice it..

surface of LTT 9779 b reach a temperature of ~2,300 K (2,000 °C). So, it is *very* hot, but it is not as hot as might be expected, given that it is only 2.5 million km above its host star's surface.

The reason why it is not even hotter than 2,300 K lies in its albedo of 0.8 (80%) so that most of the star's radiation is never absorbed by the planet. The high albedo, in turn, arises from highly reflective clouds formed from molten titanium lying high above the exoplanet's (probably not solid) surface.

Thus, whilst LTT 9779 b misses out on the 'Hottest Exoplanet' (above) award, it still deserves its slightly less obvious place amongst these 'Juicy Bits'.

FASTEST ROTATION

Few exoplanets have had their rotations measured directly so far. β Pic b[217], though, is an exception, having had its rotation determined from the broadening of a carbon monoxide spectrum line in 2015. The rotational period was determined to be ~ 8ʰ 6ᵐ. This is shorter than Jupiter's 9ʰ 56ᵐ period, but it seems unlikely to continue to be seen as extraordinarily rapid when more data becomes available.

Exoplanets which are very close to their host stars are quite likely to be tidally-locked onto the star. Their rotational and orbital periods will then be the same. Thus K2-137 b (Section '*Smallest Orbit*'), whose orbital period is ~ 4ʰ 20ᵐ, may have this same value for its rotational period.[218]

SLOWEST ROTATION

Who knows? – and anyway what do we mean by rotation? An exoplanet which is tidally locked onto its host star is not rotating with respect to that star, but it is rotating with respect to everything else in the Universe – but it is the interrelationship with the star which has the greatest effect on the exoplanet.

We can use the Universe as a whole as a reference frame for rotational motions,[219] so somewhere and somewhen there must be/has been/will be an exoplanet which is not rotating as far as the rest of the Universe is concerned. We certainly have not found that exoplanet yet though.

Anyway, non-rotation with respect to the Universe would have no significance[220] for the nature of the exoplanet or for any ETs and/or ETIs living on it – so we'll ignore this particular 'juicy bit' from now on.

RETROGRADE AND SANDSTORMS AND BLISTERINGLY HOT

WASP-17 is a hot star some 1,300 ly away from us, which lies close to Antares in the sky. Its exoplanet, WASP-17 b, was the first exoplanet system to be found (in 2009) in which the orbital motion is in the opposite direction (retrograde) with respect to the direction of rotation of its host star.

This exoplanet is also distinguished by being shown recently, via JWST observations, to have nanocrystals of silica (a.k.a., quartz, sand) floating in its atmosphere. Although the silica crystals are very tiny, they are being carried by winds moving at thousands of km/h and which have a temperature around 1,500 K (~1,200 °C); any ETI capable of surviving under those conditions could surely conquer the entire Universe one-handed before breakfast.

---

[217] A 13 M$_{Jupiter}$ hot, super-Jovian gas giant hosted by the naked eye star, β Pic and lying some 60 ly away from us.

[218] For comparison, the Earth would have to rotate every 1ʰ 20ᵐ or so, for objects on its surface and which are near to the equator to be at risk of being flung off into space (i.e., for the Earth to start breaking-up).

[219] The ability of the Universe to form an absolute reference for rotational motions, but not for linear motions is an intriguing problem which I leave for the reader who has *lots* of spare time to research for him/her self.

[220] Actually, it would probably make such an exoplanet into a renowned tourist destination and the ETI astronomers would not need drives for their telescopes; any object in the sky to which the telescope was pointed would just stay in the field of view forever, or until the telescope was deliberately moved to view another object.

OLDEST

The oldest exoplanet currently known is one of the pulsar exoplanets (Chapter 5) called PSR B1620-26 b (Table 5.2). This exoplanet's origin is put at about 12.7 to 13 billion years ago since that is the age of the globular cluster M4, of which it forms a part. Given that the age of the universe is currently estimated at ~13.7 billion years, there cannot be many older exoplanets than this one.

YOUNGEST

Since stars are still being formed throughout the Universe today; so also must exoplanets still be originating. Somewhere, therefore, there will be exoplanets in all the earliest stages of their formation. However, at these very early stages we cannot know whether *that* slight increase in the density of *that* molecular cloud over in *that* direction will disperse again, or, if it continues to contract, whether it will become a star, brown dwarf, exoplanet or a couple of dozen big boulders.

   The youngest object found to date therefore and which is fairly clearly an exoplanet is V830 Tau b.[221] Its age is estimated to be ~ 2,000,000 years. The host star is a cool T Tauri star[222] slightly brighter than the Sun, but with about the same mass as the Sun. The exoplanet has a mass of about 80% that of Jupiter and it orbits ~0.06 AU (~9,000,000 km) out from its host star once every ~5 days. T Tauri stars are generally young stars, but V830 Tau is a particularly youthful example.

SMALLEST ORBIT

Some white dwarf stars show signs that an exoplanet has crashed into them relatively recently and been absorbed into the star – and you can't get orbits any smaller than that (i.e., exoplanet's orbital radius = host star's radius).

   These signs shown by the white dwarfs are of 'pollution' in their surface layer compositions. The chemical elements making up most white dwarf stars are carbon and oxygen, oxygen and neon or oxygen, neon and magnesium.[223] Quite a high proportion, though, also have small quantities of elements such as calcium, iron, magnesium and others (heavy element 'pollutants') detectable in their surface layers. These dense elements will sooner or later settle down towards the white dwarf's core and so cannot have been present in the outer layers for very long. It is now widely accepted that these pollutants come from the core of an exoplanet which approached so closely to the white dwarf that it was broken up by tidal forces and its fragments then fell into and were absorbed by the white dwarf (Box 8.2).

---

[221] The standard designation for a variable star; making it the 830th variable star to have been found (many years ago) within Taurus.

[222] A type of variable star.

[223] A small proportion of white dwarfs may be predominantly helium, at least for their surface layers.

BOX 8.2 – ENDINGS FOR THE SOLAR SYSTEM PLANETS AND FOR EXOPLANETS (AND FOR A FEW OTHER THINGS AS WELL)

Crashing into their host stars is one way in which planets and exoplanets might meet their ends and the metal pollution of white dwarf atmospheres is one line of evidence that this can occur.

For a white dwarf to engulf an exoplanet, that exoplanet must already have survived engulfment in the red-giant precursor of the white dwarf or (much less likely) be a captured free-floating exoplanet. If there is still some gas surrounding the white dwarf, then interactions with that gas ('drag') may suffice to cause the exoplanet's orbit to approach the white dwarf even more closely. Eventually, tides will pull the exoplanet apart[98] and the fragments in their turn, probably by now within the star's outer reaches (corona), will crash into the solid surface of the white dwarf. This will undoubtedly produce a strong very high temperature flare which will probably be observable from the Earth.

Recently, x-ray emission has been detected by the Chandra spacecraft coming from the white dwarf, G29-38, which is some 60 ly away from us in Pisces. The star's normal surface temperature is ~12,000 K (~12,000 °C), but the sources from which x-rays are originating are at over 100,000 K and so it seems possible that this is direct evidence of the accretion of exoplanetary debris onto a white dwarf.[224]

If two or more exoplanets survive (Chapter 5) in orbit around the white dwarf, then their tidal interactions may cause one of them to crash into the star. Given long enough (many times the current age of the Universe), gravitational radiation[225] and other small effects will lead to all white-dwarf exoplanets crashing into their stars, unless other interactions cause them to be flung out from the system to become free-floating planets.

Similar futures, to those of white dwarf exoplanets, await the exoplanets orbiting low mass stars ($< 0.25 \, M_{Sun}$) and brown dwarfs, neither of which will evolve into red giants, but the time scale will be *much* longer.

Some of the planets orbiting the Sun and other stars with masses more than about $0.25 \, M_{Sun}$ face the same fate as some of the white dwarf exoplanets, being engulfed by their host stars, and this may well include the Earth.

But don't worry too much about it; for the Earth, feeding the Sun is some 5,000 million years into the future and may not happen anyway, although other changes may render the Earth uninhabitable before then. Recent super-computer models, for example, suggest that plate tectonics will lead to the merger of most current terrestrial land masses into a single super-continent in around 250 million years from now. Combined with the slow evolutionary rise in the solar luminosity, terrestrial surface temperatures are then likely to exceed 50 °C and lead to the extermination of most, if not all, mammalian types of life. Of course, evolution can do a lot in 250 million years, so, maybe some highly evolved varieties of humankind will still be around to bask in that extreme heat.

The possible fates of the Earth and with greater certainty, those of Mercury, Venus and of those exoplanets orbiting in the near vicinity of all except the least massive host stars, arise from the evolution[97] of those host stars.

Except for the very low mass stars, when a star's supply of hydrogen for its fusion energy generation reactions begin to come to an end, its outer regions start to swell enormously and

---

[224] In 2020, a transient event occurred some 12,000 ly away in Aquila and given the name of ZTF SLRN-2020 (but it is also known as the 'Red Nova') and which has been plausibly interpreted as a Jovian exoplanet being engulphed by its evolving host star. Two similar(ish) earlier events have also been observed.

[225] For further details of gravitational radiation / gravitational waves see, for example, the Author's *'Understanding Gravitational Waves'*, 2021, Springer or conduct an internet search for 'gravitational radiation'. In the latter case note that the term 'Gravitational Waves' is also used for very different types of waves in the Earth's atmosphere.

the star becomes a Red Giant.[226] The main sequence lifetime for a star, however, varies enormously with its mass:

Stellar mass	Main sequence lifetime
~0.4 M_{Sun} (M0)	– ~200,000 million years
~0.8 M_{Sun} (M0)	– ~15,000 million years
~1.1 M_{Sun} (G0)	– ~10,000 million years
~1.8 M_{Sun} (F0)	– ~3,000 million years
~3.0 M_{Sun} (A0)	– ~500 million years
~15 M_{Sun} (B0)	– ~10 million years
~40 M_{Sun} (O0)	– ~1 million years

Thus, the Sun (spectral class, G2) has a main sequence lifetime of about 10,000 million years, but it is already some 4.500 million years old and so it will actually start evolving towards being a red giant in the aforementioned ~5,000 million years.

What will happen to us then?

Well, humankind, or even our very distant and highly evolved descendants, are unlikely to be around then (see the Section 'Kepler 452 b' in Chapter 13 for a further discussion of this topic), although Tardigrades (or their very distant and highly evolved descendants – Chapter 12) just might still be around. The Earth itself, though, will be either just inside or just outside the outer reaches of the red-giant Sun. The Earth's surface temperature will then probably reach ~2,500 K (~2,200 °C) and it will lose all its volatile elements and compounds. The Sun's red giant phase will last for ~150 million years and it will lose some one-third of its mass during that time via an enormously more intense solar wind than the Sun has at present. During this time the Earth may fall into the Sun as it loses energy whilst moving through the Sun's outer layers and so be totally destroyed. Alternatively, it may survive as a burnt-up cinder in its present orbit or even be in a larger orbit because of the loss of mass from the Sun.

Mercury and Venus will almost certainly be completely destroyed during the Sun's red giant phase, whilst the outer planets, including Mars, will probably survive after losing some or all of their volatile materials and also being in larger orbits than at present.

Thus in, say, ~6,500 million year's time, the Solar System will comprise a white dwarf host star with five or six planets and a few asteroids.

In a similar, but sometimes shorter, sometimes longer, periods of time, most exoplanet systems around most host stars will suffer a related fate and end up in a kindred state to that forecast for the Solar System. For example, the hot Jovian exoplanet, KELT-16 b, is currently in a 23.3-hour orbit, ~3 million km out from its-rather-more-massive-than-the-Sun host star and will probably be engulfed by that star within the next 500,000 years, So, keep your eyes peeled; it is about 1,200 ly away and is to be found (using a moderate-sized telescope) in Cygnus, about 12° South of Deneb.

Although significantly less probable than colliding with the host star, exoplanets can also collide between themselves, perhaps destroying each other completely, perhaps coalescing, perhaps breaking into smaller bodies. Many theories of how the Earth-Moon system formed

[226] One of the brightest naked-eye stars, α Orionis (Betelgeuse), is a red giant. Its mass is ~18 M_{Sun} and its radius is ~4.5 AU; sufficient to engulf Mercury and Venus *plus* the Earth, Mars and almost Jupiter were it to take the place of our present Sun.

suggest that the proto-Earth and a Mars-sized proto-planet collided during the early stages of the formation of the Sun and Solar System, leaving the proto-Moon orbiting the proto-Earth. The double star, BD+20 307, is currently surrounded by a warm disc of dust and it seems possible that this could have originated recently from the collision of two rocky exoplanets. The star ASAS-SN-21qj[227] has similarly undergone a brightening and fading since 2012 which is consistent with two icy giant exoplanets colliding (the brightening) and then the resultant debris cloud transiting the host star (the fading).

Stars with masses above ~8 M_{Sun} are likely to become supernovae and destroy most of their exoplanets in the process. But unlikely as it may seem, it *is* possible for supernova remnants to have a planetary system (Chapter 5).

In the *REALLY* distant future; a million times the present age of the Universe or so, all stars will have completed their energy-emitting stages. Most will have become black dwarfs with temperatures very close to absolute zero, and some will also have further collapsed into black holes. Planets will have been absorbed by their host stars or flung out to roam the Universe alone.

In the *REALLY, REALLY* distant future (and if some physics theories are to be believed), protons may be unstable particles with half-lives estimated to lie between 10^{20} and 10^{35} years, decaying into neutral pions and positive electrons (positrons). The positrons are stable insofar as is currently known, whilst the neutral pions decay in just 10^{-16} seconds into two photons; the photons then also being stable. Stellar mass and larger black holes may evaporate through Hawking radiation[225] into positive and negative electrons and (stable) neutrinos on time scales of ~10^{100} years.

Thus, in somewhere around

10,000,000,000,000,000,000,000,000, 000,000,000,000,000,000,000,000,000,000,
000,000,000,000,000,000,000,000,000, 000,000,000,000,000,000,000,000,000,000,
000,000,000,000,000,000,000,000,000, 000,000,000,000,000,000,000,000,000,000,
000, 000,000,000,000,000 years

from today, we shall, all of us:

Humankind, Terrestrial Life-forms, ETs. ETIs, the Earth, the Solar System,
the Sun, all the stars, all the nebulae, all the black holes,
all the galaxies and anything else around,

will exist only in the form of individual positive and negative electrons, photons and neutrinos, with each such particle being separated from the next by at least the diameter of the current universe.

Then, indeed, will Wordsworth be able to:

"… wander lonely as a cloud"[228]

and after that – who knows? Certainly, *nobody* currently living on the Earth does.

---

[227] All-Sky Automated Survey for Supernovae.
[228] Wordsworth W. *Poems in Two Volumes*, 1807. I have altered the original slightly (from 'wandered' to 'wander').

The arguments for the origin of white dwarf pollution from exoplanets are sometimes reversed, so placing the first detection of an exoplanet as occurring in 1917. That date was when Adriaan van Maanen, working with the 1.5-m telescope at the Mount Wilson observatory, first observed calcium and iron spectrum lines in a white dwarf star.[229]

Amongst the normal exoplanets, the smallest semi-major axis, at the time of writing is for K2-137 b, at 0.0058 AU (~900,000 km). It is a hot, Jovian gas giant with a mass of ~ 0.5 $M_{Jupiter}$ orbiting a red dwarf star some 300 ly away from us in Virgo.

LARGEST ORBIT

The red dwarf star, TYC 9486-927-1 is close to the Southern Celestial Pole, in Octans. It is about 110 ly from the Earth and has a borderline exoplanet/brown dwarf in orbit around it at a distance of some 6,900 AU (~ 1×10^{12} km). The exoplanet's mass is ~ 13.3 $M_{Jupiter}$ and its orbital period is ~900,000 years. Its temperature is estimated at ~ 1,800 K (~ 1,500 °C), but very little of the energy for this comes from its host star. The exoplanet, unusually, is not named for its host star but has the separate name of 2MASS J2126−8140.[230] The exoplanet differs little from being a free-floating exoplanet and only a very minor gravitational disturbance would suffice to free it from its host star.[231]

MOST PROLIFIC HOST STARS

Some host stars have been found with several exoplanets and undoubtedly, as exoplanet detection methods become better, the number of such systems and the number of exoplanets within them will increase rapidly.

For now, let us have a round of applause for the award for the most prolific host star, which goes to, ….

The Sun

As we have seen (Chapter 8, Table 8.1), by the definitions adopted for this book, the Solar System has some 32 objects orbiting the Sun whose masses are at least that of Ceres.

You may feel that this is cheating, but it illustrates the point that our data on exoplanets is in its infancy and that we may expect that most host stars will have multiple exoplanet systems, some, I am sure, involving far greater numbers than the poor old Solar System.

Apart from the Sun, the next most prolific host star is Kepler-90 (Figure 8.10). This star is an analogue of the Sun but about 20% more massive and it is some 2,800 ly away from us, in Draco. It has eight identified exoplanets (Figure 8.10) at the time of writing. All the exoplanets are close to their host star, with semi-major axes ranging from 0.074 to 1.01 AU (10,000,000 to 150,000,000 km), with their sizes ranging from 30% larger than the Earth to 11 times larger than the Earth and with their masses ranging from 1.8 to 400 M_{Earth}.

---

[229] Van Maanen, at the time though, thought that he was studying a hot main sequence star and so was not surprised by the composition.

[230] 2 Micron All Star Survey.

[231] The phrase 'One flap of a butterfly's wing' comes to mind, but that probably would not be quite enough. In fact, to give a 13.3 $M_{Jupiter}$ exoplanet a velocity of 1 mm s^{-1} would require ~22 times all the energy generated by humankind in the whole of the year 2021 (~6×10^{20} J).

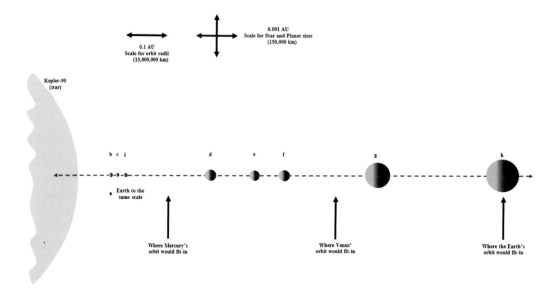

FIGURE 8.10 Kepler-90 and its exoplanetary system (N.B. note the scale difference between the orbits and the star's and exoplanets' sizes. If the orbital scale were to be made the same as the planet scale, then the page would need to be some 50 feet (~16-m) wide. Add where Jupiter's orbit would fit in and the page becomes ~250 feet wide. Include Neptune and the page becomes over a quarter of a mile wide (nearly half a kilometre). (© C. R. Kitchin 2023. Reproduced by permission.)

TATOOINES[232] AND OTHER BINARY STAR HOSTS

The majority of exoplanets discovered to date have single stars as their hosts. This may seem natural given that the (single star) Sun is host to all the Solar System planets and to all the other Solar System objects. However, when we look around the galaxy at stars, a third, perhaps even half, of them are binary stars or form a part of multiple star systems, so binary, etc. host stars should be quite common.

Now, there are possible reasons why exoplanets should be rarer amongst binary stars; the main one being that in many circumstances their orbits may be unstable, so that the exoplanet gets flung out of the system or collides with one of the stars – and that situation applies in spades to multiple star systems. Stable orbits may exist in complex systems, though, Jupiter's Trojan asteroids[233] being one example, but such orbits are few and far between.[234]

There are, though, two circumstances when binary star exoplanets can have orbits which are stable over stellar lifetimes. The first of these is the Tatooine configuration wherein the stars are close together and the exoplanet orbits them, almost as though they were a single star, from much further out. However, a Tatooine exoplanet is orbiting stars which themselves are moving around their orbits and so, when such a planet is detected via the transit method (Chapter 7), the intervals between the transits are likely to vary slightly.

---

[232] In case you have only just arrived back on Earth following your 50-year long round trip (in hibernation, of course) to Prox Cen b, Tatooine is the home planet of Luke Skywalker, a leading character in many of the *Star Wars* films and games. Tatooine is a hot(ish) version of the Earth with a predominantly desert climate. It orbits a close binary star system and its fictional name seems to be well on the way to being used as the official name for exoplanets with such binary star hosts (actually, their official name is 'circumbinary planets').

[233] The Trojan (and Greek) asteroids are a small number of asteroids which have (very roughly) the same orbit around the Sun as does Jupiter, but are (again very roughly) 60° (~ 800,000,000 km) ahead or behind the planet.

[234] At the time of writing, of the 5,580 or so known exoplanets, 40 have host stars which may be part of triple or quadruple stellar systems.

The second configuration is when the exoplanet is in a close orbit around one of the stars so that *that* star dominates its motions and the second star is then sufficiently far away that it has little effect on the exoplanet. This latter configuration has yet to appear in a *Star Wars* feature and so it does not yet have a nice, evocative designation. Some tens, maybe hundreds of this type of binary star host have now been found, in most cases though, the stars' separation is large so that they differ little in practice from single-star-host exoplanets systems. Some triple-star hosts and at least one four-star host have also been found.

Several Tatooine-type exoplanets have been detected, with Kepler-16 b being a good example (Figure 8.11). Kepler-16's two stars are less massive than our Sun and are about 0.22 AU (~30,000,000 km) apart in their 41-day orbit around each other. Kepler-16 b is a cold Jovian gas giant in a 229-day orbit, 0.7 AU (100,000,000 km) out from the binary stars.

FIGURE 8.11   Kepler-16: An artist's impression of the binary star system as might be seen from a spacecraft near the Tatooine exoplanet. (Reproduced by kind permission of NASA/JPL-Caltech/T.Pyle.)

A second, recently discovered, tatooine system is TOI[235]-1338 b and c. This latter system has a binary star comprising a slightly larger star than the Sun and a Red Dwarf separated by ~0.13 AU and with a 14.6-day orbital period. The binary system lies about 1,300 ly away and may be seen using a moderate-sized telescope in Pictor. It is host to two tatooine exoplanets: TOI-1338 b has about twice Neptune's mass and orbits the binary every 95 days at a distance of ~0.46 AU, while TOI-1338 c has almost four times Neptune's mass and orbits the binary every 216 days at a distance of ~0.8 AU.

STRANGEST COMPOSITION

Few exoplanets' bulk chemical compositions are known yet, so this particular accolade goes to PSR B1257+12's three exoplanets (Chapter 5) whose composition *may* be largely metallic.

---

[235] TOI -TESS Object of Interest

THEME 2

ETs and ETIs (Plus HUMANKIND)

PREPARATORY NOTE – THE BEST WE CAN DO AT THE MOMENT

In Theme 1 (Exoplanets) we were examining material that has been observed, measured, studied and even sampled (for solar-system objects). Thus, although it is still quite a new field of astrophysics, it has plenty of reliable data and a firm scientific basis. A lot of the subject, nonetheless, remains speculative (i.e., guesswork – see all the 'Artist's Impression' images in Chapters 3 to 8). Some speculations may be good, and some will undoubtedly be far from the mark, if not actually wrong, but they are all 'The Best We Can Do at the Moment'. Given the number of exoplanets now known (~ 5,580), we may hope, though, that the 'good' speculations will be in the majority

Moving on now to our second theme (Life in any and all of its known and unknown varieties), we know a great deal about current life on Earth and quite a fair amount about how life on Earth once was, millennia to aeons ago. Beyond that, though, it is *all* 'The Best We Can Do at the Moment'. Again, some speculations may be good, some will undoubtedly be far from the mark, and many now though will simply be wrong. The overall balance will also move away from 'good' towards 'wrong'.

So, for the next few chapters, please bear in mind that, now, fewer of science's arrows will be hitting the gold and more will be missing the target altogether – but it *is* The Best We Can Do at the Moment. – or as Søren Kierkegaard expressed it, almost two centuries ago:

> "Det er ganske sandt, hvad Philosophien siger, at Livet maa forstaaes baglaends. Men derover glemmer man den anden Saetning, at det maa leves forlaends. Hvilken Saetning, jo meer den gjennemtaenkes, netop ender med, at Livet i Timeligheden aldrig ret bliver forstaaeligt, netop fordi jeg intet Øieblik … (lever i gang) … baglaends."[236]

In short (and translated):

> Life can only be understood backwards; but it must be lived forwards.

---

[236] It is quite true, what philosophy says, that life must be understood backwards. But above that, one forgets the second part that it must be lived forwards. Which concept, the more you think about it, just ends with the fact that life at any given instant will never be quite understandable, precisely because I can never for a moment … (live going) … backwards.
Søren Kierkegaard. *Journals,* 1843

DOI: 10.1201/9781003246459-10

9 World(s) Enough, and Time[237]

INTRODUCTION

The opening words of Marvell's ode *To his Coy Miftrefs*[237] form an appropriate initiation to this book's next major theme:

> how did life originate and develop on the Earth (and perhaps elsewhere)?

The basic answer to that question is:

> if something *is* possible
> and
> you have *enough* working-space (worlds) available
> and *enough* time available,
> then it is bound to occur
> somehow, somewhere and somewhen.

The second link of the Ode to this theme's concerns, though, is less immediately apparent. The full first line of the poem reads 'Had we but World enough, and Time', and these days it is usually used as a lament by someone living such a busy life that not everything which she/he wants to do can be done.

Marvell's intention though was much more specific; he was lamenting the passage of time on the human biological clock and that his desire for (pro)creating new life with his mistress was being delayed by her refusal (coyness) to cooperate. Later in the poem he chides her with:

> 'But at my back I alwaies hear
> Times winged Chariot hurrying near:
> And yonder all before us lye
> Defarts of vaft Eternity.'[237]

and

> 'The Grave's a fine and private place,
> But none I think, do there embrace'[237]

So, almost four centuries ago, Marvell expressed the essences for life's origination (whether it be for the first time throughout the entire Universe or one of innumerable independent repetitions minutes, centuries or many, many aeons later) quite clearly as:

---

[237] Andrew Marvell. '*To his Coy Miftrefs*'. Written, possibly, in the 1650s, published posthumously 1681. Spelling as in the original.I have taken the liberty of adding an ''s' to Marvel's original 'World'; I'm sure he would have used the plural, since it scans better, had he known about all 'our' exoplanets.
Translations:
Miftrefs' – Mistress,
alwaies – always,
lye – lie,
Defarts – Deserts.
vaft – vast.

DOI: 10.1201/9781003246459-11

Copious Room and Plentiful Time.

Marvell's concerns with sex and death will come a bit later-on in our story; once we have got life itself in *some* form going.

Another rummage through the pick-and-mix of history, although not quite so far back on this occasion, brings us to Charles Darwin (Figure 9.1) and his letter of 1871 to Joseph Hooker. Most people associate Darwin with his book *On the Origin of Species*[238] (though far fewer have actually read it). Darwin, in that book, was concerned with the origin of *SPECIES* (i.e., Evolution, Natural Selection/Survival of the Fittest/Transmutation[239]) and *not* the origin of *LIFE*. We will come to the evolution of life in Chapter 10.

FIGURE 9.1   A statue of Charles Darwin sited outside the Natural History Museum, London. (Reproduced by kind permission of Pixabay and Andrew Martin.)

But Darwin did also ponder life's origin as well. Thus, in 1837, just after returning from Beagle's second voyage, we find him writing in his *Notebook C*[240]:

"The intimate relation of Life with laws of chemical combination, … render spontaneous generation,[241] not improbable. "

This remained Darwin's life-long opinion, so that three and a half decades later, he was expressing it in the far-better-known Hooker quotation[242]:

---

[238] Full title: *On the Origin of Species by Means of Natural Selection, or the Preservation of Favoured Races in the Struggle for Life*. 1859, John Murray.

[239] These are not all exact synonyms. Evolution is the *phenomenon*, Natural Selection / Survival of the Fittest / Transmutation, etc., are the names for the *mechanism* by which evolution works.

[240] http://darwin-online.org.uk/ - *Notebook C*: [Transmutation of species]. CUL-DAR122.

[241] Of a new life-form. Spontaneous generation of life is often, now, called Abiogenesis because spontaneous generation, as envisaged by Aristotle and others over the last two millennia, includes suggestions such as fleas being spontaneously generated from dust particles and maggots spontaneously generating from dead flesh, rather than today's idea that it occurred only rarely, at the molecular level and that it was billions of years ago.

[242] Letter to Joseph Hooker, 1st February 1871. Darwin Correspondence Project, https://www.darwinproject.ac.uk/.

"It is often said that all the conditions for the first production of a living organism are now present, which could ever have been present. - But if ... we could conceive in some warm little pond[243] with all sorts of ammonia & phosphoric salts, - light, heat, electricity[244] &c present, that a protein compound was chemically formed, ready to undergo still more complex changes, ..."

The quote continues with Darwin writing that nowadays such a protein would be "instantly devoured" by present-day life-forms, but the clear, though unwritten, implication is that in the *absence* of those present-day life-forms, the protein *would* have continued with its 'complex changes' and potentially could thus have become a recognisably living entity.[245]

As the philosopher/scientist, Iris Fry, similarly expresses the situation today[246]:

"The question of the origin of life is among the most difficult problems faced by science today. ... Thus, the scientific goal is to reconstruct a possible scenario, ... that could have led to the first living systems. Optimistically, as Earth science, astronomy, and other relevant fields of science gain more knowledge about conditions on the primordial Earth, the gap between the possible and the actual will become narrower. Various scenarios delineating possible mechanisms of emergence are being raised ... , but none of them has as yet gained the consensus of the origin-of-life community and none has as yet been simulated in full in the laboratory. Investigators are nevertheless convinced that the transition from chemical compounds on the primordial Earth to the first living systems was an evolutionary process."

and herself quotes Leslie Orgel[247] from half a century earlier:

" ... any 'living' system must come into existence either as a consequence of a long evolutionary process or a miracle".

So, for almost two centuries, given plenty of Space and Time, 'Spontaneous Generation' or 'Abiogenesis'[241] in a 'warm little pond', or its variations, has been credibly proposed as the way in which life on Earth could have originated. Over those two centuries, whilst much effort has been

---

[243] 'Warm little ponds' may actually have been quite common, quite large and quite hot just before life emerged on the early Earth. Over the period between about 4,100 million and 3,800 million years ago, the Earth and other inner Solar System objects may have been subject to a very intense storm of impacts by proto-asteroids and proto-comets, known as the Late Heavy Bombardment (LHB). This would have left numerous craters suitable for filling with water and, for the bigger impacts, would also have left a large buried mass of hot/molten rocks below the crater to provide energy to heat that water for a long time after the impact (c.f. today's hydrothermal vents – see below and Chapters 7 and 12).

[244] Darwin was aware that the origin of life would need energy as well as the appropriate chemicals; however, he was less specific about what might be the energy source. Likely sources would include, depending upon the local environment: Cosmic rays and other high energy particles, Energy from gravitational collapse, Exothermic chemical reactions, Lightning, Radioactive decay, Shock energy from impacts, Stored internal energy from within the exoplanet (i.e., geothermal, hydrothermal, plate tectonics, volcanoes, etc. for the Earth), Tides, UV, visible and Infrared radiation.

[245] One assumption behind Darwin's ideas and also more modern theories, which is often not spelled out, is that the laws of physics, chemistry, biochemistry, etc., are assumed to be essentially the same here and now as they were/are billions of light years away and billions of years ago.

We do know of some things which have changed; when we look at objects at great distances away from us (>10,000 million light years) and long ago, many of the heavier elements had not then been synthesised in significant quantities inside stars and distributed into the wider Universe via stellar explosions. The abundances of important life elements, such as carbon and oxygen, were therefore then much lower than they are today.

More fundamentally, Relativistic gravitational time dilation means that we see events happening much more slowly at great distances away from us than if they were quite nearby. A recent measurement suggests that time runs five times slower than for ourselves on objects observed as they were a 1,000 million years after the Big Bang. Perhaps surprisingly though, the same laws of physics, etc., as ours today, would actually be deduced by life-forms existing a 1,000 million years after the Big Bang and a second of time would have appeared to them to be the same as a second of time does to us ®. Readers puzzled by this may consult the author's book '*Understanding Gravitational Waves*', 2021, Springer or conduct an internet search for 'the Mössbauer Effect' for further explanations.

Fortunately, insofar as we can tell, the laws of physics, chemistry, biochemistry, etc., do not differ significantly from today's versions even when looking back to a time just 1,000 million years after the Big Bang.

[246] Kolb V.M. (Ed.), Section 6.5.1, *Handbook of Astrobiology*, Taylor and Francis, 2019.

[247] The *Origins of Life*. 1973, New York: Wiley

put in to verifying that idea, it still has yet to be proven to be correct by some scientists actually mixing the right non-living ingredients together in the proper way and producing something which can clearly be described as 'living'.

But What IS Life?

Like many cooking recipes (see 'First catch your hare' – Chapter 4) our discussions have run somewhat ahead of the first and most basic requirement for the topic; in this case:

> before we can decide that some process, whether in nature or in the laboratory,
> *has* created life,
> we need to be able to decide if something
> *is* living or non-living.

There is no universally accepted definition of life. In fact, there isn't even agreement on how many definitions of life have been suggested. Some recent publications, which have attempted to list all the proposed definitions of life, have come up with figures ranging from 77 to 95.

A recently developed and developing quantitative approach to defining life is advocated, *inter alia*, by Leroy Cronin and Sara Walker and is called the Assembly theory. It is based upon a simple algorithm for defining the complexity of a molecular structure, which results in a single integer number, and this is the complexity class of the structure. The higher that number, the higher the probability that the creation of the structure involved life in some way. But, even Assembly theory cannot (yet) suggest a complexity number whereby non-life changes to life.

Probably the division between life and non-life occurs somewhere in the region between viruses and prions. Most of the 77 (or 95) definitions of life, class viruses as non-life, although the definition I have adopted for this book (below) does class viruses as being living organisms.[248]

Viruses can *reproduce*, but they utilise their host's protein synthesis mechanisms to do so. However, viruses can and do *evolve*, as Covid-19, Influenza, SARS and many other examples have recently demonstrated to our sorrow.

Prions (the agents causing Creutzfeldt-Jacob disease, Mad Cow disease and Scrapie) are proteins whose structure is folded abnormally. They can cause the same protein in a living organism to change its usual benign folded structure to their harmful folded version, so leading to the above diseases. Prions are generally classed as non-living entities, although they *can* evolve.

The definition of when non-life becomes life adopted for this book is deliberately very basic and all-inclusive and is:

> The transition from non-life to life occurs when
> a single molecule or a complex of molecules
> develops which can replicate[249] itself
> but which
> makes mistakes in the replication occasionally[250]

– and that's it (see also First-Living-Molecule, below). The advantageous mistakes make further progress, while the disadvantageous mistakes do not progress, or at least progress less well, i.e., the fundamental operating mechanism behind Darwinian evolution.

---

[248] As a 'Gee-Whiz' aside at this point, if a definition of life excludes viruses, then the amount of life on Earth is reduced by a factor of ~10. The current estimated number of individual prokaryotic viruses existing on the Earth is around 10^{31} and if you were to lay them out in a line, then that line would stretch for some 40 million light years ®.

[249] Chemical reactions whose products promote the occurrence of the same reactions again, whether abiotic or biotic, are also called 'autocatalytic' reactions.

[250] Advantageous mistakes are what lead to evolution. If disadvantageous mistakes occur too frequently then they can accumulate leading to what is called the 'Error Catastrophe' when the whole evolutionary line becomes extinct.

The above is my own definition, but it concurs with NASA's almost identical operational specification[251]:

> "Life is a
> self-sustained chemical system
> capable of
> undergoing Darwinian evolution"

You might think that such broad definitions would cover everything, but a simple counter-example will quickly show that even these 'all-encompassing' definitions are inadequate (Figure 9.2).

FIGURE 9.2 A counter example to this book's and NASA's definitions of when something is alive. A mule train in the Zion National Park, Utah (about 220 km NE of Las Vegas), May 2023. Mules and hinnies are sterile; they cannot reproduce (never mind evolve) and so by both criteria of both above definitions, they are *not* alive! Their riders, though, (currently in the nearby saloon) would probably think their drinks had been spiked, if informed of this minor detail. (Reproduced by kind permission of Sue Harris and Maddie Pass.)

So, who is wrong? – NASA and I, of course.

The many alternative definitions of when something is alive though, quickly run into similar problems to these two definitions and/or soon have so many special cases and caveats as to be of little practical use.

Additionally, there will probably soon ✿✿ be the question of 'Is a self-reproducing robot a form of life?'. On the definition adopted for this book and on some of the other 77 (or 95) definitions, the answer must be 'Yes'. Further discussion of this topic and the related question of 'Can a machine be intelligent?' may be found in Chapter 13 and Appendix E. What further complications may arise when we *do* encounter real ETs and ETIs are (literally) beyond our current imaginations.

[251] Kolb V.M. (Ed.), Section 1.1.4.1, *Handbook of Astrobiology*, Taylor and Francis, 2019.

So, what can we do?

The only certainties about a precise and accurate definition of life are, that if it is to cover every possibility, then it is guaranteed to be extremely long, involved, complex and unusable in practice.

However, my Grandmother had a single answer for many problems and she would have certainly intervened by using it a lot earlier than now in this discussion and it was:

"Use your gumption"[252]

and so, the question of whether something is life or non-life is probably best decided by NASA and my definitions above – provided that you and I then also apply our common senses, intuitions, gut-feelings, … or gumptions to the answers given by those definitions.

RECENT DEVELOPMENTS

Discussions about abiogenesis have, for the last several decades, been more concerned with later stages in the process than those on which Darwin, Oparin, Haldane and this book were/are focussing.

Nonetheless, abiogenesis in Darwin's 'warm little pond' or some variant of it (see below and also the section 'The Origin of Life – A Caesarean Birth?') allied to Marvell's 'World enough, and Time' is still current science's presumptive (and most plausible) explanation for the origin of life[253] and forms the basis for further discussions of the topic in this book.

VARIATIONS ON A THEME BY DARWIN

There are some variations on Darwin's basic scenario,[254] mostly concerned with refining its details and which have arisen as more understanding of molecular biology/chemistry has been gained.

---

[252] A. Lees (uttered many thousands of times within my hearing over the mid-20th century years)

[253] The innumerable Creation Myths / Non-Evidence-Based Fables / Miraculous Explanations / etc. for the origin of life, attributing it to various gods and deities, are not given any credence here. They do not answer the question 'How did life originate?', they merely change it to 'How did the gods / deities / etc. originate?' Richard Dawkins' books, *The Blind Watchmaker*, 1986, Norton & Co and *The Selfish Gene*, 1976, Oxford University Press, may interest readers wanting to research this topic further.

I know that the believers in the various above ideas will remain believers. As the popular saying goes 'A man convinced against his will, is of his old opinion still' (to be found originally, with slightly different wording, in Samuel Butler's poem '*Hudibras*', written between 1663 and 1678). However perhaps a few such believers may be influenced to follow Bertrand Russell's hope that ' … it is undesirable to believe a proposition when there is no ground whatever for supposing it to be true' (*On the Value of Scepticism*, in *Sceptical Essays*, 1928, George Allen and Unwin).

Although rather more scientifically based, I have also chosen to reject Lamarck's evolutionary theory, wherein characteristics acquired by an organism from its activities/environment throughout its life are passed on to its offspring.

Thus, for example, Lamarck would have it that a weight-lifting athlete who has developed extreme muscularity through her/his gym training would automatically pass that muscularity on to her/his offspring. In fact, some people may indeed have a better innate ability to develop muscles than others and that ability has been acquired through mutations amongst their ancestors (i.e., Darwinian evolution), but if their offspring want those muscles, they will also have to spend many hours in the gym.

Some recent work on plants grown under stressful conditions has shown that external conditions *can* affect the way that the body *applies* the DNA's (genes') instructions (Epigenetic changes). These changes do not, though, affect the DNA itself and do not seem to be passed on to the next generation.

[254] Mary Shelley's 'Thought Experiment' (published in 1818 when Darwin was still only nine years old) and given the title '*Frankenstein; or, The Modern Prometheus*', by-passes all this 'Warm Little Pond' stuff. Her hero, Victor Frankenstein, a scientist, obtains human body parts from body snatchers and assembles them to form what he hopes will be a perfect living creature. He then brings the creature to life using galvanism (now called 'electricity'). Of course, the creature turns out to be physically monstrously ugly.

Whilst generally regarded as fantasy, *Frankenstein*, in some ways today, seems uncannily prescient of our current defibrillators and our transplants of body parts from one human to another. At the time of writing, there has also just been the announcement that synthetic embryos, including those of humans, have been built up in the laboratory from stem cells, although, at the moment, there is no intent to grow these embryos beyond the present widely applied 14-day limit. Synthetic yeast and other cells, assembled from components from other cells have been around for about a decade and recently a record of a yeast cell with over 50% of its DNA being synthetic was announced.

(Continued)

Oparin-Haldane-Eigen Hypotheses (a.k.a. Heterotrophic[255] Theory)

In 1924 Alexander Oparin proposed a theory of how life might have originated on Earth. Initially he thought that the proto-Earth's atmosphere would contain oxygen, but soon afterwards changed his mind to thinking that it would be oxygen-free (called a Reducing atmosphere[256]). J(ohn). B. S. Haldane suggested the same ideas, independently, five years later. Oparin thought that cells developed first. Eigen had similar ideas, but thought that genes were the first to develop (i.e., molecules first – the approach adopted in this book),

Now, as well as allowing us to breathe, atmospheric oxygen also protects us from some UV radiation. Oxygen, in the form of Ozone (O_3), absorbs radiation from a wavelength of about 200 nm out to about 1.1 μm. Which is to say, it absorbs right the way from the IR through the visible and into the mid UV. Fortunately, for our ability to see things,[257] the absorption is small in the visible and peaks in the UV, with the absorption in the yellow/green being ~0.01% that of the peak of the UV absorption.

Thus, three aeons or so ago, before significant levels of ozone were present in the atmosphere, high(ish) energy UV photons were penetrating down to the Earth's sea-level. There, they would have encountered, *inter alia*, hydrogen and carbon atoms, mostly linked into simple molecules. For the next couple of paragraphs, I will leave it to Haldane to tell you what might then have happened[258]:

" … we may … legitimately speculate on the origin of life on this planet. Within a few thousand years from its origin it probably cooled down so far as to develop a fairly permanent solid crust. … The primitive atmosphere probably contained little or no oxygen… almost all the carbon … (was) in the atmosphere as carbon dioxide. Probably a good deal of the nitrogen now in the air was combined with metals as nitride in the Earth's crust, so that ammonia was constantly being formed by the action of water. … chemically active ultra-violet rays from the Sun penetrated to the surface of the land and the sea …

… when ultra-violet light acts on a mixture of water, carbon dioxide, and ammonia, a vast array of organic substances are made, including sugars and … some of the materials from which proteins are built up. This fact has been demonstrated in the laboratory by Baly[259] of Liverpool … before the origin of life they must have accumulated till the primitive oceans reached the consistence of hot dilute soup. … The first precursors of life found food available in considerable quantities, and had no competitors in the struggle for existence …"

The provision to Frankenstein of body parts by body snatchers from graves also has uncomfortable echoes in Larry Niven's futuristic novels featuring 'organleggers' who commit murder wholesale in order to supply body parts for transplants – see *The Long Arm of Gil Hamilton* (Ballantine Books, 1976), for example. A century earlier H. G. Wells was envisaging the possibility of hybrids between humans and other animals 'forcing the pace of evolution' *The Island of Dr Moreau*. 1896, Heineman.

[255] Heterotrophic organisms cannot make some or all of their nutritive requirements themselves and so must depend on obtaining those missing requirements from their food intake. Humankind and other animals are heterotrophs. Autotrophs, such as green plants, by contrast, manufacture their nutritional requirements from scratch, usually via photosynthesis (Chapter 10).

[256] The oxygen in the atmosphere mostly comes from life's activities; starting around three aeons ago, cyanobacteria (Chapter 17) began photosynthesis to obtain their energy. Oxygen is an unwanted (by the cyanobacteria) by-product of that process. Oxygen abundance near to present levels (~20%) took until about 700 million years ago to be achieved (Chapter 17). 'Reduction' is the chemical term for reactions which *reduce* the oxidation levels of molecules. For example, an Iron Ore (Fe_2O_3) molecule is reduced to two Iron atoms (2Fe) by reaction with Carbon Monoxide (CO) converting the latter to Carbon Dioxide (CO_2). The Iron Oxide is <u>reduced</u> in this reaction to Iron, but simultaneously the Carbon Monoxide is <u>ox</u>idised to Carbon Dioxide, so this type of reaction is often called a Redox reaction.

However, whenever water was present in the Earth's atmosphere, there would also be small amounts of free oxygen. This would arise from the splitting of water molecules into hydrogen and oxygen by UV radiation from the Sun. After the splitting (a.k.a. dissociation or photolysis) at least some of the hydrogen would escape into space limiting the chances of the free oxygen forming water again.

[257] As we shall see in Chapter 10, this is Cart-before-the-Horse. Evolution has evolved our eyes to be sensitive to the visible rather than to the UV *because* the visible radiation gets to ground level and the UV does not (or at least it hasn't done so for most of the last 700 million years).

[258] *The Origin of Life*. The Rationalist Annual, page 3, 1929. Ed. C. A. Watts.

[259] Edward C. C. Baly, 1871 to 1948, Professor of Inorganic Chemistry at the University of Liverpool.

Thus, was the idea of the 'Primordial Soup' or 'Prebiotic Soup' as a birthplace for life invented and 'Baly of Liverpool' became the first to try his hand at being midwife to new-life originations.

Oparin suggested that the carbon compounds, such as iron carbide (Fe_3C) would combine with hydrogen to form hydrocarbons and then interact with water and ammonia to produce carbohydrates, amino acid type molecules and then even go directly to proteins. He further expected proto-cells to form from his organic molecules when they were concentrated within the tiny droplets (known as coacervates) in colloid liquids which had developed from the 'primordial soup'.

In all of this, whilst reasonably sensible speculation about the *constituents* of the Earth's early atmosphere can be made, the *atmospheric pressure* is also likely to be important, if only because the boiling point of water varies with it – and we have much less idea how that has varied over time. Various current suggestions range from ¼ to 5 times the current atmospheric pressure at various times in the past. Water boils at a temperature of 64 °C for an atmospheric pressure of ¼ the present value and at 150 °C at 5 times the present value. Thus the 'warm little pond' could have been an 'exceedingly scaldingly-hot little pond' or a 'just-a-bit more than tepid little pond' (for further discussions around this possibility see Chapter 12 'Thermophiles').

Of course, if life-forms originated to be independent of water (Chapter 13) their 'birth temperature' and 'living temperature' would not be constrained by water's freezing and boiling points – however, whilst there are SF stories about life-forms on the Sun or at temperatures close to 0 K, I do think that these go beyond the bounds of what is reasonable (Chapter 13) and that water or some other liquid is required for life's origin(s).

Most recent work has been concerned with how *terrestrial* life originated and so places the transition point to life at the times when RNA-type and even DNA-type molecules are evolving, not concerning itself with other and different evolutionary possibilities (i.e., ETs). The two main lines of thought about terrestrial abiogenesis are briefly outlined below – but remember they are specialised and parochial in their concerns – they are called the Genes-First Hypothesis[260] and the Metabolism-First Hypothesis.

Genes-First Hypothesis

By some process, nucleic acids (Box 9.1) have appeared in the environment.

Ribozymes are types of RNA molecules which also act as catalysts. It is possible for a ribozyme to catalyse the chemical reactions which produce a duplicate of itself, so that a self-replicating molecule is formed. Such molecules have been produced in the laboratory.[261] Molecules simpler than RNA might also be capable of similar processes. The development of metabolic functions evolves subsequently.

Metabolism-First Hypothesis

This is basically the Oparin-Haldane Hypothesis with possible actual chemical pathways being suggested. Natural processes produce a stock of 'food' for some non-self-reproducing compound to utilise and the self-replication develops later. Natural catalysts such as minerals and clays may aid the formation of the simpler molecules and the complex molecules, such as nucleic acids, then follow-on. Recently it has been suggested that, on the early Earth, iron particles from meteorites might act as catalysts to convert carbon monoxide and hydrogen into hydrocarbons[262], a reaction used commercially today and known as the Fischer-Tropsch process.

THE ORIGIN OF LIFE – A NATURAL BIRTH?

As the Introduction to this Chapter concluded, the answer to the question 'How did Life Originate?' is, basically:

---

[260] a.k.a. 'the RNA world hypothesis'.

[261] See, for example, *A self-replicating ligase ribozyme*, Paul N., Joyce G. PNAS, 12733, **99**, 2002. Another study has shown that in the presence of basaltic glass (plentiful near volcanoes) nucleosides (Chapter 12) can spontaneously link into RNA-type chains 100 to 200 units long.

[262] Peters S. *et al*. Nature Scientific Reports, **13**, 6843, 2023.

'We don't quite know, but we do have some
strong suspicions about how it could have happened'.

Before going further into the topic, we need to be clear about quite what we are thinking of when considering life (see above) at this watershed point between life and non-life. We only have one example to work from: the Earth (again), so this means that we are essentially concerned with 'How did Life Originate *on Earth*?'[263]

OK – WE'RE CLOSING-IN ON THE NON-LIFE TO LIFE WATERSHED, BUT WE'RE NOT QUITE THERE YET

Life, when it originated, was not an incredibly complex organism, such as a human being, nor even a single-celled organism like an amoeba (Figure 9.3), it was a single molecule (Figure 9.3). That molecule may have been quite large and it would have been able to generate replicas of itself, but a single molecule is *all* that it was.[264] We will, henceforth, call it the '**First-Living-Molecule**'[265] .

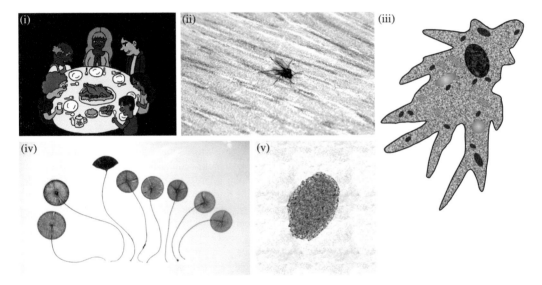

FIGURE 9.3 CONTINUED

---

[263] Panspermia, the idea that life originated elsewhere than on the Earth and then somehow made its way here is discussed in Chapter 9.

[264] The study of how non-biological processes could lead to life-forms is often called 'Pre-Biotic Chemistry'. The term 'Abiotic-to-Biotic Transition' may also be encountered.

[265] The term 'Last Universal Common Ancestor' (LUCA) is also in common use. This is the last organism from which *all* current terrestrial life-forms can trace their descents, i.e., today's different branches of terrestrial life started appearing *after* LUCA. LUCA maybe the same as the First-Living-Molecule, but I suspect not.

A single First-Living-Molecule must almost certainly have produced many variants, some of which later died out, or First-Living-Molecule events may have happened many times on the early Earth. The oldest fossils found to date (cyano-bacteria) are estimated to have existed some 3,500 million years ago and LUCA is generally estimated to have been living 3,800 million to 3,500 million years ago. Since cyanobacteria must have already been evolving for a long time by 3,500 million years ago, if they were able to leave identifiable fossils, the First-Living-Molecule could well have appeared soon after 4,000 million years ago (when the Earth probably was cool enough for liquid water to exist – see below).

I have thus chosen to use the term 'First-Living-Molecule' because it is unambiguous and I have preferred it to 'First Universal Common Ancestor' for reasons which may be obvious to afficionados of acronyms.

Many workers in this field place the origin of life at more complex levels than a single molecule – even quite complex cells being required before they regard such structures as being living entities. My inclination is that non-life changes to life at quite an early stage. But wherever <u>you</u> feel non-life changes to life, the First-Living-Molecule is the name given to that organism in this book. All subsequent developments are much the same in principle, whatsoever may be the details after that key moment.

(vi)

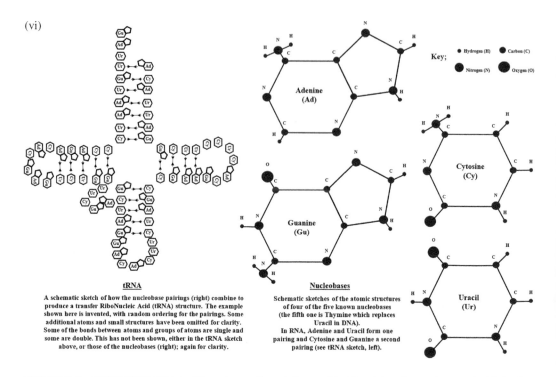

tRNA

A schematic sketch of how the nucleobase pairings (right) combine to produce a transfer RiboNucleic Acid (tRNA) structure. The example shown here is invented, with random ordering for the pairings. Some additional atoms and small structures have been omitted for clarity. Some of the bonds between atoms and groups of atoms are single and some are double. This has not been shown, either in the tRNA sketch above, or those of the nucleobases (right); again for clarity.

Nucleobases

Schematic sketches of the atomic structures of four of the five known nucleobases (the fifth one is Thymine which replaces Uracil in DNA).
In RNA, Adenine and Uracil form one pairing and Cytosine and Guanine a second pairing (see tRNA sketch, left).

FIGURE 9.3   <u>QUESTION</u>: Which of these (currently-existing-on-Earth) entities, shown above and with further details below, is a *simple* life-form? (i) – Human Beings – A large multi-cellular form of intelligent life. Sexual reproduction. Scale of image ~ 5 metres. (Reproduced by kind permission of Pixabay and LillyCantabile.) (ii) Flies – A medium-sized multi-cellular form of sentient life. Sexual reproduction. Scale of image ~ 10 millimetres. (© C. R. Kitchin 2023. Reproduced by permission.) (iii) Amoebae – A uni-cellular, eukaryotic (with a nucleus) form of animal life responsive to external stimuli. Reproduction by fission (division/budding). Scale of image ~ 10 to 100 micrometres. (© C. R. Kitchin 2023. Reproduced by permission.) (iv) Algae – A uni-cellular, eukaryotic form of plant life. Reproduction by fission or spore production. Scale of image ~ a few to 100s of millimetres[266]. (Reproduced by kind permission of Pixabay and WikimediaImages.) (v) Archaea (*Pyrodictium occultum*[267]). A uni-cellular, prokaryotic (without a nucleus) form of life. Reproduction by fission. Scale of image ~ 1 micrometre. (© C. R. Kitchin 2023. Reproduced by permission.) (vi) tRNA[268] molecule. A molecule composed of some tens of nucleobases (Box 9.1) which is involved in the synthesis of proteins and amino acids.[269] Scale of image ~ 10 nanometres. (© C. R. Kitchin 2023. Reproduced by permission.) <u>ANSWER</u>: None of them – Entities (i) to (v) are *all* highly complex, highly evolved forms of life. Entity (vi) is not a form of life, but the proteins and amino acids produced via tRNA are essential components of any form of terrestrial life known today, so it is a part of life at any level. The First-Living-Molecule was almost certainly *not* as complex ® as tRNA,[270] though it may have been as complex as the nucleobases and/or their linked pairs.

---

[266] Astonishing as it may seem (at least, it seems astonishing to me) these algae (*Acetabularia ryukyuensis*, a.k.a. Mermaid's Wineglasses) are 10 to 20 mm in size and yet are a single cell with just a single nucleus.

[267] Found living in (and presumably enjoying) the 100 °C temperatures around ocean-bed hydrothermal vents.

[268] Transfer RiboNucleic Acid.

[269] Amino acids are organic molecules formed primarily from carbon, hydrogen, nitrogen and oxygen. They are the building blocks of proteins and other biological molecules. The alpha amino acids found in terrestrial life are based upon a central carbon atom bound to an amine group (NH_2) and a carboxyl group (COOH) plus a side chain (called the R group) which varies from one amino acid variety to another. Glycine, wherein the R group is just a single hydrogen atom, is the simplest. Hundreds of amino acids are known, of which some 20 or so are necessary to human life. Nine of these necessary amino acids are called 'Essential' because they are not synthesized synthesised by human metabolism and so must be provided in the diet.

Polymers of amino acids are called peptides and they link, in their turn, to form proteins.

[270] Organisms still found today, which are still vastly more complex than the First-Living-Molecule would have been, but which form the biological class into which the First-Living-Molecule would probably also have fitted, are called the anaerobic chemoautotrophic thermophiles (Chapter 8). These live at temperatures near 100 °C and so as the Earth cooled, they would have been the first to have any opportunity to produce a First-Living-Molecule (unless life-forms can exist above 100 °C).

Given that most current life-forms are primarily based around the elements, Carbon and Hydrogen (Box 9.1), those elements may well have also dominated within the First-Living-Molecule, but we cannot be certain.

BOX 9.1 – LIFE'S ATOMS[271] AND LIFE'S MOLECULES

The terminology of the molecules and compounds involved with the chemistry of life can be confusing at the outset, so here is a brief introduction:

LIFE'S ATOMS:

Hydrogen is the lightest and simplest of all atoms/elements, comprising a single proton as its nucleus, together with a single electron.[272] It is the commonest element in the Universe (73.9% by mass, 93% by number) and comprises about 10% by mass of the average human body (mostly in the form of water).

Carbon is the 6th element (6 protons in the nucleus). It forms 0.5% of the matter in the Universe by mass (0.05% by number) and comprises about 23% by mass of the average human body. It is probably the single most important element for life, at least in its terrestrial form, because of its ability to form large, complex molecules based upon long backbones and/or polygons of linked carbon atoms.

Nitrogen is the 7th element (7 protons in the nucleus). It forms 0.1% of the matter in the Universe by mass (0.009% by number) and comprises about 2.6% by mass of the average human body.

Oxygen is the 8th element (8 protons in the nucleus). It forms 1.0% of the matter in the Universe by mass (0.08% by number) and comprises about 61% by mass of the average human body (mostly in the form of water).

Sodium is the 11th element (11 protons in the nucleus). It forms 0.003% of the matter in the Universe by mass (0.000 2% by number) and comprises about 0.14% by mass of the average human body.

Phosphorous is the 15th element (15 protons in the nucleus). It forms 0.0007% of the matter in the Universe by mass (0.000 03 % by number) and comprises about 0.14% by mass of the average human body.

Sulphur is the 16th element (16 protons in the nucleus). It forms 0.04% of the matter in the Universe by mass (0.0014 % by number) and comprises about 0.2% by mass of the average human body.

Chlorine is the 17th element (17 protons in the nucleus). It forms 0.001% of the matter in the Universe by mass (0.000 04% by number) and comprises about 0.14% by mass of the average human body.

Potassium is the 19th element (19 protons in the nucleus). It forms 0.0003% of the matter in the Universe by mass (0.000 008% by number) and comprises about 0.13% by mass of the average human body.

---

[271] The ten listed elements are often called biogenic elements because of their omnipresence in terrestrial life-forms.

[272] All atoms can exist in different forms known as isotopes, though not all are found to exist naturally. All the atoms forming one type of chemical element will have the same number of protons and electrons (one of each for hydrogen). However, the number of neutrons in the nucleus can vary. Thus, the main form of hydrogen does not have any neutrons in its nucleus, but other isotopes have one neutron as well as the proton (called deuterium or heavy hydrogen) or two neutrons as well as the proton (tritium).

Calcium is the 20th element (20 protons in the nucleus). It forms 0.006% of the matter in the Universe by mass (0.000 2% by number) and comprises about 1.7% by mass of the average human body.

From very early on in quantitative chemistry[273] it was recognised that the most abundant elements which could be identified as present in life-forms were not the same as the most abundant elements to be found more generally (compare the abundances by mass in the above list). The above ten elements are the ones which are the most abundant in terrestrial life-forms and therefore probably the most essential to the origin of terrestrial life (at least).[274]

LIFE'S MOLECULES

Organic molecules contain carbon atoms bonded to other carbon atoms and/or carbon atoms bonded to hydrogen atoms, as well as atoms from any of the other chemical elements. Example: Methyl Alcohol has one carbon, one oxygen and four hydrogen atoms (see also footnote [38]).

Confusion arises because the term 'organic' does not *necessarily* mean that the molecules have been produced by life-forms. Organic molecules *are* produced by life-forms, but they are *also* produced inorganically as well, i.e., in non-biological chemical reactions.

In fact, it is perhaps because some inorganically produced organic molecules can be identical to the same organically produced molecules, that the transition from inorganic to organic (i.e., to the First-Living-Molecule) can be made.

Hydrocarbons are organic molecules which contain just hydrogen and carbon. Example: Methane with four hydrogen and one carbon atoms. Polycyclic Aromatic Hydrocarbons (PAHs) are plentiful amongst the molecules found in the ISM (Interstellar Medium – Box 9.2) and are assemblages of several, often many, benzene rings.[275] Hydrocarbons divide into aliphatic and aromatic varieties. Aliphatic hydrocarbons, such as have been found on Ceres, have a chain structure, whilst aromatic compounds, found in many places, are based upon rings.

Carbohydrates – this term is easily confused with hydrocarbons, but the 'hydrate' term derives from water (hydro-) not, directly, from hydrogen. These molecules contain hydrogen, carbon and oxygen atoms. Many carbohydrates are found in human food-stuffs. Example: Glucose with twelve hydrogen, six carbon and six oxygen atoms.

DNA[276] – The 'N' and 'A' in DNA stands for Nucleic Acid and nucleic acids are formed from Nucleotides.

Nucleotides are built up from carbon, hydrogen, nitrogen, oxygen and phosphorous atoms (at least) and are each built upon the more widely known nucleobases (Figure 9.3): Adenine, Cytosine, Guanine and Thymine (for DNA) with Uracil replacing Thymine for RNA (often just labelled as 'A', 'C', 'G', 'T' or 'U'). The nucleobase combines with sugar and phosphate compounds to form the nucleotide.

---

[273] Antoine Lavoisier is generally credited with the birth of quantitative chemistry and bio-chemistry.

[274] The lack of active plate tectonics, such as is the current state of the surface layers of Mars and Venus, is called 'Static Lid Tectonics'. Effectively the whole surface is a single plate, so that although convection in the mantle of the planet still occurs, producing isolated volcanoes, etc., the surface material is not dragged back into the planet's interior. This probably reduces the availability at the surface of the essential chemical elements needed by life-forms, but not to the point where life becomes impossible. Plate tectonics on the Earth seems to have started about 3,300 million years ago and the first life-forms some few hundred million years earlier.

A third possible case recently been proposed which may apply to a few currently known exoplanets and that is called the Eggshell crust. The planet's crust is just ~ 1 km thick (c.f. 40 km for the Earth) and the resulting surface may appear like Venus' lowland areas. Eggshell exoplanets are probably by far too volatile to support life at present.

[275] A benzene ring comprises a hexagon of six linked carbon atoms each with an attached hydrogen atom.

[276] DeoxyriboNucleic Acid – Most readers will be aware that these molecules are fundamental in many ways to present-day life on Earth.

Because, as Darwin makes clear, present life-forms would 'instantly devour' (see above) any new First-Living-Molecules forming today, we should not now expect to find any naturally produced First-Living-Molecules (and we have yet to produce any in the laboratory).

The nearest that we can get today to a molecule which might be related to the First-Living-Molecule may be one of the molecular building blocks of today's life-forms such as a nucleobase and/or a nucleotide (Box 9.1, Figure 9.3). Nonetheless, we can infer some things about its basic nature.

The First-Living-Molecule's most fundamental requirement, by my usage (above), is that it "will replicate itself" or in NASA's words "… (be) capable of undergoing Darwinian evolution". The two wordings are equivalent because any replication process will not be perfect and there will therefore be errors in some of the replications. Evolution will then ensure that the advantageous mistakes make further progress, whilst the disadvantageous mistakes do not progress, or at least progress less well (see comment at the start of Chapter 10, though). This is the fundamental operating mechanism of evolution (Chapter 10) and the 'Mistakes' are more usually labelled as 'Mutations'. As John Wyndham puts it in his SF book *The Chrysalids*[277]:

" … life is change, that is how it differs from the rocks, change is its very nature."

Replication at the level of the First-Living-Molecule is unlikely to be an active process. The 'first' First-Living-Molecule, if it originated in a Darwinian 'warm little pond', would be immersed in a solution of the types of atoms and molecules from which it, itself, was formed (otherwise it would never *have* formed). A possible replication process might then occur via a 'right' atom or simple molecule colliding at the 'right' place on the First-Living-Molecule and being captured there. This is followed by other captures until another First-Living-Molecule has formed and the two split apart, under their continual bombardments from all the other molecules in their vicinity, to go their own ways; though soon to be joined by another two First-Living-Molecules, then another four, then another eight, then sixteen, then thirty-two … and the first population explosion of a life-form is well under way.

Of course, the First-Living-Molecule may replicate via a quite different manner (or several different manners), but for this book's purposes, that does not matter; *replication* is the crucial factor, and the details of how it happens are important, but secondary.

As human beings we probably tend to favour 'warm little ponds', or their equivalents (Figure 9.4), because they sound like nice, comfortable habitats for ourselves. But does the First-Living-Molecule' have to originate in one such? The observational evidence seems to suggest not.

We have now discovered that many of the important molecules, which we know to be fundamental to terrestrial life-forms, occur within interstellar gas and dust clouds (Box 9.2), within meteorites, on asteroids and comets, on the surfaces of ice crystals and even in distant galaxies and quasars. So perhaps 'we' actually began on a small ice crystal which itself had just frozen from the thin frigid gases of a GMC (Giant Molecular Cloud – Box 9.2) or within the remote reaches of the proto-planetary disc of a newly forming star or deep in a tiny crevice in a far distant cometary nucleus. We could even, at the other extreme, have had our First-Living-Molecule being produced in seething environment around a deep-ocean hydrothermal vent, although, in that case, we might like the smell of bad eggs (sulphides) a great deal more than we actually do.

---

[277] Michael Joseph, 1955.

FIGURE 9.4   Do we enjoy warm sunny beaches so much because our very, very remote ancestors started in a 'warm little pond'? (Reproduced by kind permission of Pixabay and StockSnap.)

BOX 9.2 – THE INTERSTELLAR MEDIUM AND GIANT MOLECULAR CLOUDS (GMCS)

The vast regions of space between the stars look to be completely empty. These vast volumes of near-nothingness are called the ISM (Interstellar Medium) and, in fact, on average every cubic metre of that space contains very approximately[278] 1,000,000 atoms, most of which are in the form of molecular hydrogen (two hydrogen atoms linked together – see also Box 9.1).

If that sounds like a lot, then for comparison, a cubic metre of the Earth's atmosphere at sea-level contains $\sim 3 \times 10^{25}$ molecules (equivalent to $\sim 8 \times 10^{26}$ hydrogen atoms), making the Earth's atmosphere around 800,000,000,000,000,000,000 times denser ® than the ISM.[279]

Most of the ISM, if anything, is somewhat less dense than just discussed. This is because we have been considering its average density, and there are large volumes of the ISM with much higher densities. These regions are known as Globules, Nebulae, Interstellar clouds, GMCs, etc.

The GMCs are enormous and mostly hidden from (visual) sight. They often, though, embrace even denser portions wherein proto-stars and proto-planets are currently collapsing under gravity to form stars and planets. These star-forming regions are usually easily spotted because they already contain stars which started life a little earlier and these stars illuminate the denser gas, causing bright and glowing emission nebulae to become visible. The Orion Nebula (M42) is one such star-forming region and it is a small part of Orion GMC A within the Orion Molecular Cloud Complex (Figure 9.5). The latter, in the sky, covers an area about four times larger than that covered by the principal stars of Orion.

---

[278] The figure can range from 1,000 to 10,000,000,000 atoms per cubic metre, the latter figure being for GMCs.

[279] If that isn't sufficiently mind-boggling for you, then the average matter density of the Universe (i.e., what you would get if you broke up all the stars, etc., and spread their material evenly across the whole Universe) is just six hydrogen atoms per cubic metre. So, with 1,000,000 atoms per cubic metre between the stars ... don't ever talk about 'empty space' again – and the Earth's sea-level atmosphere is thus $\sim 100,000,000,000,000,000,000,000,000$ times denser ® than the average across the whole Universe.

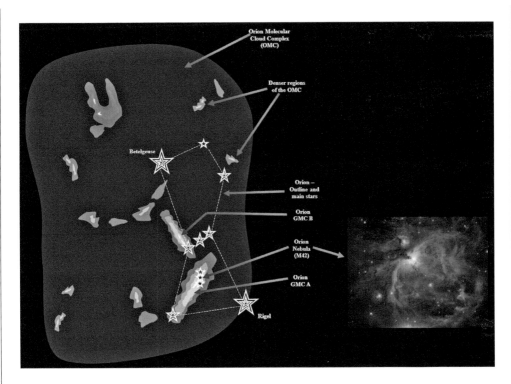

FIGURE 9.5   A sketch of the Orion Molecular Cloud Complex. The whole structure is some 300 ly across and it is about 1,200 ly away from us. The principal stars forming Orion are, of course, seen visibly. Most of the cloud complex, though, can only be detected in the IR and micro-wave parts of the spectrum, but the Orion Nebula is one of the most magnificent (visible) objects in the night sky (right). Orion GMC A and Orion GMC B are each estimated to have masses of around 100,000 M_{Sun}, so the total mass of the entire structure could be in the region of 500,000 M_{Sun}. (© C. R. Kitchin 2023. Reproduced by permission. Orion nebula image reproduced by kind permission of ESO/G. Beccari.)

Smaller scale, often opaque, nebulae, such as Bok globules and dark nebulae with sizes of a few light years or less are sometimes found within larger hotter nebulae and elsewhere in the ISM. They have masses of a few to a few hundreds of solar masses and their central temperatures can fall as low as 10 K (−260 °C). Some simple molecules, such as carbon monoxide and ammonia, have been detected within these objects.

So, Perhaps, It Was Not a 'Warm Little Pond' After-all?

The First-Living-Molecule might have originated in a 'warm little pond', but then again, perhaps 'warm little ponds' are not suitable for life's first step after all – or perhaps there are other environments which can do the job just as well. Until we discover in the laboratory how the First-Living-Molecule could have originated, we should not neglect any other plausible scenarios – and *they* seem to be abundant. Organic molecules, some very complex but short of being the First-Living-Molecule, seem to be able to originate in many different situations and under many different conditions, which bodes well for the origin of life being able to occur under a similarly wide set of situations and conditions.

So, before returning to the First-Living-Molecule, let us examine the evidence we already have for its possible precursors.

Organic Molecules: Here, There and Everywhere?

Between the Stars

At the time of writing around 250 varieties of molecules have been detected[280] as existing in the interstellar medium, nebulae, GMCs, etc. (Box 9.2), though 13 of these have yet to be confirmed.

Recently for example, in the IC348 star-forming region of the Perseus GMC, the carbon-bearing molecules, acetylene (C_2H_2), benzene (C_6H_6), cyanoacetylene (HC_3N), cyanobutadiyne (HC_5N), diacetylene (C_4H_2), ethane (C_2H_6), hexatriyne (C_6H_2) and hydrogen cyanide (HCN) plus water, carbon dioxide and ammonia, many of which can be built up into amino acids and the like, have been detected. The detection of the essential amino acid,[281] tryptophan, in IC348 has also just been announced, based on its IR radio emissions.

Acetaldehyde (CH_3CHO), ethanol (C_2H_6O), formaldehyde (CH_2O), formic acid (HCOOH), methane (CH_4), methanol (CH_3OH) and silicates have similarly been observed by the JWST[282] around the proto-star IRAS 15398-3359.

Perhaps surprisingly, many moderately complex molecules are also to be found in the hot, UV-filled environments comprising planetary nebulae. Since the material of a planetary nebula is being ejected from its central, white dwarf, star, these molecules will feed into the ISM in due course and there they may provide the raw material for the formation of more complex molecules.

Many of these molecules are organic, but created non-biologically. Discounting the fullerenes, which contain only carbon atoms, the most complex molecule so far detected has 19 atoms, with the chemical formula $C_{10}H_7CN$ and the name 2-cyanonapthlene.

None of the nucleobases have been found in the ISM to date (Figure 9.3 (vi) – but see below). However, Indene (C_9H_8), which has a basic hexagon-pentagon structure related to that of the nucleobases, adenine and guanine (Figure 9.3), *has* been found.

Perhaps of more interest to wine-loving readers, ethyl alcohol (C_2H_6O) occurs at a rate of around one molecule per cubic metre in some parts of the ISM and some GMCs. That's not a lot of alcohol until you realise just how many cubic metres there are in a GMC. The Orion GMC A has an actual length of ~300 ly and if we take it to be basically cylindrical in shape with a diameter of 20 ly, then we get its volume as ~90,000 cubic light years. But a light year is ~10^{16} m in length so the total volume of the Orion GMC A is ~ 8×10^{52} m^3. At one alcohol molecule per cubic metre, that's almost 6 $\times 10^{27}$ kg of pure alcohol. This would suffice to make 1,000 pure alcohol exoplanets, each with the mass of the Earth or to be enough for ~10^{29} bottles ® of champagne.[283]

But how would molecules form in the ISM and/or GMCs with their very low gas densities (Box 9.2)? Even within GMCs where the gas density may reach ~10,000,000,000 atoms per cubic metre and the separations of the atoms or molecules average ~0.5 mm, the distances apart of *collisions*

---

[280] In almost all cases, the detection is via the molecules' IR and micro-wave emission or absorption spectrum lines (Box 6.1). These usually arise from the molecules' vibrations or rotations rather than the transitions of electrons. This, though, is taking us well outside the remit of this book and so the interested/puzzled reader will need to research the topic for him/herself. Two of the author's other books – *Astrophysical Techniques* 7th Edn., CRC Press, 2020, and *Optical Astronomical Spectroscopy*, Institute of Physics, 1995, may help or an internet search for 'Molecular vibrational transitions' or 'Molecular rotational transitions' will bring up many suitable articles.

[281] Some 20 amino acids are required by mammals for their full health, including ourselves. Human metabolism, however, cannot generate nine of these and so they must be obtained in the diet. These nine amino acids are called 'Essential' although this is slightly misleading, in that all 20 amino acids are essential, but the nine which are not internally produced are *essential* in the *diet*.

[282] The presence of the main constituents of the Earth's atmosphere: nitrogen, oxygen, argon, water vapour and carbon dioxide in exoplanets' atmospheres are difficult to detect using Earth-based instruments because the earthly gases' strong absorptions mask the trace signals from the exoplanets. The JWST, HST and other spacecraft observatories can make much better observations because they are above the atmosphere and so the Earth's atmospheric absorptions no longer affect the results. Recently the JWST *has* detected carbon dioxide in the atmosphere of the hot Jovian WASP-39 b exoplanet.

[283] About 100,000,000,000,000,000 bottles of champagne each for every human currently living on the Earth plus all those who have lived in the past, going back to Homo habilis.

between any of the particles (their mean-free-path) is ~1,000,000 km ® and the time intervals between such collisions are around 15 days. Given typical impact speeds of ~ 1 km/s, most particles will not stick together during a collision anyway, but will simply bounce apart.

The number of 'bouncing apart' to 'sticking' collisions' is a matter of guesswork, though it seems likely to be quite high. Calling it for the moment X, to produce Indene with its 17 atoms (above) via random collisions of this sort, assuming every appropriate collision was successful and the partially built construct never broke-up during a subsequent impact, and also assuming that for every one collision between the partial construct and a carbon atom, there are 1860 with hydrogen atoms and 139 with other atoms (Box 9.1) it will take a total of ~18,000 X collisions to produce one indene molecule. Given the 15-day intervals between collisions and taking a value of 'X' of 999 (i.e., only one in a thousand collisions is a sticking one), we get a guestimate of ~270,000,000 days (~750,000 years) for the minimum formation time of an Indene molecule in a dense GMC.

Of course, 'X' could well be much larger than just assumed, not every appropriate collision may be successful and the partially formed molecule could be broken-up many times so that the entire age of the Universe starts to look none-too-long to produce an Indene molecule in practice.

On the plus side, there may be short cuts in the building process, such as two or three sub-units of the molecule forming independently and then joining up to form the whole molecule in less than 750,000 years. Recent observations also suggest that dust particles in the gas may act as catalysts for the formation of complex organic molecules.

Also on the plus side, a GMC *does* contain an *awful* lot of atoms and molecules, around 10^{62} ® for the Orion GMC A cloud, for example (and that is a number large enough for even astronomers to think it to be a bit sizeable). So, some Indene is likely to get produced even via this low probability route, and the origin of many other organic molecules including, maybe, the First-Living-Molecule could thus be inside the denser parts of Giant Molecular Cloud complexes (Figure 9.5).

If collisions between the particles in a thin gas seem to be a slow way of building up complex molecules, then can the process be speeded-up? The slowness arises from the long intervals between collisions and the high kinetic energies of the particles. So, speeding things up involves reducing the intervals between collisions and reducing the kinetic energies involved in the collisions – and that brings us to ice.

The ISM and GMCs, with the exception of regions heated by stars, like the Orion Nebula (Figure 9.5), are very cold, typically around 10 to 20 K (~ −260 ºC). So, most gases, except for helium will freeze out and form tiny ice crystals within the GMC. Other molecules may then form thin coatings on the surfaces of the ice crystals where they will be in close proximity to each other and almost stationary with respect to each other. Recently, also, a molecule called ortho-benzyne (C_6H_4)[284] has been detected in the Taurus Molecular cloud. This is a highly reactive molecule (i.e., it has a low activation energy – Chapter 10) which can easily build into much more complex organic molecules. If we factor in an energy supply of UV radiation from newly formed stars, then we may have factories for producing organic molecules.

Similar processes may well occur on the surfaces of dust particles. UV radiation also produces another highly reactive 'starter' molecule, H_3^+, known as Hydron or the Trihydrogen Cation. This molecule is formed by the ionisation of a normal (H_2) hydrogen molecule, forming the ion H_2^+ which then captures a single hydrogen atom. H_3^+ easily reacts with other simple molecules to produce ions such as the Formyl cation (HCO^+).

Such processes cannot be observed currently out in the 'real' Universe, but they can be and have been simulated in laboratories (see also the section 'The Origin of Life – A Caesarean Birth?'). One recent such experiment,[285] for example, found that not only complex organic molecules could originate via this process, but the actual building blocks of terrestrial life, the nucleobases Adenine,

---

[284] Basically, it is a benzine (C_6H_6) molecule which is missing two hydrogen atoms.
[285] *Nucleobase synthesis in interstellar ices.* Y. Oba *et al.* Nature Communications, 4413, **10**, 2019.

Cytosine, Thymine and Uracil[286] (Figure 9.3 (vi)) were generated on the surfaces of water, carbon monoxide, ammonia and methanol ice crystals. So there seems to be the distinct possibility of the First-Living-Molecule originating in a 'very cold little niche' in some part of the space between the stars,[287] instead of a 'warm little pond'.

Microwave receivers and telescopes in the last decade or two have become sufficiently sensitive to observe objects out towards the edges of the visible Universe. So, we now know that molecules are present in the interstellar media of galaxies billions of ly away and even in a quasar at a distance of 11,000,000,000 ly. The Cloverleaf quasar (H1413+117) is gravitationally lensed (Figure 7.7) by an intervening galaxy (or perhaps by two such galaxies) along our line of sight to it and so we actually see four 'images' of it. The quasar's millimetre and microwave spectra show lines from at least 14 different molecules including hydrogen cyanide (HCN), hydrogen sulphide (H_2S) and water. Time and distance in these matters are closely linked so we therefore also know that interstellar molecules were being formed some 2,500,000,000 years after the Big Bang, as well as more recently. The JWST has recently detected PAHs in an even more distant galaxy, SPT0418-47, which is another 1,000 million ly further away from us than the Cloverleaf. Whilst closer to home (~26,000 ly) and at longer wavelengths the first isomer (Chapter 11), of glycine, has been detected in an interstellar gas cloud which is near to the centre of the Milky Way galaxy.

Closer to Home

Within the Solar System there seems to be a good deal of evidence that organic molecules of many varieties can be formed in many different environments. Thus, we have:

(i) In comets, acetaldehyde, benzaldehyde, benzoic acid, ethylene glycol, ethanol, naphthalene, thioformaldehyde and so on (~ 25 compounds in total to date) have been observed in or on comets. For Comet 67P/Churyumov-Gerasimenko, visited by the Rosetta spacecraft in 2014, it was found that some 45% by mass of the analysed samples were organic[288] in their nature.

(ii) Tholins (a.k.a. 'Sticky Browny-Reddish Crud' or 'Star Tar'): The name was invented by Carl Sagan and Bishun Khare as a label for the mess of complex brownish organic compounds with which they were often left during their laboratory experiments involving irradiating various naturally occurring gases, such as methane and ammonia, with UV light or when passing sparks through them, etc. The name, tholin, is derived from the Greek for 'Muddy' or 'Blurred' (Tholos – θολὸσ) – and the crud has proved difficult to analyse because of its complexity and the variety of individual compounds (Sagan and Khare estimated there could be hundreds of thousands of different molecules in their 'crud'). Nonetheless, it is certainly 'organic' in the Box 9.1 sense, and amino acids, at least, amongst the known requirements for terrestrial life, have been isolated from that 'crud'.

The relevance of tholins to this discussion lies in their browny-reddish colour. From the laboratory experiments it seems likely the tholins could be produced within the Solar System wherever there is UV radiation (from the Sun) or lightning (in an atmosphere) and where appropriate precursor molecules, such as ammonia, carbon dioxide, hydrogen sulphide, methane, nitrogen, water, etc., are available (also in an atmosphere). Tholins produced in such a manner could give a reddish colour to the surface of a planet, satellite, asteroid, etc., or produce reddish liquids or reddish gases.

---

[286] Uracil *has* recently been found in the samples from the asteroid Ryugu brought back to the Earth by the Hayabusa 2 spacecraft. Niacin (a.k.a. vitamin B_3) was also found in the samples – see earlier discussions.

[287] It would then be incorporated into the protoplanet during its final stages of formation and thus would be a type of panspermia (see below).

[288] Remember (Box 9.1) 'organic' essentially means 'contains carbon' not that is living material or even material derived from living processes. Organic material can be and is produced abiologically.

FIGURE 9.6 Combined New Horizons' images of Pluto and Charon in their natural colours and correct relative sizes, clearly showing their large reddish patches which may be due to the presence of tholins on their surfaces. (Reproduced by kind permission of NASA/JHUAPL/SwRI.)

As many of you will know, Pluto's surface and to a lesser extent that of its satellite, Charon, when observed by the New Horizons spacecraft[289] in 2015, were seen to be covered in very large red patches (Figure 9.6) and it is widely expected that tholins will be found to be the cause of the colour of the patches.

Similar red colorations, which are also expected to be due to tholins (although the tholins may vary in their compositions between different sites), have been found to date on and/or in:[290]

- Atmospheres of the outer planets
- Carbonaceous chondrite meteorites
- Ceres
- Comets including 'Rosetta's comet'; 67P/Churyumov–Gerasimenko
- Europa
- Icy bodies in the outer Solar System including KBOs (Kuiper Belt Objects) and Centaur asteroids
- Makemake
- Rhea
- The main belt asteroids, 24 Themis, 203 Pompeja and 269 Justitia

---

[289] Launched January 2006, flyby of Pluto July 2015.

[290] They are no longer to be found on the Earth, except in laboratories. However artificial tholins have proved to be very palatable to terrestrial soil bacteria.

- Titan – both on the surface and in the atmosphere
- Triton
- The interstellar medium
- The system of the star HR 4796A. The star is young, hottish and embedded in a dust ring. It is about 235 ly away in Centaurus and forms a binary with a red dwarf some 560 AU distant from it. The dust in its ring has been observed by the HST to have a reddish tinge (i.e., tholins – perhaps).

Thus, if current expectations and the results of laboratory experiments are correct, then large parts of the outer part and even the middle part of the Solar System may house megatonnes,[291] even giga-tonnes ®, of pre-biotic organic material – which is just awaiting the opportunity to become life (an excellent reason for making future astronauts, after they have spent an exciting day exploring Pluto or Triton, wipe their feet well before coming back into their spacecraft – Chapter 13).

(iii)   In meteorites, there is an entire group of meteorites: the Carbonaceous Chondrites, which are rich (up to 3% of the whole meteorite) in carbon and carbon compounds, including abiologically produced organic molecules such as the alcohols, amines and sugars. Also recently discovered from a meteorite is the pre-biotic compound hexamethylenetetramine (HMT – $(CH_2)_6N_4$), which, in the laboratory and in industry, is the feedstock for a large range of more complex organic molecules.

With meteorites, since they, by definition, are found on the surface of the Earth, there is always the possibility of contamination by terrestrial biologies. The meteorites found on the surface of the Antarctic ice cap though (where they show up like 'Rudolph's' renowned 'Red Nose') generally have little such contamination.

The carbonaceous chondritic meteorites are thought to have originated very early in the formation of the Solar System, perhaps predating the formation of the Earth by 100 million years, or so and to be largely unchanged since their formation. Their compositions therefore reflect that of the earliest stages of the Solar System and even of the pre-solar GMC. Over 80 amino acids have been detected in the Murchison meteorite, whose total mass is over 100 kg, which fell in Australia in 1969 and was quickly retrieved, so limiting any terrestrial contamination. Many of these amino acids are racemic mixtures (Chapter 11) suggesting an abiotic origin. Alcohols, hydrocarbons, sugars, purines, pyrimidines and many other organic molecules have been detected in the Murchison and other carbonaceous meteorites. The presence of the amino acids, including non-protein varieties, etc., so early in the formation of the Solar System argues strongly for their presence already in the material condensing from the molecular cloud.

Around 300 known meteorites are thought to have been blasted off the surface of Mars during major cratering impacts there and these are available for direct laboratory analysis (see also 'Sampling', below). Two in particular have attracted attention for the presence of organic material within them: the ALH 84001 and Tissint meteorites.

ALH 84001 (see also Chapter 14) is a ~2 kg meteorite found in the Allen Hills (Antarctica) in 1984 but which probably fell to Earth about 13,000 years ago. Analysis of the meteorite revealed organic molecules associated with small globular structures which, from the high abundance of Carbon-13,[272] are thought to be terrestrial contaminants. But other organics with lower carbon-13 abundances are likely to be similar to the extraterrestrial but abiotically produced organic matter in carbonaceous chondrites such as polycyclic aromatic hydrocarbons[292] (PAHs).

---

[291] Looking at Figure 9.6, some 1/3 of the visible hemisphere is covered by the reddish deposits. Even if that is all, then the deposits cover an area on Pluto of about 3,000,000 km². At a thickness for the deposits of 1 µm and a density of 1,000 kg m⁻³ for the tholins, that amounts to ~3 megatonnes. If the deposit's thickness is 1 mm, then its mass is ~3 gigatonnes – and that's just on the hemisphere of Pluto which we can see.

[292] Components of, e.g., coal and oil etc., but can also be generated abiotically (Box 9.1).Some recent work has shown that for ethane (C_2H_6), molecules, wherein both carbon atoms are the carbon-13 isotope, are significantly more abundant in biologically produced ethane than in abiologically produced molecules.

The Tissint Martian meteorite fragmented during its fall, but some 12 kg of it have been retrieved. The fall, in Morocco in 2011, was observed and so its fragments were retrieved quickly before much contamination could occur although much of it was then sold off to private collectors and has never been scientifically examined.[293] Nonetheless, PAHs, organic compounds such as aldehydes, organic-magnesium compounds and more, all of which are probably of abiotic, extraterrestrial origin have been identified in this meteorite.

In a similar manner to the Martian meteorites, there are about 60 meteorites which are thought to have originated from the Moon. Organic molecules, clearly of non-terrestrial origin, have yet to be found in them. Indeed, it is sometimes suggested that the very small quantities of organic material detectable in actual lunar samples (see 'Sampling' below) originate from 'outer space' meteorites such as carbonaceous chondrites which have hit the Moon in the past.

(iv) On the Earth, we are too late for any remaining signs of organic molecules which are clearly *not* the result of current living organisms' activities (such signs have all been, as Darwin suggests, 'instantly devoured').

(v)   Sampling – It is all very well deciding that reddish colours on Pluto arise from organic molecules, but whatever the remote sensing evidence may be, nothing beats having a bit of Pluto in the laboratory and subjecting it to the full and detailed attentions and equipment of Analytical Chemistry. It will be a while[294] before we do have a bit of Pluto in a terrestrial laboratory, but in the meantime, there are other Solar System objects where samples have already been brought back to Earth and/or been subjected to chemical analysis by other means.

Unless a fully equipped analytical chemistry laboratory, were to be built and launched to land on Mars, or Ceres, or Titan, or wherever, together with some analytical chemists – and that seems unlikely – samples of those objects or others, need to be brought back to Earth. So, one system for sampling in detail the objects within the Solar System is to send a (manned or un-manned) spacecraft to the object, land on it, robotically dig up a few shovels-full of the surface and bring it back to the Earth. This is called a Sample-Return mission and just five objects, to date, have been examined in this way: the Moon, Comet Wild-2 and asteroids 101955 Bennu, 25143 Itokawa and 162173 Ryugu. Current and near-future missions may add Mars, Phobos and Deimos to the list within the next decade or so.

Alternatively, while a fully-equipped and manned analytical chemistry laboratory cannot currently be sent out, even to the Moon, smaller, less comprehensive but automatic or remote controlled, laboratories can be and have been used in this way – starting with Viking 1 in 1975 whose lander carried a small laboratory which examined Martian soil samples for evidence of life. This type of approach is called an In-Situ Sampling mission.

Since sampling is the most hopeful way of finding life (whether terrestrial in nature or not) these two approaches deserve more in-depth examinations.

Sample-Return Missions

The sampling of outer space objects is, so far, totally dominated by the Apollo Manned Moon-missions, which brought back, in total, about 382 kgs of lunar material to the Earth.

Additionally, Russia's (robotic) Luna 16, 20 and 24 Moon landers returned a total of about 320 g and China's Chang'e 5 retrieved about 1.73 kg of lunar material. NASA's Stardust sampled the

---

[293] At the time of writing, asking prices on the internet for small Tissint fragments are around $1,000 per gram (~ 17 times the price of gold).

[294] It took the New Horizons spacecraft about nine years to get from the Earth to Pluto. If the decision were to be made today to send a sample-return mission to Pluto, it would be around 35 years before the samples could be got back here: 10 years design, planning and constructing, 12 years journey outwards (longer than New Horizons because it would have to slow down at Pluto), 1 year exploring and sampling, 12 years for the return journey. Feasible, but improbable.

material in Comet Wild-2's tail and captured many tiny particles with a total mass of ~1 mg. Japan's Hayabusa and Hayabusa-2 spacecraft visited the asteroids Itokawa and Ryugu respectively, returning < 1 mg of sampled material in the first case and ~5.4 g in the second case. NASA's OSIRIS-REx has visited and sampled asteroid 101955 Bennu and has returned around 0.12 kg of material to the Earth. That material is still being unloaded from its container at the time of writing. Whilst in December 2022 NASA's Mars Perseverance Rover started caching its samples for collection by a future Mars mission.

The results of all of this activity for the subject matter of this book have been less than might have been hoped. Low levels of amino acids were about all that was found of relevance in the Apollo lunar samples. If not due to contamination directly by terrestrial sources, these could have been produced by the landers' rockets' deposits[295] or from molecules carried in the Solar Wind and left on the Lunar surface. The most hopeful possibility would be if it was material carried to the Moon from elsewhere within the Solar System, such as via carbonaceous chondrite meteorites (Section (iii) Meteorites, above). The Russian Luna samples showed no signs of hydrocarbons or organic molecules, while Chang'e 5's samples showed only low levels of water in addition to the expected rocky/silicate minerals.

The Stardust mission obtained its samples by trapping rapidly moving inter-planetary dust particles in aerogel[296] (Figure 9.7). The aerogel has sufficient strength to resist the penetration of a dust particle and slow it down to a rest gradually without damaging the particle. The particle then remains trapped in the aerogel for later examination. A wide range of organic compounds were

FIGURE 9.7   Aerogel. (Reproduced by kind permission of NASA/JPL-Caltech.)

---

[295] The rockets used methyl hydrazine (H_3CNHNH_2), dimethyl hydrazine (($H_3C)_2NNH_2$) and nitrogen tetroxide (N_2O_4) as their fuel.

[296] An aerogel is *not* a gel. It is the remnants of a gel that has had all the 'jelly' bits removed. What remains is an extremely lightweight (~ 0.001 x density of water) foam-like substance which has sufficient structural strength to retain its shape and even to be handled gently.

found in Stardust's trapped particles including the amino acid, glycine and many varieties of hydrocarbons (Box 9.1). Also, some 45 particles (out of ~ one million) were (probably) of interstellar origin.

Hayabusa's sampling method was to catch material ejected from the asteroid's surface by the impact of a small 'bullet'. It returned only a tiny quantity of material (~1,500 dust particles) from Itokawa, but these particles have been very intensively analysed. Some carbon-based material was found in the samples including graphite and disordered aromatic compounds (see PAHs, Box 9.1). Hayabusa-2's much larger sample from asteroid Ryugu revealed the presence of two of terrestrial life's basic building blocks, Uracil (Box 9.1) and Niacin (a.k.a. Vitamin B_3) plus prebiotic compounds including glycine and alanine,[297] eight more types of amino acids, PAHs and carboxylic acid.

In-Situ Missions

Almost all soft-landing Solar System space missions, except the very earliest ones, have included some means of conducting a chemical analysis of the surface on to which they have landed. Soft landings have successfully been made on[298]:

- the planets: Mars and Venus
- the satellites: the Moon and Titan
- the asteroids: Bennu, Eros, Itokawa and Ryugu

and

- the comet 67P/Churyumov-Gerasimenko.

Out of this list, we have yet to consider Eros, Mars, Titan and Venus. Mars is considered separately (Chapter 13) because of humankind's great expectations about Martian life and the many strenuous efforts which have been expended on looking for evidence of life (past or present) on that planet. Here, we briefly look at what we know about Eros, Titan and Venus and the prospects for life on them.

The spacecraft, NEAR-Shoemaker[299] touched-down on Eros in February 2001. A landing had not initially been a part of the mission, so only the γ-ray spectrometer was available to probe the asteroid's surface. This, however, studied the elements (atoms) in the surface layers, not the molecular compounds and so no data relevant to this book was obtained.

Titan has had only one visit, in 2005 by the Huygens lander of the Cassini[300] mission to Saturn. The 'lander' was primarily designed to examine Titan's atmosphere but it did survive the landing and then operated for another 90 minutes. Like the Earth, Titan's atmosphere is primarily nitrogen, but methane is then the next most abundant constituent. As a result, Titan has an abundance of lakes and seas of liquid methane[301] (plus ethane) which are predominantly located towards the North pole.

---

[297] Amino acids involved in cell growth and protein production.

[298] The Galileo spacecraft carried a probe for Jupiter's atmosphere. This successfully entered the atmosphere and, *inter alia*, analysed the composition of the atmosphere, but the probe was destroyed by the pressure long before it reached any solid surface of Jupiter (if, indeed, there is one).

[299] Launched February 1996, last contact February 2001.

[300] Named for Giovanni Cassini who, in 1675, discovered the gap in Saturn's rings which is named after him and for Christiaan Huygens who discovered Titan in 1655. The Cassini spacecraft was launched in October 1997 and disposed of by deliberately crashing it into Saturn's atmosphere in September 2017.

[301] Headline writers have made much play of the quantities of hydrocarbons present on Titan, with phrases like '300 times the entire oil reserves on Earth' being bandied around – obviously they anticipate vast new energy reserves becoming available. They rather seem to miss the point that for hydrocarbons to be used as fuel they have to combine with oxygen (i.e., be 'burnt') and Titan's oxygen reserves are minuscule.

Concept studies for future Titan Sample-Return missions do, though, envisage electrolysing water ice from Titan to produce oxygen, which, with the methane, would then supply the spacecraft's fuel for the return journey.

Huygen's scientific instruments included a gas chromatograph and mass spectrometer,[302] but these were for analysing the atmosphere, not the surface.

When Huygens was constructed, the prevailing opinions about Titan were that it was covered with a hydrocarbon ocean and so the probe was mainly designed to fall into and analyse liquids. However, it actually landed on a solid surface comprised of ice-crystal 'sand' and so the instruments designed for the liquid landing (such as liquid refractometers) went unused. While those designed for a dry landing, did not include anything to analyse the surface composition.

Titan's atmosphere and surface have a reddish colour as already discussed above (Section (ii) – Tholins) and we may expect an abundance of organic molecules of numerous varieties to be present on and in its surface and atmosphere. The mass spectrometer on board the orbiting part of the mission (Cassini), however, did not have sufficient mass resolution to identify individual types of prebiotic/organic molecules present in the atmosphere. Nonetheless, Cassini was able to detect the presence of molecules with masses a hundred or more times that of the hydrogen atom – and those are highly probably organic in nature.

Not including the several missions which have used Venus for a gravity assist on their way to somewhere else, there have been 22 successful or partially successful spacecraft missions to Venus. Of these 14 were landers or included landers which formed a part of the mission. However, none of these missions are still operating at the time of writing. Indeed, the last to operate was back in June 1985.[303] Then, Vega-2 was in action on Venus' surface for 56 minutes. If that seems a very short time after all the effort it takes to get a spacecraft to Venus, then just remember that the Vega-2 lander had to operate at a temperature of ~740 K (~470 °C – a bit hotter than the molten zinc in a hot-dip galvanising bath) and under a pressure 90 times greater than the Earth's sea-level pressure ® (about the pressure 900 m below an ocean's surface).

With Venus' surface temperature hotter than molten zinc, it should come as little surprise to find that no organic molecules have been detected by the landers.

What may come as a surprise though, is the number and frequency of serious suggestions that Venus could have organic molecules, even life-forms, high in its atmosphere. At an altitude of 50 to 60 km, Venus' atmospheric temperature and pressure are similar to those of the Earth near to sea level and so this location (after the Earth) is the most Earth-like environment in the whole Solar System ®. Indeed, for photosynthesising organisms, Venus' atmosphere could be *better* ® than the Earth, since photosynthesis could continue throughout the 1,400-hour-long night, utilising the thermal radiation from Venus' surface. The atmosphere's composition of 97% nitrogen and 3% carbon dioxide with clouds of sulphuric acid[304] but almost no water vapour, however, may keep the tourists away for a while.

Several space missions to Venus have already probed Venus' atmosphere using sounders which fell slowly through the atmosphere to the planet's surface[305] whilst supported by parachutes or which were held aloft by balloons circulating through the atmosphere, so making for longer sampling times and more varied sampling regions. Several future missions to Venus are in the early proposal stages, some, such as VAMP (Venus Atmospheric Maneuverable Platform – launch, 2029??), may

---

[302] Details of these instruments may be found in the Author's book '*Remote and Robotic Investigations of the Solar System*', 2018, Taylor and Francis, or via an internet search.

[303] Orbiting missions, though, have visited Venus since that date: Magellan (launched May 1989, mission concluded October 1994), Venus Express (launched November 2005, mission concluded January 2015) and Akatsuki (launched May 2010, mission still continuing at the time of writing).

There are several proposals for future Venus landers, with possible launch dates from the late 2020s onwards, but none seem to be very certain of fulfilment at the time of writing.

[304] Phosphine (PH_3), an extremely toxic gas for humankind, has recently been claimed as a trace component of Venus' atmosphere. Later work seems to contradict this claim. However, if phosphine is present, it could be a biomarker (Chapter 17), i.e., a sign that life-forms exist within Venus' atmosphere. This is because, on Earth, despite phosphine's potential lethality for humankind, it is produced in tiny quantities by some terrestrial microbes.

[305] E.g. Venera 4 (parachute) and Vega 1 (balloon).

carry balloon/aircraft hybrids to explore those parts of Venus' atmosphere where the previously postulated life forms might just possibly exist.

THE ORIGIN OF LIFE – A CAESAREAN BIRTH?

It should, by now, be clear that organic molecules (in the Box 9.1 sense) form naturally, in a great many varieties, in a great many places throughout the Universe and under a very great many differing environmental conditions. Except for terrestrial life, though, we have yet to find any evidence of the First-Living-Molecule in all that data.

Perhaps the First-Living-Molecule or its current manifestation(s) is/are extremely rare; some might even say that it only occurred once, on Earth – and that is, of course, a possibility. However, if we adopt that latter viewpoint, then there's little more to do towards solving the question of the Origin of Life; it happened once and that's it – Finish.

Science and Scientists, on the whole though, are optimistic. To quote Helen Keller (who needed optimism more than most):

> "No pessimist ever discovered the secret of the stars … or opened a new heaven
> to the human spirit"[306]

and

> "Optimism is the faith that leads to achievement."[307]

So – we will go for it – and perhaps, in due course, we will achieve something.

'Going for it' in this context means:

> … if we can't find the *!!*@**!$'!* stuff naturally, then we'll find it artificially.

In other words, we'll go down the road of the laboratory approach of Sagan and Khare (Chapter 9, Section (ii) – Tholins).

Rolling Your Own[308]

Sagan and Khare's laboratory experiments in the 1970s and 1980s were not the first attempts to 'Roll their own' First-Living-Molecule (or at least to see what would happen if they simulated the conditions which might have led to the First-Living-Molecule). That honour belongs to Stanley Miller and Harold Urey working 20 years earlier (if not to 'Baly of Liverpool' – see above).

The Miller-Urey experiments aimed to re-create the atmosphere of the Earth around 4 aeons ago. There is, however, still discussion going on/disagreement occurring over the nature of the Earth's atmosphere immediately following its formation. What seems (fairly) likely is that:

- The Proto-Earth comes into existence about 4,600 million years ago
- Initially the Earth was too hot, perhaps even having a molten surface, for any significant atmosphere at all to be present.
- As the surface cooled, gases emitted from the intense volcanic activity which was then still occurring, gradually accumulated to build up the first atmosphere. Those gases would have been mainly carbon dioxide, hydrogen sulphide, methane, nitrogen,[309] sulphur dioxide and water vapour.

---

[306] Keller, H. *Optimism*, T.Y. Crowell Co. New York, 1903
[307] Ibid.
[308] In these vaping days, there may be readers who do not realise that this means making your own cigarettes from loose tobacco and paper, or in this case, making your *own* First-Living Molecules.
[309] Reactions between the nitrogen, methane and hydrogen sulphide would then add ammonia to the mix.

- Carbon dioxide may have been the principal constituent (as it is today on Mars and Venus).
- Some 500 million years after its formation[310] the Earth's surface cooled to the point at which water could condense (see above).
- Huge oceans formed on the surface of the Earth into which most of the atmospheric carbon dioxide dissolved,[311] leaving an atmosphere largely formed from nitrogen (which is an unreactive element and so tends to accumulate) plus ammonia, hydrogen sulphide, methane sulphur dioxide and (some) water vapour, plus new compounds produced from these initial constituents.
- This would now be a reducing atmosphere.
- There is still debate, at the time of writing, over just how intensely reducing the atmosphere might have been at the start of the Archaeon Eon, some workers suggesting compositions containing up to 40% molecular hydrogen, others that there might even have been some free oxygen molecules roaming around – although this latter claim has been disputed recently. In terms of the end product (organic molecules) of Miller-Urey-type experiments (below), though, so long as free oxygen is *not* available, organic molecules are produced for a wide range of mixes of the basic 'starter' compositions.

Miller and Urey's Work

In 1952 Miller had just graduated with his BSc from the University of California, Berkeley, and was searching for a PhD topic. After some preliminary false starts he suggested to Urey (at the University of Chicago) the idea of studying the products of electrical discharges (sparks – to imitate lightning in nature) in various gas mixtures. With Urey as his supervisor, he then built and operated an apparatus to do just this experiment.

The experimental set-up was relatively straightforward:

- a reservoir for the liquid ingredients
- an inlet for the gaseous ingredients
- a heat source for the reservoir to evaporate all the ingredients and to drive them around to a chamber where the spark was produced
- a cooling system to condense the products
- a pipe to return the products to the reservoir to be cycled around the closed-circuit system again and again and which also had a trap with an outlet for the products so that they could extracted for chemical analysis when they had been 'cooked' for long enough.

The experimental procedure was:

- to sterilise / clean the apparatus so that it was free of any organic molecules and the like.
- to load the ingredients (Miller and Urey used ammonia, hydrogen, methane and water to begin with).
- to heat the reservoir
- to turn on the spark.
- to leave the apparatus running for a time ranging from a day to a week.
- to collect the product – by now ranging in colour from pink to dark brown (see tholins above), into sterile/clean containers.
- to analyse the product and identify the molecules from which it was now composed.

---

[310] This is the start of the Archaean Geological Eon which lasted for ~1,500 million years until 2,500 million years ago. The Earth's first 500 million years are known as the Hadean Eon after Hades, the Greek god of the dead and of the underworld (which seems somewhat premature, since life hadn't even started then).

[311] Later becoming locked more permanently into carbonate rocks.

The chemicals in the product of the experiment were identified by a technique which was then only ten years old, paper chromatography (Box 9.3). Miller found glycine and alanine in the experiment's product plus, less certainly, a few other organic molecules. Preserved samples from Miller and Urey's experiment, though, when analysed with far superior techniques some 50 years later, showed that the experiment had actually produced at least 20 amino acids. Other recent work has shown that they used borosilicate glass to construct their apparatus and this provided silica and catalysts which aided the formation of their organic molecules greatly. Nonetheless, their experiments were still highly significant in the context of this book since rocks within a 'warm little pond' would also act in similar ways to the silica, etc.

BOX 9.3 – PAPER CHROMATOGRAPHY

Paper chromatography[312] separates out different compounds using their rates of travel through/along a sheet of absorbent paper when dragged by a suitable liquid (the solvent) which is also traveling through/along the paper.

The paper strip will often be suspended above a bath of the solvent with the bottom edge of the paper dipping into the solvent, although other arrangements are possible.

The compounds in the sample to be tested are adsorbed (i.e., are adhering weakly – this is not a mis-spelling of 'absorbed') on to the surface of the paper substrate (or sometimes to a second solvent such as water with which the paper is already soaked) and also have an affinity with the primary solvent. At the start of a test a drop of the sample will be placed close to the bottom of the paper strip and the strip immersed into the solvent but with the sample drop higher than the surface of the solvent. The solvent then, through capillary forces, gradually moves up the paper strip. When the solvent passes the sample, the compounds' molecules experience two forces – a tendency to stick to the paper and a tendency to move with the solvent. The balance between these forces will vary from one type of compound to another. If the balance is in favour of the solvent, then that compound's molecule will ascend almost as fast as the speed of the solvent's movement. If the balance lies with the paper substrate, then that compound's molecule will move upwards much more slowly.

Thus, after an interval of time, the faster-moving molecules will have reached the top of the paper strip, while the slower-moving molecules are still near the bottom of the strip, i.e., the two molecules will have been physically separated from each other.

All the molecules of a particular type of chemical compound will move at roughly the same speed, but molecules of a different chemical compound will generally move at a different speed.

With a wide mix of molecules such as that obtained by Miller and Urey, at the end of the test, the various compounds will be strung out along the paper strip. If the compounds are coloured, then the result will be seen immediately as bands of colour running horizontally across the paper strip. If the compounds, or some of them, are clear or are white in colour, then a dye can be applied in order for them to be seen.

All that remains in order to identify the components of the original sample, is to measure the speeds of the molecules over the paper strip and compare these with the reference list of already known speeds of pure compounds.

If some differing compounds have the same speeds, then a different solvent can be tried which may separate them out. This can sometimes be done on the same sample already tested by turning the paper strip through 90° and repeating the process with the different solvent, thus giving a 2-dimensional distribution for the chemical compounds from the original sample.

---

[312] There are now other types of chromatography, such as gas, liquid and thin layer, but Miller and Urey used paper chromatography.

Miller and Urey repeated their experiment many times using different ingredients and/or changing other conditions. Other workers have also tried the same experiment, sometimes using UV light[313] in place of the spark as the energy source, right up to the present day[314]. A recent example[315] used laser-driven plasma shock waves as well as sparks in a carbon monoxide, ammonia and water reducing atmosphere and produced formamide (an amide[316] derived from formic acid) and RNA nucleobases (Figure 9.3, Box 9.1).

Laboratory simulations of the conditions in Titan's atmosphere today (carbon monoxide, methane and nitrogen, but *no* water) have also shown that all five nucleotide bases plus the amino acids, glycine and alanine, can be synthesised.

On a slightly different tack, Manfred Eigen showed in the 1970s that the nucleotides (Box 9.1) in an aqueous solution along with suitable protein catalysts for the reactions and a small amount of 'seed' RNA, would self-assemble into RNA molecules in an hour or two. The process was quicker and more consistent with the seed RNA, but it also occurred without its help. He found that for the nucleotide solutions without the seed RNA, the first RNA molecule to be assembled by random processes was then rapidly replicated up to some 10^{14} copies. Only one type of RNA was produced in each experiment, but the types could differ from one experiment to the another; clearly the first RNA variant to be 'born' by the random processes replicated so quickly that any later-comers could not get a look-in.

Leslie Orgel had conducted similar but inverse experiments to Eigen's and showed that nucleotides would assemble into RNA in the presence of the seed RNA but without the protein catalysts.

Thus, once the First-Living-Molecule evolves to the stage of producing nucleotides, it seems possible that progression to forms of biochemistry and life, basically similar to today's terrestrial life-forms, might be almost automatic; maybe even unstoppable.

THE ORIGIN OF LIFE – "IT WEREN'T ME GUV. – HONEST"

This section examines a theory which does not explain the origin of life at all; it merely puts the blame firmly somewhere else.

The theory is called Panspermia and, in fact, it is not a theory because we know that it has actually happened – and We, Humankind, were the instrument and guilty party of that happening.

Panspermia may be defined as the transportation of life-forms from a place where they already exist to a separate, previously life-less, place. Panspermia is generally intended to mean that the life-forms inhabiting a planet, or some other environment, did *not* originate on that planet. Instead, they originated somewhere else and were transported by some means to their new home. It does not, therefore, explain the actual origin of life, but it does mean that, if panspermia works, life need only have originated once (perhaps on Earth?) somewhere and somewhen within the entire Universe – and that, if it is then found elsewhere, then it is because it was been transported there.

Humankind's involvement with panspermia dates back to the Surveyor and Apollo Moon landing programmes (1966 to 1968 and 1961 to 1972, respectively), although I am sure that similar incidents must have occurred many times, with many other space missions.

---

[313] Many years ago, I gave an MSc student this experiment to duplicate as a part of his MSc project. Unfortunately, the UV lamp burnt out (and we did not have the money to replace it) before any definitive results could be obtained. Happily, though, he still graduated with his MSc.

[314] Only a few examples are mentioned here since, although the details of the experiments vary, their generic resemblance is marked. For further discussions of the topic see, for example, Kolb V.M. (Ed.), *Handbook of Astrobiology*, Taylor and Francis, 2019.

[315] *Formation of nucleobases in a Miller-Urey reducing atmosphere.* Ferus M. *et al*, PNAS 4306, **114**, 2017

[316] Amides are products of reactions between amines and other organic molecules.

Surveyor-3 soft-landed on the Moon on 20th April 1967 within the Oceanus Procellarum and a part of its scientific payload was a TV camera. Subsequently (and deliberately), the manned Apollo-12 spacecraft landed less than 200-m away from Surveyor-3 (Figure 9.8).

FIGURE 9.8   Surveyor-3 with Apollo-12 in the background, plus Charles Conrad and a plentiful supply of (unseen) *Streptococcus mitis*. (Reproduced by kind permission of NASA.)

On their return to Earth, the Apollo-12 astronauts brought that TV camera from Surveyor-3 back with them, as well as some other items.[317] When examined on Earth, the TV camera was found to be harbouring specimens of the common terrestrial bacterium, *Streptococcus mitis*.

Although there have been subsequent claims that the bacterium contaminated the camera *after* it was returned to the Earth, there is no proof of this. Thus, there is a distinct possibility that the bacterium *was* transported from Earth to the Moon on the Surveyor-3 spacecraft and existed there for 2 ½ years. There is, therefore, also the possibility that similar or other bacteria are still, to this day, living in niches and crannies aboard Surveyor-3, i.e., a possible start to a panspermia 'infection' of the Moon.

So, what we might call 'Inadvertent Directed Panspermia' may first have been instigated by Surveyor-3. I am quite sure, though, that inadvertent directed panspermia has, by now, also occurred

---

[317] The opposite process also happens and the Apollo astronauts undoubtedly brought back some lunar regolith dust on their space suits. In fact, this dust may be a major problem during future long-term missions like NASA's Lunar Gateway. Eugene Cernan (Apollo 17), during his debriefing on return to Earth said: "… dust is probably one of our greatest inhibitors to a nominal operation on the moon" (https://ntrs.nasa.gov/citations/20210026411). The dust is very abrasive, clings to surfaces electrostatically, clogs up, wears and damages moving parts and causes 'Lunar Dust Hay Fever' in humans. The dust, however, is not a living organism so, unless it is carrying micro-organisms which *are* living, then the problems which it causes are 'only' physical/engineering ones.

for most (all?) Solar System bodies which have had contact with any of humankind's mechanisms and I am *certain* (sorry Neil) that following any visit by actual humans, even though encased in spacesuits, that contamination of Solar System bodies by terrestrial organisms has already occurred.

In making this last statement, I am not apportioning any blame; the contamination is inevitable given that spacecraft sit atop their rockets for hours exposed to the normal terrestrial environment before their launch. Also, on manned missions[318] when inside their spacecraft, the astronauts remove their helmets at least, so that micro-organisms from the humans will pervade the interior of the spacecraft and in due course get onto the exteriors of the space suits and then be taken outside. As John Grunsfeld of NASA's Science Mission Directorate expressed it in 2015[319]:

"We know there's life on Mars already because we sent it there"

Heating up an entire spacecraft throughout its complete volume to ~130 °C,[320] after it is in space, might sterilise it completely, but that is clearly impractical for manned spaceflight and almost so for just about all other types of missions – and in any case it is not attempted.

Sterilisation, as best as can be managed,[321] is now the general practice for most missions intended to land on another Solar System body, but even if that is successful, there have been many missions in the past where little or no anti-contamination measures were attempted at all.

Many people involved with spacecraft and space missions argue that the harsh environment of space – a vacuum, plus extremely low or extremely high temperatures, plus γ-rays, x-rays, UV radiation and high energy charged particles, will do the sterilisation for us – and that may well be the case. But a *Streptococcus mitis* bacterium is only about 700 nm in size (0.01 x the width of a human hair) and so it doesn't need much room in a well-protected little corner of a TV camera to survive all those perils.

In conclusion, we must now live with the fact that any life-forms found on any Solar System bodies previously visited by humankind, by whatever means and in whatever fashion, could be the result of contamination brought by those earlier visitations and/or by the mission presently examining that body (Box 9.4). Inadvertent Directed Panspermia, or terrestrial contamination, will therefore be a major consideration and problem, if/when we get to the stage of wondering if *these* organic molecules found on *that* asteroid are actually of genuine ET origin.

BOX 9.4 – WHEN IS A RESULT 100% CERTAIN? – IF YOU'VE GOT THE NERVE, THAT'S A CLUE TO THE ANSWER.[322]

No scientific measurement, procedure or result (indeed no measurement, procedure or result of any kind) is *ever* absolutely certain or completely accurate. Uncertainty (perhaps together with death, taxes and leaking flat roofs) is about the only certain thing in life – but please don't ask me if I'm sure about that.

Almost any reputable piece of scientific work will be accompanied by an estimate of just how convinced the scientists involved in getting the result are that the result is correct/accurate. Incidentally, that estimate itself will be uncertain – and if you estimate the uncertainty

---

[318] So far, this only applies to the Moon.

[319] Reported in the *New Yorker*, 8th October 2015.

[320] The prion proteins which cause Creutzfeldt-Jakob disease (CJD, Mad Cow Disease and Scrapie, although these are not truly living entities, require emersion in caustic soda heated to 121 °C for 30 minutes in order for them 'probably' to be de-activated. So, even 130 °C, might not be hot enough just by itself.

[321] The Mars rover, Perseverance, for example, is currently stashing its soil samples onto the surface of Mars to be collected and returned to the Earth in a later mission (see also Chapters 3 and 13). It is currently planned for the samples to be transferred into a spacecraft orbiting Mars and there to be placed into a container which would be sterilised on-board the spacecraft and then packed into a second sealed container.

[322] Hint to the clue: Try an anagram.

in that estimate, the new estimate will also be uncertain – and so on ad infinitum. We surely do live on shifting sands.

One of many ways of evaluating the certainty/uncertainty of a result is via its 'Sigma value'. Sigma (or σ, a.k.a., Standard Deviation) is a statistical parameter which I will leave to you to research if you wish to know more about it.[323] It will suffice here to note that if a result is measured as being X with a σ value of s, then there is a 68.27% probability of the *true* value lying between (X – s) and (X + s), i.e., in the range X ± s.

For example, for a measured value of 100 and for s = 1.5, there is a 68.27% chance that the true value of the quantity will lie between 98.5 and 101.5. So, the chances that the true value does *not* lie within the range would be 31.73%.

100 ± 1.5, would be called a one-sigma, or 1-σ, result. Extending this to better levels of certainty gives:

- a 1-σ result would be 100 ± 1.5, with a probability of the true value being inside that range of 68.27% and a chance that the true value is not in the given range of 31.73%.
- a 2-σ result would be 100 ± 3, with a probability of the true value being inside that range of 95.45% and a chance that the true value is not in the given range of 4.55%.
- a 2.5-σ result would be 100 ± 3.75, with a probability of the true value being inside that range of 98.76% and a chance that the true value is not in the given range of 1.24%.
- a 3-σ result would be 100 ± 4.5, with a probability of the true value being inside that range of 99.73% and a chance that the true value is not in the given range of 0.27%.
- a 4-σ result would be 100 ± 6, with a probability of the true value being inside that range of 99.9937% and a chance that the true value is not in the given range of 0.0063%.
- a 5-σ result would be 100 ± 7.5, with a probability of the true value being inside that range of 99.999943% and a chance that the true value is not in the given range of 0.000 057%.
- a 6-σ result would be 100 ± 9, with a probability of the true value being inside that range of 99.999 999 80% and a chance that the true value is not in the given range of 0.000 000 20%.

For routine work a 2.5-σ or 3-σ result will generally suffice, but for critical work a 5-σ result is needed, although not always attained. Despite the '… *chance that the true value is not in the given range* …' values in the above list decreasing rapidly, though, they *never* reach zero. So, the answer to the question posed in the title for this box is also 'never' – and once there is the remotest suspicion that a Solar System body could have been contaminated by ourselves, we can *never* be *certain* that any life-forms discovered on that body are genuine ETs.

Directed Panspermia is little different from Inadvertent Directed Panspermia save that the entities involved in spreading the life-forms were/are doing so intentionally and in full knowledge of what they are doing. It is simply the deliberate introduction of life-forms into a life-less environment, generally with the hope/intention that those life-forms should colonise and establish themselves permanently within that environment.

Given that the Earth was formed about 9,200 million years after the Big Bang and that the first evidence of life on Earth occurs some 800 million years after that, it would be quite credible that ETs and ETIs should have evolved elsewhere in the Universe *before* any forms of life were present on the Earth. Rather, therefore, than ourselves originating from some First-Living-Molecule which came into being on the Early Earth by chance, perhaps some of those early ETIs seeded replicas of themselves in their own directed panspermia programme onto the Earth before any native-born

---

[323] The author's book *Telescopes and Techniques* 3rd Edition, Springer, 2013 will provide further details of error/uncertainty estimation and an internet search for 'Sigma Measurement Confidence' will give you plenty of information.

Earth life could come into being. In this way, we may find surprising resemblances to ourselves sometime in the future when we finally meet up with ETIs and so be able to greet them with:

'Great-to-the-100,000,000th Grandfather, I presume?'.

The scenario in the last paragraph is not actually that fanciful. We have, in just the last few decades, come up with reasonably practical systems for sending swarms of micro-spacecraft out to other stellar system light years distant from us (Chapter 18). Already, then, debates are underway about perhaps using those swarms to carry terrestrial micro-organisms out to nearby exoplanets and so to seed *them* with *our* versions of life. Such programmes raise significant ethical questions and we may decide not to go along that route, but in a few decades or so, we will be *able* to do so, if we do *decide* to do so ✿.

Panspermia 'proper' is the introduction of life-forms into a life-less environment by natural processes. In many ways, it will not differ from Inadvertent Directed Panspermia except that the transport mechanism is just some normal process arising from non-biological processes.

Suggestions that individual micro-organisms might be carried to the outer edges of an exoplanet's atmosphere and then pushed by radiation pressure[324] from the host star out into space may actually be correct. But, such unprotected organisms, however tough and hardy they might be (c.f. Tardigrades and *Deinococcus radiodurans* – Chapter 12) will only survive very briefly in space. Certainly, they will not remain viable for even the few tens of hours required to travel from the Earth to the Moon, never mind the hundreds of thousands of years needed to travel to a different planetary system.[325]

The situation, though, changes once the micro-organisms are within some sort of protective environment. We know that meteorites can travel from both the Moon and from Mars to the Earth (Chapter 9) and we also know that most other meteorites come from asteroids or even date back to the origin of the Earth (Carbonaceous Chondrites – Chapter 9). We also know that organic compounds of many types are to be found within many of those meteorites.

Thus, if organic compounds can survive intact within meteorites and meteorites can travel from the asteroids, Mars and the Moon to the Earth, it is but a short step to having meteorites containing spores and other forms of life travelling from Earth to the asteroids, Mars and the Moon and vice versa – and perhaps also throughout the entire Solar System.

Terrestrial-origin meteorites will require a large meteorite to hit the Earth first and so send some fragments of Earth out into space with velocities higher than the Earth's escape velocity (11.2 km / s, 25,200 mph). The 'Dinosaur-Killer' (66 million years ago) would do the job, but so also, would smaller, more recent impacts, such as the Bosumtwi impact (1,100 000 years ago, near what is now Ashanti in Ghana and which produced a 10 km diameter crater).

If we find life-forms within the Solar System, elsewhere than on the Earth and they have a similar biochemistry to terrestrial organisms (i.e., the same amino acids, proteins, nucleobases, etc. – Chapter 9), then we now have a choice between six possible different interpretations:

(i)   They have originated completely independently of terrestrial organisms.
(ii) They are the result of panspermia transferring terrestrial organisms to other Solar System objects.
(iii) They are the result of panspermia transferring other Solar System objects' life-forms to the Earth.

---

[324] Although normally imperceptible, all forms of e-m radiation, from radio waves to γ rays, *do* exert pressure on the objects which they illuminate ®. The curved dust tail of a comet is thus produced by the radiation pressure of sunlight on tiny solid particles sprayed out from the comet's nucleus (Chapter 18).

[325] In the last quarter of the 20th century, though, Chandra Wickramasinghe and Fred Hoyle proposed that some viral diseases, such as the 1918 influenza outbreak, resulted from the viruses involved being transported to the Earth via dust particles from comets. These ideas have found very little support from the rest of the scientific community.

(iv) They are the result of contamination/inadvertent directed panspermia (see above) by one or more space missions sent out from Earth.

(v) They are the result of inter-stellar panspermia (natural or directed) transferring non-Solar System life-forms to the Earth.

(vi) They result from processes as yet unknown to humankind.

Now, if these hypothetical Solar System ETs do *not* have a similar biochemistry to terrestrial life-forms (Chapter 11), then out of the above options we have to conclude that either (i) or (iv) must be the case and that we *do* now have evidence of genuine ETs, unrelated to ourselves – and so life must have evolved at least twice within the Universe.

Before concluding this part of our studies, though, we need to consider 'interstellar panspermia' a little further.

Directed interstellar panspermia will presumably be well designed and planned by its ETI origi-nators and so could be sent out at relativistic speeds. Nonetheless, even at (say) 0.1 c, the journey time over 100 light years will be 1,000 years. It would seem credible, though, that the ETI origina-tors of the directed panspermia 'units' (whatsoever those might be) would design them appropri-ately to allow for this, send them towards a likely prospect of a planetary system and probably have terminal guidance so that the units arrive at a planet (or planets) with good chances of success in their mission.

Natural (proper) panspermia will be *many* orders of magnitude less likely to succeed, but it may occur *many* orders of magnitude more frequently than directed interstellar panspermia. The vec-tors for natural interstellar panspermia would probably need to be much larger than the small (ish) meteorites considered for Solar-System panspermia, simply in order to provide better protection over the much longer travel times involved with natural panspermia. 'Oumuamua (Chapter 8, Figure 8.9) would possibly be a suitable object for natural interstellar panspermia's transport; its speed through galactic space relative to the Solar System was ~23 km/s, so it would take ~ 1.3 million years to travel 100 light years (and then it missed when it got here). Current models for the origin of the Solar System suggest that large numbers of planetesimals[326] are flung out into interstellar space by gravitational encounters with other early Solar System objects and presumably similar occurrences will happen during the formations of some/many/most other stars.[327] Planetesimals might well be able to protect lifeforms adequately for an interstellar journey, but whether life-forms would have had time to form before the initial storm of ejected planetesimals from a proto-star died away is another matter.

Thus, panspermia *is* possible and it probably *does* happen. Whether or not it happens *success-fully* (i.e., new colonies of life are established) is an entirely different and presently totally unknown affair. It would seem reasonable, nevertheless, that if an ETI civilisation (and/or humankind) were to attempt directed interstellar panspermia, they would only do so with a reasonable expectation of succeeding.

Natural interstellar panspermia would seem likely to be occurring, but its success rate is prob-ably very tiny indeed (perhaps it equals zero *exactly*).

You can make your own mind up about these possibilities – but I would not bet the farm on any of them.

ALTERNATIVE LIFE-FORMS: PERHAPS POSSIBLE, PERHAPS IMPROBABLE OR PERHAPS IMPOSSIBLE?

This section is where *your* imagination is allowed to run completely free. But, if you want to start off with some speculations from other people, read-on.

---

[326] The small bodies which form intermediate stages in the progression between the gas of a collapsing interstellar nebula and a fully formed planet, sizes ~10 m to ~100 km.

[327] Perhaps this is where 'Oumuamua came from?

Why Hydrogen, Carbon, Nitrogen and Oxygen (Plus a Few Bits of Calcium, Phosphorous, Potassium and Silicon)?

Perforce, most discussions of the possible chemical ingredients for life-forms are based upon our only known examples, those inhabiting the Earth – and that implies those chemical elements listed in the title of this sub-section (see also Box 9.1). However, we have also seen (Chapter 2) that silicon might be a substitute for carbon and ammonia or methyl alcohol substitutes for water in other forms of life. We shall also see later that for much of the time whilst life was originating and evolving on the Earth, little oxygen was available and life-forms were anaerobic (Box 13.1).

So, are we hydrocarbon-water agglomerations with high opinions of ourselves, the only chemical basis possible for a life-form?

Well, the answer *could* be *Yes* – for all that we *know* at the moment.

In Your Dreams (or Nightmares)

Other combinations of chemical elements from those leading to terrestrial life-forms, which might lead to alternative life-forms, have been envisaged. As we have seen within this chapter, no one has yet succeeded in making terrestrial life from scratch, so success with different chemistries is hardly to be expected and insofar as I can find, no experiments in this area are being attempted anyway. However, for what it is worth, some of the suggestions for alternative biochemistries are:

(i) Swap isomeric biological molecules for their inverse forms. ✿✿✿✿✿

Isomerism (a.k.a., handedness, chirality) is discussed in more detail in Chapter 11. Here, however, we will simply regard it as the ability of some molecules to exist in two physical versions: left-hand-type-shapes and right-hand-type-shapes (a.k.a. mirror images of each other[328]). As will be discussed further in Chapter 11, many of the biological molecules making up ourselves and other terrestrial life-forms can exist in both forms, *but*, only *one* of those forms takes part in the operation of the living being.

Thus, all terrestrial life-forms depend in various fashions on just *one* physical form of many of the molecules which go into our make-up, although *two* physical forms are possible.

If we somehow were to exchange (say) all the left-hand glucose molecules for right-hand glucose molecules in our bodies, much/most biochemical processes keeping us alive would be severely disrupted. We would become very sick almost instantly and almost certainly be dead within seconds or minutes.

However, suppose we exchanged *all* the left-handed molecules in our bodies for their right-handed versions and *all* the right-handed molecules for their left-handed versions, simultaneously and instantaneously.[329] Would we then still function normally? There would seem to be a possibility that we might do so, or at least that it might be possible that some ETs could evolve from their First-Living-Molecules and develop viable living forms with all *their* left and right-handed molecules in the inverse forms to those which we possess.

Whilst this all sounds rather complex, it is probably the least biochemically-different-from-us-ET that is possible, should the altered biochemical processes actually be able to support the existence of a living being.

---

[328] The molecules are not actually shaped like hands, but they are asymmetrical and one form is the mirror image of the other.

[329] I am *not* volunteering to be the subject for this experiment.

(ii) Replace carbon. 🐚🐚🐚🐚🐚🐚

Silicon has been mentioned in Chapter 2 as a possible replacement for carbon in liv-ing beings because the structure of its outermost electrons (which are the ones involved in chemical reactions) is similar to that of carbon. In the chemical periodic table, they are therefore both in Group 14 with silicon following carbon. The heavier elements in that same group, which have the same outer electron structure as carbon and silicon are: Germanium, Tin, Lead and Flerovium.[330] Grouping within the periodic table was based originally upon the elements within a single group having similar chemical behaviours. But 'similar' does not mean 'identical' and so, whilst carbon is the basis for terrestrial life-forms and silicon can form some similar molecular structures[49] to those of carbon,[49] the heavier elements in the group almost certainly cannot be carbon replacements in living beings. It would, though, be interesting to meet, albeit very briefly and at a good distance, a flerovium-based ETI, since almost all the isotopes of this radioactive synthetic element have half-lives measured in milli-seconds.

Sulphur can form chain molecules, but as a replacement for carbon in a life-form, that is about as far as it goes.

(iii) Replace water. 🐚🐚🐚🐚

Silicon dioxide (silica, quartz) in its molten form behaves in many ways similarly to water and might be able to take the place of water for life-forms on very high temperature exoplanets (Chapters 8 and 13). Perhaps this might be in conjunction with oxygen and alu-minium, but this is probably a 🐚🐚🐚🐚 suggestion at least, since silicon dioxide's melting point is in the region of 2,000 K (~1,700°C).

Other liquids, which have been suggested as perhaps being possible replacements for water within living beings include: ammonia, hydrogen fluoride, hydrogen sulphide, meth-ane and methyl alcohol.

(iv) Other suggestions. 🐚🐚🐚🐚
   a. Replacement reaction cycles for photosynthesis
   b. Replacement chemical elements for some of the chemical elements in terrestrial nucleic acids
   c. Arsenic as a replacement for phosphorous
   d. Boron and nitrogen compounds as replacements for hydrocarbons
   e. Space is left here for your own contributions (dreams, nightmares)

....

....

NEARING SUCCESS? – 5/10 ONLY, SO FAR

Thus, to date, the First-Living-Molecule has yet to reveal itself to us, either through Natural or Caesarean processes and the situation on its origin still remains, as it did at the start of this Chapter:

'We don't quite know,
but we do have some strong suspicions about
how it could have happened'.

One way or another, though, it would seem that success in this endeavour could be 'just around the corner' or still be some decades or be even a century or two away from us.

---

[330] A transuranium element.

Despite this somewhat downbeat conclusion, the work on finding the First-Living-Molecule has led to an extremely important result which one day may enable us to decide if ETs and/or ETIs are from the same stock as we are or if they have arisen from a completely separate First-Living-Molecule moment and maybe to decide if some interesting little collection of organic molecules is non-life or life. This result is slightly technical (but do not worry – if you have got this far, you can cope) as well as being very significant, so the next two chapters will be devoted to it.

10 Evolution

The origin of life and the evolution of life are often treated as though they are much the same thing. Here, a clear distinction is made between them, because they are not the same thing. Without evolution, the First-Living-Molecule (Chapter 9) would just stay the First-Living-Molecule. Without the First-Living-Molecule, evolution would never get going. So, the origin and the evolution of life are *linked* topics, but, especially when we are speculating about ET life-forms, we need to deal with them separately.

We have examined possible modes for the origins of life in Chapter 9, here, therefore, we turn to examine the evolutionary side of the coin.

EVOLUTION, OR DID THE FIRST-LIVING-MOLECULE LIVE HAPPILY-EVER-AFTER?

Let us get rid of two common false impressions immediately:

(i) Evolution, although it affects living entities, is *not in any way animate itself.* Evolution has:
no agenda, no awareness, no biases, no bigotry, no conscience, no consciousness,
no fears, no favourites, no hopes, no ideas, no intelligence, no mission, no motives,
no objectives, no opinions, no prejudices, no purposes, no sensitivities, no reasons,
no thoughts, no wishes …
nor is it
awake or asleep, depressed or excited, ethical or non-ethical, kind or unkind,
just or unjust, politically correct or incorrect, right or wrong,
woke or not woke …
it is *only* a *chemical process* acting entirely automatically according to the laws of physics and chemistry and the natures of the interactions between atoms and molecules.

(ii) Statements along the lines of
'The Earth is an ideal home for Humankind'
put the *cart before the horse.* Humankind (and all other terrestrial life-forms) have evolved to suit their own 'bits' of the Earth's environment.[331]

The effect of evolution is to change the descendants of an organism so that they are better fitted to live in their environment than were their parent or parents. Thus, the *life-forms evolve to suit their environment,* NOT *the environment to suit the life-forms.*

In Chapter 9 the requirements for the First-Living-Molecule were listed as the ability to replicate itself and then to undergo evolutionary development. Evolution arises when the replication process

---

[331] Life undeniably does change the Earth's surface and near-surface environments; which is why, for example, we now have oxygen in our atmosphere; the oxygen being a by-product of photosynthesis.[258]

However, the resultant changes may make the life-forms *less* well suited to their environment than before.

Evolution's (successful) mutations generally make the life-forms *more* suited to their environment than before.

Humankind and a lot of other terrestrial life-forms are currently experiencing the former type of changes right now, from the effects of climate change, etc.

Left to evolution, the life-forms surviving such environmental changes would adapt to these new conditions, but that could take millions of years. Larry Niven's science fantasy Hanville Svetz stories explore *inter alia* how humankind (and dogs) might evolve to live with levels of industrial pollution hundreds or thousands of times worse than those we have at present – see *The Flight of the Horse*, 1975, Futura Publications, for example.

DOI: 10.1201/9781003246459-12

is not perfect and the new molecule is not an exact replica of the original molecule. Those mistakes, we normally call 'Mutations'.

A mutation may be fatal, i.e., the new molecule cannot reproduce itself at all – in which case we can forget about it. Alternatively, it may be beneficial, neutral or detrimental; meaning the new molecule reproduces itself more effectively than the older one, that it is much the same as the older one or that it reproduces itself less effectively than the older one.

Now Evolution has the alternative names[239] of 'Survival of the fittest' and 'Natural Selection', but to the author, at least, these terms seem to carry an implication that the entities with a beneficial mutation somehow actively kill-off their predecessors without that mutation. That is not normally the case – they simply outbreed them (see below) and mathematics takes care of the rest.

Of course, lions do kill their prey and that process influences the evolution of the lions and of the prey. The prey evolves towards being able to run away faster (say) from the lions and the lions evolve also; towards being faster and/or to being better camouflaged and/or to becoming craftier in their ambushing (i.e., more intelligent) and so on. But these changes are not the result of conscious decisions by anybody or anything; evolution is just pushing both organisms towards becoming better suited to their own changing environments.

If the environment changes (the prey runs faster), those lions with mutations which enable them also to run faster will feed better and so breed more successfully. They will thus gradually replace the slower lions, as the latter starve or die off at the ends of the natural lives. The faster lions do not deliberately kill off the slower lions – in fact there's no reason why the slower lions might not be better at fighting than the faster ones – so if the slower lions win more mating fights, the issue could become considerably confused, but almost certainly will sort itself given enough of Marvell's second marvellous ingredient (Time).

Once a First-Living-Molecule originates, it is possible that it will take over its local environment very quickly (see Eigen's experiments – Chapter 9). But, inevitably, at some moment, a mutation will occur. As already mentioned, this could be 'beneficial, neutral or detrimental'. We need only concern ourselves with beneficial mutations from now onwards, and in this context 'beneficial' means 'Better Replication'.

'Better replication' usually will mean faster replication but, especially with the more complex organisms, it could also mean:

- more accurate replication (fewer mutations),
- more efficient replication (getting more copies of itself out of the same amount of available material),
- using less energy/food to continue living (when the life-forms get to the stage of needing energy) and so on.

But all this, in the end, just comes down to there being more of the mutated molecules/organisms in existence than would have been the case without the mutation.

We can illustrate the dramatic effect which a beneficial mutation can have – and so, see why evolution 'works', with a simple case study.

Eigen's molecules multiplied from one, or a very few, up to 10^{14} in a few hours and for this would need around 50 cycles of the replication process, with a doubling of the numbers for each cycle.[332] So, for our case study, we will assume that our First-Living-Molecule replicates itself every 10 minutes. After 100 minutes there should be 1,024 First-Living-Molecules in the 'warm, little pond' (or whatever). But at this point, let us suppose a mutation occurs and that the new, mutant molecule can replicate every 5 minutes. So, we actually will have 1,023 First-Living-Molecules and 1 mutant molecule. The progress after that is shown in Table 10.1.

---

[332] For the mathematically minded, this is exponential growth.

TABLE 10.1

Replications of the First-Living-Molecule and Its Mutant

Time (minutes)	First-Living-Molecule numbers	Mutant numbers	% First-Living-Molecule	% Mutant Molecule
0	1	0	100	0
50	32	0	100	0
100	1,023	1	100	0
150	32,736	1,024	97	3
200	1,047,552	1,048,576	50	50
250	3.35×10^7	1.07×10^9	3	97
300	1.07×10^9	1.10×10^{12}	0	100
350	3.43×10^{10}	1.13×10^{15}	0	100
400	1.10×10^{12}	$1.15. \times 10^{18}$	0	100
450	3.52×10^{13}	1.18×10^{21}	0	100
500	1.12×10^{15}	1.21×10^{24}	0	100

From columns 2 and 3 in Table 10.1 we may see that the new ('better') mutation rapidly out-paces the original First-Living-Molecules; reaching equality by 200 minutes, despite the mutant's later start. Thereafter the mutant takes an unassailable lead in its numbers over those of the First-Living-Molecules. However, the First-Living-Molecules *do* continue to exist, and to increase; they have not been killed off by the mutant molecules.

If we were to continue the progression, then in less than a day (actually ~21⅔ hours), *all* the normal matter in the entire Universe ® would be in the form of these two molecules – and almost all of it would be the mutant version.

Clearly this is somewhat unlikely.

So, now let us bring in a limit of (say) 10^{20} as the maximum number[333] of First-Living-Molecules and/or Mutant Molecules which biochemist's flask/'warm little pond' can support. Also let us factor in a lifetime for both type of molecule of (say) ten hours, after which the molecules disintegrate back to their original constituents, i.e., undergo a form of death. The new results are shown in Table 10.2.

The numbers of both molecules increase as in Table 10.1 until we get to 430 minutes. Before the First-Living-Molecules can double again (at 440 minutes) the mutant molecules would have doubled to 1.18×10^{21} (at 435 minutes) – except that the capacity limit of 10^{20} kicks in (i.e., the 'food' runs out) and the replication of both molecules ceases. The figure for the population of the First-Living-Molecules is thus fixed at 8.79×10^{12} and that for the mutant molecules[334] is fixed at $9.99999912 \times 10^{19}$ giving the total of both as 10^{20} ®.

From 450 to 600 minutes, the numbers are constant because no deaths or replications are occurring. From 610 minutes onwards, the first of the First-Living-Molecules will start disintegrating. The molecules' constituents will immediately become available to form new First-Living-Molecules and new mutant molecules.

However, because the mutant molecules replicate twice as fast as the First-Living-Molecules and there are 11,000,000 mutant molecules for every First-Living-Molecule, the constituents will almost all go into forming new mutant molecules. N.B.: The figures in the table do not change at the 650th minute because only three significant figures are being shown. If the full 21 significant figures were

---

[333] This may seem a lot, but it translates to about 0.000 02 kg (~20 mg) of the relevant molecules and is thus safely some way from equalling the 'entire mass of all the normal matter in the Universe'.

[334] This is the accurate figure, which rounds up, with the three-significant figure display used in Table 10.2, to the 1.00×10^{20} shown there.

TABLE 10.2

Replications of the First-Living-Molecule and its Mutant within a limited sample

Time (minutes)	First-Living-Molecule numbers	Mutant numbers	% First-Living-Molecule	% Mutant Molecule
0	1	0	100	0
50	32	0	100	0
100	1,023	1	100	0
150	32,736	1,024	97	3
200	1,047,552	1,048,576	50	50
250	3.35×10^7	1.07×10^9	3	97
300	1.07×10^9	1.10×10^{12}	0	100
350	3.43×10^{10}	1.13×10^{15}	0	100
400	1.10×10^{12}	1.15×10^{18}	0	100
430	8.79×10^{12}	7.38×10^{19}	0	100
450	8.79×10^{12}	1.00×10^{20}	0	100
500	8.79×10^{12}	1.00×10^{20}	0	100
550	8.79×10^{12}	1.00×10^{20}	0	100
600	8.79×10^{12}	1.00×10^{20}	0	100
650	8.79×10^{12}	1.00×10^{20}	0	100
700	8.79×10^{12}	1.00×10^{20}	0	100
750	8.79×10^{12}	1.00×10^{20}	0	100
800	8.79×10^{12}	1.00×10^{20}	0	100
850	8.79×10^{12}	1.00×10^{20}	0	100
900	8.79×10^{12}	1.00×10^{20}	0	100
950	8.72×10^{12}	1.00×10^{20}	0	100
1000	6.59×10^{12}	1.00×10^{20}	0	100
1010	4.39×10^{12}	1.00×10^{20}	0	100
1020	0	1.00×10^{20}	0	100
1050	0	1.00×10^{20}	0	100
1100	and so on	and so on	and so on	and so on

to be shown, then the figures would be changing; the First-Living-Molecule numbers decreasing and with the mutant molecules' numbers just below 10^{20}, but slowly increasing towards it.

By 950 minutes, the decrease in the First-Living-Molecule numbers is starting to show in the figures and by 1040 minutes the First-Living-Molecules have completely disappeared (i.e., the '0' in the table is really zero, not something rounded down to zero).

Thus, the effect of there being a limit to the numbers of the molecules in the sample is as you might expect; once the numbers reach 10^{20}, they stay there.

The effect of imposing a lifetime on *both* types of molecules is perhaps unexpected; the First-Living-Molecules die out completely by 1040 minutes.

This is not because the mutant molecules have destroyed them directly, it is because once the natural disintegrations of the oldest molecules started, the mutant molecules, being quicker at their replications and vastly in the majority, utilised the released molecule components in producing more mutant molecules, before the remaining First-Living-Molecules had any opportunities to produce any replicas of themselves.

Thus, the older strain (First-Living-Molecules) is completely replaced by the newer strain (mutant molecules) in a very short time – just through the natural processes of evolution and *without* there being any active killing of the older strain by the newer strain. If one or two First-Living-Molecules do manage to replicate at this stage, then they will disappear in another ten hours (their lifetimes).

This is quite a simple model for the workings of evolution, but it serves its purpose of demonstrating the basics of how evolution works. If we took a (perhaps) more realistic estimate of 5% for the advantage which the mutant molecules have over the First-Living-Molecules (instead of 100%), the end result would be unchanged, it would just take longer. Thus, if we take the First-Living-Molecules' replication time as 10 minutes, as before, and make the mutant molecules' replication time 5% shorter than that, i.e., 9 ½ minutes, then we find that the time to produce 10^{20} molecules is:

First-Living-Molecules:	11 hours 10 minutes
Mutant molecules:	10 hours 36.5 minutes

---

and that, when the mutant molecules have reached 10^{20} in numbers, the First-Living-Molecules have only reached 1.8×10^{19}, a ratio of ~5.5:1. So, as might be expected, the mutant molecules take longer to swamp the First-Living-Molecules, but they are still well on their way to becoming totally dominant in less than a day.

Of course, most mutations will not give the organism an advantage. In that case, the mutations resulting in a disadvantage will die out, whilst those which are neutral in their effects will probably just live side-by-side with the original organisms.

We see, therefore, in outline, how with even a small advantage in the replication rate, a natural limit to the size of the population and death in some form, the natural processes of evolution lead to one organism being completely supplanted by another which is better suited to the then prevailing environmental conditions. We have also covered three of Marvell's concerns (Chapter 9, opening quotation), but have yet to see where sex comes into the equation.

Now we shall remedy that omission.

ALL ABOUT SEX AND ENERGY

SEX

In fact, sex is *not* essential for the operation of evolution. The First-Living-Molecule and its immediate descendants plus plenty of organisms living today manage without it. Their (non-sexual) replication/reproduction processes are via:

- Fission (a.k.a. cloning): the organism first duplicates its genetic material (usually DNA today) and then splits into two, with each half having a complete copy of the genetic material.
- Budding: basically, this is the same as fission, but with the duplicate developing as a smaller copy of the original. Perhaps, also, from a special part of the original organism which is devoted to the budding process.
- Fragmentation: similar to fission, but with numerous versions of the original being produced instead of two.
- Spores: a single cell containing all the genetic material of the original organism which can be released by the original organism to make its own way in the world. They differ from seeds in not having a food supply and in being asexually produced. Produced in millions by, for example, fungi, as a method of reproduction, spores may also be produced by organisms, which normally reproduce sexually, when they are experiencing harsh environmental conditions, as a 'safety net', since spores can be very durable and long-lasting.
- Parthenogenesis: reproduction solely from an egg without it being fertilised by a sperm, although the latter can also occur. A somewhat related process can occur in some plants and is given the name apomixis.

- Vegetative propagation: when a fragment of a plant, whose main purpose is not reproductive, grows to be an entirely new plant. This will be familiar to most gardeners who use it as a means of propagating their best specimens via 'cuttings', i.e., small twigs, bits of stem, suckers or even just leaves which can be persuaded to develop roots and eventually grow into a new plant.

All these asexual reproductive/replicative methods can involve imperfect copying during their reproductive processes and so can evolve over time. So why did sexual reproduction evolve?

At the time of writing, there are at least 20 theories regarding why sexual reproduction began at all. However, they mostly come down to one single reason; sexual reproduction *speeds up evolution*.[335] Under rapidly changing environmental conditions, asexual reproduction has to await the 'right' mistake being made in order for the organisms to evolve to be a better fit for the new conditions; and that could take so long that the organisms are killed off by the changed conditions first.

Sexual reproduction requires two organisms to cooperate in some way in order to produce a new organism; and that new organism will generally be different from either of its parents because it has a mix of their attributes. The new organism may then be somewhat better suited to the changed conditions and the time required for the improvement is just one reproductive cycle.

Sexual reproduction is not independent of mutations; in fact, the reason why the parents *do* differ in some way is almost certainly due to mutations. In the parents' offspring, many mutations are likely to be mixed together and much more quickly than would be the case if the mutations had to occur one-after-another as in an asexually reproducing organism. Unless the population of the sexually reproducing organisms is very small (i.e., they are 'in-bred'), then the offspring and the great-offspring and the great-great-offspring and so on are likely to have a much greater range of adaptabilities to the new environmental conditions than any asexually reproducing organisms could achieve in the same time. Exactly in the same manner, though, as happens with the asexually reproducing organisms (Tables 10.1 and 10.2), the best fitted for the new environmental conditions amongst the sexually reproducing organisms will quickly take over from their less fortunate cousins.

Estimates of just when sexual reproduction did evolve range from 1,200 million to 2,000 million years back from today. So, if the First-Living-Molecule occurred about 3,800 million years ago (see above), then either it is extremely difficult to get sexual reproduction working, or its advantage over asexual reproduction is not as great as it might appear. Indeed, sexual reproduction does not guarantee successful adaptation to rapidly changing environmental conditions; just ask the next Tyrannosaurus rex which you meet for her/his opinion of large meteorites.[336]

Furthermore, sexual reproduction as well as asexual reproduction may lead to an evolutionary trend which has only a temporary advantage and not to long-term survival. Thus, humankind may well be close to the brink of finding out that the evolution of high intelligence is not, after all, a long-term survival trait (see also the Fermi Paradox – Chapter 16).

Looking on the bright side, though, we may summarise evolution via a slight modification of a well-known aphorism[337]:

The masons of evolution build slowly. Yet they build exceeding tall.

---

[335] Thus, Marvell's underlying motive behind his appeal to his *Coy Miſtreſs* was probably not quite what he was thinking it was.

[336] It does seem likely that some small relatives of *Tyrannosaurus rex* (Theropods) did survive the meteor strike of 66 million years ago. The theropods in due course evolved into today's birds. It seems improbable though, at least to the author, that the Tyrannosaurus rex themselves would have seen this development as being a welcome one.

[337] The mills of God grind slowly, yet they grind exceeding fine.

All this though is more than we need for the purposes of this book. Insofar as divining what forms ETs and ETIs might take on exoplanets very different from the Earth, the First-Living-Molecule appropriate for those conditions, followed by evolution to adapt the ETs to survive optimally for those same conditions, will suffice to get us going.

Unless we should discover a 99.9999% twin-Earth exoplanet (which given the results of exoplanet searches to date is quite unlikely) the actual individual forms of life now living here on Earth will not be very relevant. As we have seen above, evolution is based upon random events and so we will not get life-forms sufficiently identical to terrestrial life-forms for interbreeding to occur.

Even on Earth, if the First-Living-Molecule had occurred in a different warm, little pond or a few years (maybe even a few seconds) earlier or later or to a slightly different molecule from where, when and how it actually did occur, then we would not now have *exactly* the same types of the aforesaid life-forms living on the Earth today.

So, as has been said before (in relation to history[338]), Evolution does *not repeat* itself, but Evolutionary situations *can recur.*

What this means is that similar environmental conditions may lead to evolution producing life-forms which resemble each other, but which are still not at interbreed-able level of similarity to each other, a process called Convergent Evolution.

A terrestrial example may clarify what this means. Marine life is immersed in quite a dense medium (water). This provides buoyancy so that its weight is very low, zero or even negative (i.e., tending to float upwards). Water (or another liquid) also, though, produces resistance to movement through itself.

Thus, when a fast marine predator life-form evolves, there is a premium on its shape being as streamlined as possible – and there is only one basic shape which does that optimally; an elongated tear-drop. Thus barracuda, black-marlin,[339] dolphins, ichthyosaurs, narwal, orca, pike, sharks and tuna, all have that basic tear-drop shape (Figure 10.1) despite having very different evolutionary histories; indeed, dolphins and orca are mammals that evolved to return to live in the water having once been land animals.

FIGURE 10.1 Streamlined shapes which have evolved for terrestrial life-forms living in water or air. Left – Barracuda.[340] (Reproduced by kind permission of Pixabay and Ahmet Cakir) Right – Peregrine Falcon. (Reproduced by kind permission of Pixabay and Sven Lachmann.)

---

[338] 'Given that history does not repeat itself, but that historical situations may recur.' Black, H.D. *Moving Frontiers*, The Australian Quarterly, 123, **19**, 1947. That is to say; problems may arise in a modern context which may be similar to, but not identical with, problems occurring in the past. Lessons may therefore be learned from past, sometimes what *to do* now, but also perhaps, what *not to do* now.

[339] The world speed holder at an incredible 129 km/h (80 mph).

[340] This shape is a little more pointed at the front than the ideal – but the fish does need something to grab its prey with, when its speed (58 km/h, 36 mph) has allowed it to get within range.

Birds experience evolutionary pressures similar to those on fish from the atmospheric resistance to movement, although this is weaker than that for marine life (and additionally, they have to have wings). Nonetheless, a peregrine falcon stooping at 389 km/h ® (242 mph[341]) with its wings folded back has exactly the same elongated tear-drop shape (Figure 10.1) as any of the fast fish mentioned above.

Thus, when we find ETs which exist by chasing fast prey through liquid methane oceans on a super-Earth exoplanet with a (dry-land) surface gravity five times that of the Earth, we may confidently expect them to have tear-drop(ish) shapes like Black-marlin[342] and Peregrine falcons.

Similarly extraterrestrial plant life existing under hot desert conditions anywhere in the Universe is likely to develop tough waterproof outer layers similar to those which are characteristic of terrestrial cacti and succulents. Unless, though, there are herbivore animals around, the ET-Cacti probably will *not* develop the cactus' spines. Amongst land animals, the Wolf and the Tasmanian Tiger (a.k.a. Thylacine and now extinct), both medium-sized carnivores with similar sizes, body shapes and types of diet, are another example of convergent evolution.

Further and more detailed discussions of the potential shapes ETs might adopt to suit their environments will be found in Chapter 13.

Energy

Whilst sex may not be essential to life, energy is. Even the formation and then the reproduction of the First-Living-Molecule must have involved some energy changes amongst the atoms and molecules involved. Generally chemical reactions take place because, after they have occurred, the energy stored in the product(s) is less than the energy which was stored in the ingredient(s) prior to the reaction. The released energy converts into the kinetic energy of the atoms and molecules and/or into the emission of electromagnetic radiation, etc. This process is simply a demonstration of the tendency of 'things' in general to adopt the most stable position/situation[343] which is available under the circumstances prevailing:

- a pencil balanced vertically on its point may be stable for a while, but the slightest disturbance will cause it to fall onto its side where it is in a *much* more stable position.
- water can remain a liquid, sometimes, even when its temperature is below its freezing point. It is then called supercooled water. A slight disturbance though is then likely to cause it to solidify very rapidly and completely into the more stable form (at that temperature) of ice.

In these two above examples, the pencil balanced on its point and the supercooled water are examples of unstable equilibrium. The pencil lying on its side and the ice are examples of stable equilibrium. But, the two unstable equilibrium situations required a nudge in order to change from their unstable equilibrium positions to their more stable equilibrium positions.

Many chemical and biochemical reactions occur because the atoms and molecules end up being a in a more stable state, but the 'nudge' required to get the reaction going, may need to be considerably stronger than a 'slight disturbance' and it is then generally known as the 'Activation Energy' for the reaction.

Thus, quite independently of the energy needed to dig the garden, press the button on the remote control or pursue the fleeing prey, life-forms, just to exist, need an energy input in order to keep

---

[341] The highest measured speed for a stooping peregrine falcon at the time of writing
[342] The colour, black, may not have such a strong evolutionary importance as the shape, so they could be Pink and Blue with Green-Polka-Dots Exo-Marlins, if that colour scheme happens to provide better camouflage.
[343] This is a manifestation of entropy increasing with time and the second law of thermodynamics, but I leave it up to you to research *those* topics any further.

their biochemical reactions going. If the chemical reactions leave the atoms and molecules in higher energy states after they have finished, then more, perhaps a lot more, energy input will be needed.

For many terrestrial life-forms, the primary initial source of that energy is photosynthesis, i.e., utilising the light from the Sun. There *are* other sources of energy and we encounter some of them elsewhere in this book (see Chemoautotrophs, etc., Chapter 12, for example); however, photosynthesis is by far the most important set of such reactions.

"OK" you're perhaps now saying: "that's fine for plants, but *I'm* not in the least bit green anywhere, (except, perhaps, after the occasional over-extravagant celebration), so *I* don't need photosynthesis".

I'm sure that's entirely true, but it's because you are a parasite (and so am I and all the rest of humankind and all other animal life and a great many microbes and fungi and fish and …). Plant life obtains its energy via photosynthesis and we, the parasites, obtain our energy by eating plants in one way or another.[255] Some of you may now be looking at the lettuce leaf on the plate in front of you and feeling guilty, whilst others may be looking at a piece of steak and feeling virtuous. But, even the steak-eaters get their energy from plants and photosynthesis, albeit, at second or third-hand. The animal who provided the steak would have had a diet based on plants or on insects, other animals, lichen, fungi, etc. and those latter would have dined on plants in their turns and so on.

Thus, except for the chemoautotrophs, all terrestrial life-forms ultimately depend upon photosynthesis. Photosynthesis, for its part and as we now know it, uses the molecule, chlorophyll, so we also need to examine what we know about chlorophyll.

The various types of chlorophyll[344] currently to be found on Earth are pigments which do *not* absorb green light – contrary to many popular misconceptions (see also Chapter 13). Since the green light is not absorbed, it must be reflected and therefore we are accustomed to seeing chlorophyll-using life-forms as being green in colour (Figure 10.2).

FIGURE 10.2 Plants: Lawn grass and Black lilyturf (*Ophiopogon planiscapus* 'Nigrescens'). The lawn grass is the expected green colour, but the lilyturf leaves are almost black. Nonetheless both plants use chlorophyll for their photosynthesis reactions. In the lilyturf, dark red/purple anthocyanins overwhelm the chlorophyll's reflected green colour; but it *is* still there. (© C. R. Kitchin 2023. Reproduced by permission.)

---

[344] Main types of chlorophyll – *Chlorophyll a* (most widespread variety, $C_{55}H_{72}O_5N_4Mg$), *Chlorophyll b* (plants), *Chlorophyll c_1* (algae), *Chlorophyll c_2* (algae), *Chlorophyll d* (cyanobacteria) and *Chlorophyll f* (cyanobacteria).

The effective absorption of light-energy[345] by chlorophyll involves several stages, but the overall effect[346] is that the light energy is used to convert carbon dioxide and water into glucose (a carbohydrate) and oxygen via the net reaction:

six carbon dioxide molecules (6 CO_2) and six water molecules (6 H_2O) plus photon energy produces one glucose molecule ($C_6H_{12}O_6$) and six oxygen molecules (6 O_2).

Thus, not only is the immediate product of photosynthesis glucose, a food for us which we can consume without further processing, but the glucose is then further processed by plants, etc. and forms the basic feed-stock of pretty well all other nutriments for all forms of terrestrial life. *Additionally,* photosynthesis provides us with the oxygen[347] which we breathe.

After the Earth's First-Living-Molecule moment (around 3,800 million years ago – Chapter 9) the most crucial step in the development of (most) modern terrestrial lie-forms was therefore the evolution of photosynthesis and chlorophyll and this may have happened around 500 million years after the First-Living-Molecule moment. Those first varieties of chlorophyll would, of course, have been much simpler than the ones now dominant on Earth. Recently an unusual bacterium found near some hot springs in Iceland and called *Heliobacterium modesticaldum* has been found to utilise a very stripped-down form of chlorophyll (bacteriochlorophyll g). This type of chlorophyll cannot duplicate all the functions of modern chlorophylls. In particular it cannot use carbon dioxide directly and oxygen is extremely toxic to it. *H. modesticaldum*'s chlorophyll may therefore indicate the sort of forms which might have been taken by photosynthesis in its very early stages.

Oxygen has been present in the Earth's atmosphere at pressures above about 1% of our present atmospheric pressure since about 2,500 million years ago. Prior to that, the Earth's atmosphere was effectively anoxic and all life-forms must have been anaerobic. So, although photosynthesis had probably started around 3,000 to 3,500 million years ago, it took a billion years for it to become significant on a planetary scale; or perhaps the early forms of photosynthesis did not produce oxygen (see previous paragraph), or produced it much less efficiently than happens today.

Starting ~2,500 million years ago, though, Earth's atmospheric oxygen levels increased to around 2% to 4% of the whole atmosphere. Photosynthesis, mostly by marine life-forms, was by then producing large quantities of oxygen, some of this went into the atmosphere, but for ~1,500 million years, most of the oxygen was absorbed by the oceans and land surfaces.

But then, with all the drama of a Prima Donna Soprano making her first stage entrance of the evening at La Scala, Oxygen suddenly takes its rightful place in the story of the Earth and its life-forms.

The time is ~850 million years ago and the sea and land reservoirs for oxygen are now at full capacity and so all the photosynthesised oxygen goes straight into the atmosphere. In an 'instant' (actually ~500 million years, but then this *is* Geology) the atmospheric oxygen levels rise by some 700% to 800% to peak at 20% more than we have today. In the 300 million years since that extraordinary climax, the oxygen levels have fallen back to a 21% abundance within the whole of the Earth's lower atmosphere.[348]

This sudden massive[349] efflux of oxygen into the atmosphere is called the GOE (Great Oxygen Event) and it should come as no surprise that its peak coincides (roughly) with the appearances on land of the first plants and animals.[350]

---

[345] Very early light absorption processes in bacteria and single-celled organisms (Archaea[310]) may have involved the molecule Rhodopsin (a.k.a. Visual Purple) – and this is still the basis of the black-and-white vision of our eyes today.

[346] Readers interested in further details about photosynthesis can easily find them by searching the internet, for example, search for 'The biochemistry of photosynthesis' and or 'Krebs Cycle'.

[347] Oxygen is actually poisonous to some life-forms (Chapter 12).

[348] The other main atmospheric constituents are nitrogen (78%), water vapour (1% to 4%) and argon (1%).

[349] The Earth's atmosphere now contains around 1,000,000,000,000,000 tonnes of oxygen ®.

[350] Other workers come up with different results, as you might well expect for this type of research – I leave it up to you to research the topic in more detail, should you wish to do so.

HUMANKIND

Once it is established, we have seen how terrestrial life can and has evolved to exist over most parts of the Earth's surface and even to occupy some of it parts a bit below and a bit above the surface. Humankind, at first sight, might seem to be a prime example of this fanning-out over the Earth. Indeed, humankind does occupy and live in almost everywhere that other terrestrial life-forms can inhabit. However, our wide dispersal is not the result of physical evolutionary changes which enable differing varieties of 'us' to exist from the Equator to the Poles, but our use of tools (in the widest sense) which enable us to exist in environments to which we have not become adapted.

Humankind and its predecessors, such as Homo habilis, originated in nice warm parts of sub-Saharan Africa. There, we had no need of clothes. Indeed, we even had no need of body hair (such as that on chimpanzees). Hence, following a mutation variously estimated to have occurred between 1 million and 8 million years ago, our ancestors and ourselves became the *Naked Apes*.[351] In the warm climate of sub-Saharan Africa, some hundreds of thousands to millions of years ago, the lack of body hair did not matter; it may even have had an evolutionary advantage in terms of helping to control body temperatures during the hotter seasons and within the hotter regions.

With our evolving intelligence we could devise means (hunting, gathering and later farming) of feeding ourselves. We also soon learnt to protect ourselves from bad weather with shelters (perhaps even including caves) and from the predators which were better than ourselves (e.g., lions – see Chapter 1).

For some reason, probably population pressure, around 2 million years ago, some humans (Homo erectus *et seq.*) started to colonise parts of Europe and Asia where the climate was less clement than the Eden to be found around the Olduvai gorge. By a million years or so ago human stone tools[352] (cleavers, hand axes) were being used throughout most of Africa, up to Southern England, across Arabia and India to Korea and Indonesia.

Now the climates at the outer edges of this expansion would be much less well suited to bare-skinned hominids than sub-Saharan Africa. Hence, as various glaciations came and went, the hominids who had moved away from Africa would have found the need to keep warm somehow. Deliberately produced fires used for cooking and almost certainly used to provide warmth as well can thus be dated back to about one million years ago, at about the time of the Gunz[353] glacial stage.

The first evidence of clothes being worn (probably, by that time, by Homo sapiens) comes from genetic differences between human head lice and human clothes lice and suggests a time of about 170,000 years ago. These first clothes would certainly have been animal hides[354] in the beginning, so that the cartoonists are near to being right (Figure 10.3). The first *hand-made* clothing was probably some type of felt, since to make felt, fibres are simply matted together and this can be done by pounding fibrous material between two rocks.

Woven and knitted textiles followed quite sometime later. Estimates for the dates when hand-made clothing started to be made suggest that it was perhaps ~27,000 years ago, although flax fibres dating back to 36,000 and needles dating back to 50,000 years ago have been found by archaeologists. It is probably not a coincidence that the first adoption of animal-hide clothes, ~170,000 years ago, more-or-less coincides with the Riss (~300,000 to 130,000 years ago) and the Würm (115,000 to 11,700 years ago) glacial stages.

---

[351] I owe thanks to Desmond Morris and his famous book for this phrase; Johnathan Cape, 1967.

[352] The oldest currently known stone tools are some 3.3 million years old, predating Homo habilis by over 500,000 years.

[353] The names of glacial stages mentioned here are those for Europe. Other names (and dates) may be encountered for glaciations in other parts of the world.

[354] The aforementioned lice would not have been from the animal fur itself. Since hominids now lacked (much) body hair, the human body/clothes lice could not evolve until clothes *were* being worn; hence giving the human body lice a pleasant warm sheltered environment with lots of food (human blood).

FIGURE 10.3 Cartoon Cavemen, Cavewomen and Cavekids. (Reproduced by Kind permission of Pixabay and OpenClipart-Vectors.)

Thus, Humankind occupies (almost) all environments across the Earth by virtue of our 'tools', Fire and Clothing[355] – and *not* by virtue of our evolution to *fit* all those various environments.

Humankind has evolved somewhat since the first Homo sapiens was born, for example, the occupants of the high Andes generally have larger lung capacities than most of humankind in order to cope with the lower atmospheric pressures and oxygen levels at those high altitudes (there are permanent human settlements up to 5,000 m altitude).

However, civilisation, especially medical science and balanced and adequate diets for (much of) humankind, mean that our evolution is now much more controlled by the artificial environment,[356] towns, pollution, sedentary lifestyles, climate change, etc., created by ourselves, than by the 'natural causes' of yesteryear.

---

[355] Plus, air conditioning, boats, books, bricks, cars, central heating, clocks, combine harvesters, computers, deep freezers, doctors, drains, electricity, farms, hospitals, houses, ovens, ploughs, potatoes, radios, rice, roads, sanitation, spades, schools, telescopes, tinned food, tractors, TVs, warehouses, wheat and so on … and so on …).

[356] Since most of the artificial environment is of recent origin, there is little direct evidence yet, of its human evolutionary effects. However, as one example, there is a well-documented genetic difference between the alcohol tolerance of Oriental and Caucasian human populations, with Caucasians generally being the more tolerant of alcohol.

One possible reason for this difference is that it arose because much of the water for drinking in the Middle Ages and earlier was polluted by sewage-derived and other bacteria and so would frequently lead to the development of fatal diseases (e.g., cholera, polio, typhoid, etc.).

In the Orient, tea was widely available and drunk instead of water – and to make tea, the water had to be boiled first and so was therefore sterilised.

In the Occident, tea did not become widely available until the 17th century. The processes involved in brewing though, did produce drinks, such as beer, small beer, wine, etc., which were largely free from bacteria. Thus, alcohol replaced tea as being the healthy thing to drink throughout most of Europe. Furthermore, the more brewed drinks that you drank, the less likely you were to die of the various water-borne diseases (though you might die from alcoholism, of course). For occidental peoples, therefore, evolution drove the development of an increased tolerance of alcohol, because more people survived to produce offspring by avoiding early deaths from waterborne diseases than died from alcoholism (or at least the latter took time to become fatal, so reproduction could occur first; cholera, though, could/can be fatal in less than 24 hours).

A BRIEF HISTORY OF DOGS

The first dogs probably evolved naturally alongside grey wolves from a now extinct common ancestor some 40,000 years or so ago. They and Homo sapiens hunter-gatherers probably began mutually beneficial associations (almost certainly many times over and in many different places) around 30,000 years ago and dogs became humankind's first domestic animal at about that same time.

Since then, dogs have evolved by incredible amounts; no ETI visiting the Earth would credit Chihuahuas and St Bernards (Figure 10.4) as belonging to the same species (*Canis familiaris*). The

FIGURE 10.4 Dog Breeds – Shown approximately in their relative physical sizes. Top – A modern Coyote (*Canis latrans*); quite possibly this resembles the first domesticated dogs. (Reproduced by kind permission of Pixabay and Brigitte Werner.) Centre – Chihuahua. (Reproduced by kind permission of Pixabay and Ann-Marie.) Bottom – St Bernard. (Reproduced by kind permission of Pixabay and Bernell MacDonald.)

evolution from a coyote-type early dog to Chihuahuas and St Bernards, though, is not due to *natural* evolution. Indeed, we normally call the process by another name entirely: Breeding.

Breeding though is still a form of evolution. Indeed, Darwin discusses the breeding of sheep and pigeons in his *On the Origin of Species*,[238] as an illustration of how natural selection might accomplish similar changes. We should, therefore perhaps give breeding the name Directed Selection.[239]

However, since evolution ensures that organisms change in ways to increase their suitability for their environment and for dogs and many other organisms (plants and animals) that environment is now dominated by the desires and needs of humankind, breeding/directed selection (or whatsoever you wish to call it), *is* thus, still actually a part of *natural evolution*.

11 The Left-Hand of Darkness and the Right-Hand of Light[357]

We have ventured, in Chapter 9, into some quite abstruse aspects of biochemistry without mentioning one very important phenomenon associated with many of the chemical substances which we encountered there. The name given to the phenomenon is Optical Activity and it involves the way in which light and materials interact with each other. This topic is quite complicated, but fortunately we can concentrate on one aspect of it.

However, to provide some motivation and to re-assure you that we *are* still heading in the right direction, George Wald (1967 Nobel Laureate in Physiology or Medicine) rated its importance as:

"No other chemical characteristic is as distinctive of living organisms as is optical activity."[358]

and I would add that the 'living organisms' may well include ETs and ETIs, although Wald was probably only thinking of terrestrial life when he wrote the comment.

First, then, let us look into those properties of light and of material substances which in combination lead to Optical Activity.

POLARISED LIGHT

Now light and all other forms of electromagnetic waves, such as radio waves and x-rays are a complex type of wave motion.[359] They are made up of an electrical field and a magnetic field, both of whose intensities simultaneously vary in a wave-like manner, with the direction of the magnetic field perpendicular to that of the magnetic field. We can follow what happens with a light wave by looking at what happens to either or both of these waves and it is usually the electric field which is chosen for this purpose.

---

[357]   "Light is the left-hand of darkness and darkness the right-hand of light.
Two are one, life and death, lying together like lovers …"
Le Guin. Ursula K. *Tormer's Lay. The Left-hand of Darkness.* MacDonald & Co. Ltd. 1969

[358] *The Origin of Optical Activity.* Wald. G. Ann. N. Y. Acad. Sci. 352, **69**, 1957.

[359] Electromagnetic waves, including light, are a very perplexing phenomenon indeed and they sometimes behave as waves and they sometimes behave as particles. The particles, in the latter situation, are called photons or quanta. If you find this somewhat puzzling, don't worry – you will be in the company of most physicists. For the purposes of this book, it is best just to accept the situation, but if you *really* want to know more, then an internet search for 'Lightwave-Particle Duality' will swamp you with information. A well-known joke amongst physicists, which is actually too close to the truth to be really funny, may comfort, if not help, you:

"On Mondays, Wednesdays and Fridays, we believe light is a wave.
On Tuesdays, Thursdays and Saturdays, we believe light is a particle.
On Sundays we hide our heads under the bedclothes and
hope it will all have gone away by the next day - but it never does."

DOI: 10.1201/9781003246459-13

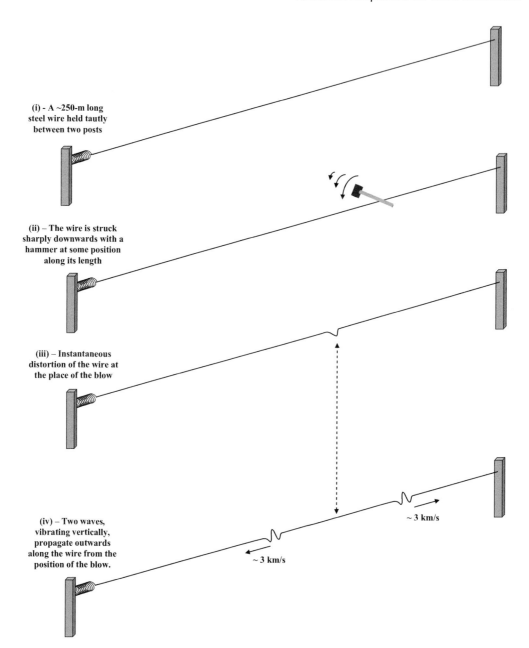

(i) - A ~250-m long steel wire held tautly between two posts

(ii) – The wire is struck sharply downwards with a hammer at some position along its length

(iii) – Instantaneous distortion of the wire at the place of the blow

(iv) – Two waves, vibrating vertically, propagate outwards along the wire from the position of the blow.

~ 3 km/s

~ 3 km/s

FIGURE 11.1 Schematic illustration of vibrations in a long, tensioned wire. (© C. R. Kitchin 2023. Reproduced by permission).

Thus, taking only the electric field, we may envisage a light wave as being analogous to the vibrations in a tensioned wire[360] (Figure 11.1). The deformations to the wire's shape, following the (downward) blow from the hammer, are in the vertical direction and are 'wavy'[361] in shape. They

---

[360] You may have seen illustrations of the plucking of a guitar string (etc.), in which the whole guitar string becomes bowed and moves up and down following the pluck (= hammer blow). This is the same physical situation as that in Figure 11.1 – but there is no contradiction between the two viewpoints. The speed of sound in steel is around 3 km/s. The length of a guitar string is perhaps 1 metre. It thus takes only ~300 μsec for the effect of the pluck to influence the whole guitar string. In Figure 11.1 we are considering a much longer string/wire and so it takes time for the disturbance to the wire, caused by the hammer blow, to propagate along the wire. In effect the 'Guitar string' picture is just that of the very peak of the disturbance to the wire as seen from the much wider viewpoint of Figure 11.1.

[361] Sinusoidal, or nearly so, if you are mathematically inclined.

also move out along the wire in both directions from the hammer's point of impact onto the wire and at the speed of sound (~3 km/s in steel). Since the movement of the wire (wave/vibrations) in this example is vertical, the wave motion is said to be 'vertically polarised'.[362] Had we hit the wire with the hammer from the side, the movements of the wire would be from side to side (horizontal) and we would call the wave 'horizontally polarised'. More generally, we just speak of 'linear polarisation'.

Normal light, such as that by which you are almost certainly reading this book, is composed of trillions upon trillions of individual light waves, each of which is linearly polarised in some direction, but those directions across the whole light beam are completely randomly distributed. We call a light beam 'linearly polarised' when all, or at least most, of its constituent light waves are vibrating in the same direction.

Light waves can be vertically or horizontally polarised and they can also be circularly or elliptically polarised. The latter states though are not relevant to this discussion. They are mentioned here, because if you are already acquainted with 'circular polarisation', then you should not confuse it with 'optical activity', even though both phenomena do involve rotation of the direction of the polarisation of light waves.

Quite when the phenomenon of the polarisation of light was first discovered is uncertain, although the fact that a crystal of good quality calcite (a.k.a. Iceland Spar) would show two images of a single object when that object was seen through the crystal (Figure 11.2) has certainly been known since the seventeenth century. Later, it would be found that the two emerging light beams were linearly polarised at right angles to each other. By 1828 William Nicol, by combining together two appropriately cut crystals of calcite, was able to separate the two linear polarised beams sufficiently that when an ordinary beam of light was fed into his 'Nicol Prism' only one linearly polarised beam emerged. The Nicol prism was thus the first polariser to be produced and it allows polarised light to be manipulated and experimented with. However, François Arago almost two decades earlier (1811) had made the first fundamental discovery which underlies optical activity, i.e., that the direction of the polarisation of a beam of linearly polarised light rotated when he passed the beam through a crystal of quartz. This rotation of the plane of polarisation of linearly polarised light is what is meant by the term 'Optical Activity'.

(i) (ii)

FIGURE 11.2 A single object (i) has two images when that object is viewed through a calcite crystal (ii). (© C. R. Kitchin 2023. Reproduced by permission)

The second fundamental discovery which underlies optical activity was made by Jean Baptiste Biot a couple of years later when he showed rotation of the direction of the polarisation also occurred when passing a linearly polarised beam of light through some organic liquids and some solutions of organic materials.

---

[362] If this should remind you about polarised sun glasses – you would be dead right.

The third fundamental discovery which underlies optical activity was made by John Herschel in 1820 and it was that quartz crystals came in two varieties and when you looked down the beam towards the light source through one of the crystals, the direction of rotation was (say) clockwise, called Dextrorotatory, or '+', but with the other type of the crystal, the rotation was anticlockwise, called Levorotatory, or '−'.

Now, even stretching one's imagination to the limit, it seems unlikely that we shall find ETs made up from quartz sometime soon. Thus, it is Biot's observation that (terrestrial) organic materials could produce optical activity which is of further interest here, especially when combined with Herschel's discovery that in two different physical forms of the same substance, one could be dextrorotatory and the other levorotatory.

Now therefore, we must look at how the same substance *can* have two physically different forms.

ISOMERS

Herschel, then, discovered that quartz crystals came in two varieties with different crystal structures and an analogy with left and right hands and gloves may be useful in envisaging the physical structures involved here. Your left and right hands are very similar in appearance, but in fact they are mirror images of each other. They are also asymmetric in their three-dimensional structure, meaning that you cannot lay your left hand over your right hand so that every point making up your left hand overlies exactly the equivalent point in your right hand. You can try this for yourself by attempting to put a right-hand glove onto your left hand (Figure 11.3), and failing. This property of hands, that they are non-superimposable, either directly or one hand and its mirror image, is fundamental to the later discussions in this chapter.[363]

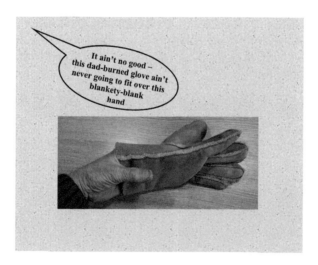

FIGURE 11.3 A right-hand glove *not* fitting a left hand. (© C. R. Kitchin 2023. Reproduced by permission)

However, isomerism is primarily an atomic/molecular-scale state of affairs. Its basic definition is that there may be:

> Two or more *differing* chemical compounds, each of whose molecules are made up from the *same* set of atoms. The differences, therefore, lie in and arise from the way that the atoms are arranged in space.

---

[363] This topic has a lot of synonyms and near-synonyms in common use. As well as the terms utilised here, the reader may encounter stereo-isomers, enantiomers, chirality, homochirality, handedness, sinistral and dextral, cis and trans, L and D and many other usages. I can only recommend that, when in doubt, you conduct an internet search to check what is being described.

To see what this means for a real molecule, let us start with a Tetrahedron (Figure 11.4 (i)) and suppose we have a carbon atom and four hydrogen atoms. With the carbon atom at the centre of the tetrahedron bonded to the four hydrogen atoms, which are at each corner of the tetrahedron, we have a familiar molecule, Methane (Figure 11.4 (ii)). This molecule is completely symmetric and all its molecules have identical structures whatsoever their orientation in space may be. Thus, methane is *not* an isomeric compound.

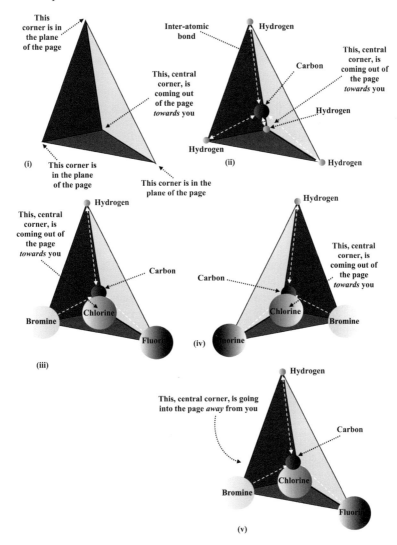

FIGURE 11.4 Configurations of tetrahedral molecules. (i) The basic tetrahedron shape (central corner coming out of the page towards you). (ii) Methane – CH_4 – The Carbon atom in the centre is bonded to four hydrogen atoms, one at each corner of the tetrahedron. (iii) Bromochlorofluoro-methane – $CHBrClF$[364] – The Carbon atom in the centre bonded to four other, differing, atoms – Hydrogen (top), Bromine (left), Chlorine (centre, coming out of the page towards you) and Fluorine (right). (iv) Bromochlorofluoro-methane – Mirror image – Hydrogen (top), Fluorine (left), Chlorine (centre, coming out of the page towards you) and Bromine (right). (v) Bromochlorofluoro-methane – mirror image rotated 180° – Hydrogen (top), Bromine (left), Chlorine (centre, going into the page away from you) and Fluorine (right).

---

[364] The same name is sometimes used for a similar molecule in which a second fluorine atom replaces the hydrogen atom ($CBrClF_2$). The two fluorine atoms make this molecule symmetric and so it, like methane, is not an isometric compound.

Now let us change three of the hydrogen atoms in the methane molecule for one each of Fluorine, Chlorine and Bromine (Figure 11.4 (iii)). This molecule goes by the jaw-breaking name of bromochlorofluoro-methane.[365] Unlike methane, though, *this* molecule is completely asymmetric. If you compare the structure of the molecule in space with its mirror image (Figures 10.3 (iii) and 10.3 (iv)), then, with both chlorine atoms coming out of the page towards you, the positions in space of the Fluorine and Bromine atoms have been exchanged. You could not, therefore, superimpose the original version of the bromochlorofluoro-methane molecule onto its mirror image so that every point on the original was aligned with every point on its mirror image.

If you try to move the mirror molecule around in space in order to get point-to-point matching you will never succeed. One obvious attempt to get such matching might be to rotate the mirror molecule through 180° (Figure 11.4 (v)) and, at first, it looks as though you have succeeded; the atoms have the same pattern on the page in Figure 11.4 (v) as those in Figure 11.4 (iii). A more careful check, though, will show the chlorine atom in Figure 11.4 (iii) is coming out of the page towards you, whilst in Figure 11.4 (v) it is going into the page, away from you. So, again, you could not superimpose the original version of the bromochlorofluoro-methane molecule so that every point on the original was aligned with every point of this version.

In fact, unless you have the ability to move around and rotate your molecule in a fourth dimension of space,[366] you will *never* be able to get a point-to-point match between the original bromochlorofluoro-methane molecule and its mirror image.

We have therefore arrived at our first isomers (see above), two *differing* chemical compounds, each of whose molecules is made up of the *same* set of atoms.

There are many other types of isomers arising from different variations between molecules' atomic structures. For example, structural isomers have differences in the bonds between the atoms within the molecule, say a double bond in place of a single bond. Many isomers also have quite different physical and/or chemical properties for their two variants. Thus n-butane and iso-butane molecules (structural isomers) are both built up from 4 carbon atoms and 10 hydrogen atoms, but n-butane melts at a temperature of 134.8 K (−138.4 °C), while isobutane melts at 113.6 K (−159.6 °C).

Isomers like bromochlorofluoro-methane, whose mirror image cannot be matched point-to-point with the original molecule (non-superimposable mirror image), differ in their optical activities (see above). However, they mostly have no other way of being distinguished.[367] This makes it very difficult to separate the two types of molecules into pure samples of each from a mixture of both. When, however, you *do* have pure samples of the two types of molecules making up an optical isomer, then, other things being equal, the clockwise (dextrorotation) rotation angle observed for one of the samples equals in magnitude the anti-clockwise (levorotation) rotation angle of the other. When you do have a mixture[368] of equal quantities of the two types of molecules, then the dextro and levo rotations equal each other and cancel out and there is then no optical activity to be seen.

With this last paragraph we have arrived, finally, at the reason for all this background material for our investigations of life-forms, terrestrial and non-terrestrial (when we find them).

IS THAT BLOB OF GUNK UNDER THAT MARTIAN ROCK, LIFE OR NON-LIFE?

The reason behind the preamble on isomers is that many of the organic molecules forming ourselves and other terrestrial life-forms *are* optical isomers. Furthermore, we have just noted (above) that it may be very difficult, if you have a (racemic) mixture of the two types of isomeric molecules, to

---

[365] Read it slowly and you'll find that it *can* be spoken out loud.

[366] Time is often called the fourth dimension, thanks to Einstein, but what is being referred to here, is an additional dimension of *Space*. If you find this a difficult concept to imagine (as I do – although, mathematically it can be dealt with quite easily), then it may, or may not, comfort you to know that some theoretical physicists working on String Theory envisage 10 or even 11 spatial dimensions ® being needed to 'understand' the Universe properly. Further research into this topic, though, I leave entirely to the reader.

[367] Sometimes optical isomers will differ in their chemical interactions *if* the molecules with which they are reacting are also isomers.

[368] Known as a Racemic mixture.

separate them out into pure samples of each type of molecule. Thus, when Miller and Urey and other workers (Chapter 9) were conducting their experiments and analysing the products, some of those products were known to be optically active – such as alanine[369] (Chapter 9) – but they did not show any optical activity. In other words, Miller and Urey *et al.* had synthesised alanine, but as a racemic mixture. Samples of alanine taken from living sources, however, *do* show optical activity (they are dextrorotatory[370]) and so must be pure samples of one of the molecule types.

In summary:

- Optically active organic molecules originating from non-living sources are *not* optically active.
- Optically active organic molecules originating from living sources *are* optically active.[371]

and so, we have a simple(ish) way of distinguishing between a blob of gunk under Martian rock which is a living organism (or which came from a living organism) and a similar blob of gunk that came about by the inter-reactions of a few simpler molecules, aided by UV radiation and/or lightning (say).

Additionally, the 'optical activity' test may allow terrestrial life-forms to be distinguished from ETs. Thus, for example, if sometime in the future, samples from the sub-surface oceans[372] of (say) Europa are returned to Earth, analysed and it is found that their alanine is levorotatory (the alanine in terrestrial life-forms is dextrorotatory, see above), then it will be a very good indication indeed that Earth's life-forms and Europa's life-forms originated in two quite separate First-Living-Molecule events.

Unfortunately, if in the above scenario, Europa's life-forms turn out to have dextrorotatory alanine, then it gives us far less information. It could, in this case, be that:

- Europa's life-forms originated in two quite separate First-Living-Molecule events.
- The Europa samples have been contaminated by terrestrial organisms.
- Life-forms from Europa were transported to the early Earth and started life here (see panspermia, Chapter 9).
- Earth and Europa were both seeded with life from the same external source (see, again, panspermia, Chapter 9).
- Life can *only* be built up utilising the dextrorotatory form of alanine – and there is a distinct possibility that this last of these options may be the correct one.

You may now, I hope, see why Wald's quotation at the start of this chapter:

"No other chemical characteristic is as distinctive of living organisms as is optical activity"

is so apposite and why the presence or absence of optical activity in specimens will be so useful when we *do* find life of some sort somewhere other than on the Earth.

---

[369] Alanine is only being used here as being reasonably representative of a great many organic molecules important to terrestrial life.

[370] RNA and many of the components of DNA in terrestrial life-forms are dextrorotatory. One recent suggestion for how this imbalance may have arisen is that there is a pre-cursor molecule of RNA called RAO (Ribo-Amino Oxazoline). This molecule, in the presence of magnetic fields (say, from magnetite crystals), will form *only* its dextrorotatory *or* its laevorotatory structure. Once formed, that molecule then only reacts with other molecules of the same structure and so any original imbalance is preserved.

[371] This is a bit of a sweeping statement; not every potentially optically active compound originating from a living source is necessarily actually optically active. However, finding one or more compounds which *are* optically active in a specimen is a strong indication that that specimen *is living* or that it has come *from a living source*.

[372] Not actually proven to exist yet, but with quite a bit of evidence that they are there to be found – Chapter 13. Worlds with sub-surface oceans are sometimes called IWOWs (Interior Water Ocean World).

12 The Extreme Limits

Over the last three chapters we have seen how life may originate, how it may evolve, how we might be able to distinguish non-life from life, and maybe, how we might be able to distinguish terrestrial-type life from extra-terrestrial-types of life.

We are almost ready to start looking into extra-terrestrial-type life, but we have to look to terrestrial life just one more time first.

The vast bulk of living organisms inhabiting the Earth will be sufficiently familiar to the reader not to require further description here and the basic requirements of current terrestrial life should anyway be apparent by now.

However, there are some specialist forms of terrestrial life which are so different from the norm that they might as well be ETs. So, we will examine these extreme forms of terrestrial life in a little more detail with high hopes in due course of identifying the exoplanets where terrestrial-life-forms might, potentially, evolve and/or be potential sites for colonies of ourselves or other Earth-life-forms (Chapter 18).

EXTREMOPHILES

Terrestrial life-forms which inhabit environments where humankind would perish rapidly and probably unpleasantly are called Extremophiles (or variations on that term).

Probably the best known of these Extremophiles are the organisms which survive in temperatures close to 100 °C (see also note about atmospheric pressure in Chapter 9). These are called Thermophiles and we have already encountered one example, *Pyrodictium occultum* (Figure 9.3 (v)). This bacterium (thermophiles can also be fungi) has an optimum temperature for growth between 80 °C and 105 °C. Its survival at such high temperatures appears to be due to differences in its tRNA nucleosides[373] from those of 'normal' life terrestrial life-forms. *Pyrodictium occultum* is an example which was discovered in 1979 inhabiting deep oceanic hydrothermal vents where the temperatures can reach 400 °C in places. A related species, *Geogemma barossii* (a.k.a. Strain 121), is even hardier – it still *reproduces* at 121 °C.

Though not quite as hot as the hydrothermal vents, hot springs, such as the Grand Prismatic Spring in Yellowstone National Park (70 °C – Figure 12.1), support thermophiles in great abundance and these produce the springs' vivid colours.

Some thermophiles (but also other, non-thermophilic organisms, such as nitrogen-fixing soil bacteria) are also Chemoautotrophs[374] (see also Chapter 7 and anaerobic organisms, below). These do not use free oxygen from the air for their living processes. Instead, they obtain their energy from chemical reactions with reduced compounds such as ammonia, hydrogen sulphide, iron, methane and sulphur.

At the other end of the temperature spectrum, Cryophiles, such as the lichen, *Xanthoria elegans*, can still be active at –25 °C. Some recently discovered microorganism cryophiles are also chemoautotrophs and alkaliphiles.[375] The water in a cold spring located on Axel Heiberg Island (Lost Hammer Spring, ~10° South of the North Pole and ~ 400 km West of Greenland) is so salty that it is

---

[373] See Figure 9.3 (vi) – nucleosides are nucleobase molecules which have combined with a sugar molecule (Ribose; c.f. RNA and DNA).

[374] a.k.a. Chemotrophs. Plants, etc., which obtain their energy via photosynthesis are classed as photoautotrophs.

[375] Able to tolerate alkalis such as sodium hydroxide.

DOI: 10.1201/9781003246459-14

FIGURE 12.1 The Grand Prismatic Hot spring in the Yellowstone National Park, Wyoming. (Reproduced by kind permission of Pixabay and Mike Goad.)

still liquid at –5 °C and contains less than 1 ppm[376] of oxygen. Similar conditions may exist at some times and in some places on Mars today. Over 100 different microbes which survive without using oxygen or pre-existing organic material and which can utilise carbon dioxide and nitrogen directly from the air were detected in that spring.

Acidophiles and alkaliphiles[377] thrive under extremely acidic or alkalinic[378] conditions. For example, *Ferroplasma acidarmanus* and its relatives can even live in solutions of extreme acidity, a pH[379] of 0.0, and they do this by using the oxidation of iron sulphate to maintain a pH of around 7 inside their cells. Alkaliphiles, such as the bacterium, *Natronomonas pharaonis*, inhabit highly saline soda lakes where the pH may reach 11 (Figure 12.2).

Radiophiles can tolerate extraordinarily high levels of x- and γ rays, up to 3,000 times the fatal dose for a human. They appear to sustain the damage from the radiation but are then to be able to repair it. *Deinococcus radiodurans* was discovered in 1956 because of its radiation resistance; it had survived the γ ray sterilization process of some tinned meat, a process which was then being tested for commercial applications.

---

[376] Parts per million. For fish to survive, a minimum of 5 ppm of oxygen dissolved in water is required and 10 ppm or more is needed for fish to thrive.

[377] a.k.a. Halophiles.

[378] a.k.a. Basic conditions.

[379] pH (potential of Hydrogen) is a scale for the levels of acidity and alkalinity of aqueous solutions and is based upon the balance between positive hydrogen ions (H^+) and negative hydroxide (OH^-) ions in the solution. Neutral acidity/alkalinity (pure water) has a pH of 7.0 and the scale runs from 0.0 to 14.0, although values outside this range are possible. The scale values differ from each other by a factor of ten in the H^+/OH^- balance relative to that of water. Thus milk (say) with a pH of 6.0 has 10 times the H^+ concentration of water, while lemon juice, with a pH of 2.0 has 100,000 times the H^+ concentration of water. Sulphuric acid at a concentration of 33% (battery acid) has a pH of 0.5, while at the alkaline end of the scale, Sodium Hydroxide (Lye – used to make soap) has a pH of 13 to 14.

FIGURE 12.2   Lake Nakuru (Kenya), a soda lake with a pH value of 10.5. Such lakes are extremely salty so the alkaliphiles are also adapted to those conditions (Halophiles). The Flamingos are feeding on blue-green cyanobacteria which are toxic to most animals, so they too are actually a form of extremophile. (Reproduced by kind permission of Pixabay and JimboChan.)

High-pressure environments such as the ocean depths and rocks deep below the Earth's surface are the habitats for Barophiles or Piezophiles and some of these cannot exist without the high pressure. The *Halomonas salaria* bacterium is an example of the latter, requiring pressures of at least 1,000 atmospheres to survive.

Anaerobic organisms live without oxygen; indeed, the most extreme forms (obligate anaerobes) are poisoned by oxygen. The obligate anaerobes obtain their energy via fermentation or by using a replacement for the oxygen such as sulphur or a nitrate. Anaerobes are quite abundant; there are probably many billions occupying your mucous membranes right now. They can cause many types of (human) infections, some very serious in nature such as botulism and gas gangrene.

Not all extremophiles are microscopic bacteria, fungi, etc., although most are. So, while it is an aerobe, the Naked Mole Rat (a.k.a. Sand Puppy, *Heterocephalus glaber*), which is some 100 mm in length and is found in Eastern African countries, can survive without oxygen for long(ish) periods, ~20 minutes with no oxygen at all and several hours at oxygen levels which would kill a human in minutes.[380]

Tardigrades are so weird that you could easily believe that ETs are already living with us on Earth. They also go by the names Water Bears and Moss Piglets and are usually less than 0.5 mm in size. They are not strictly extremophiles since they survive extreme situations rather than choose to live in them. Their earliest forms may have emerged over 500,000,000 years ago.

Tardigrades have eight legs, segmented bodies and resemble, if you don't count the legs and on a microscopic scale, short, plump rhinoceros without horns. They have light-sensitive zones but no proper eyes, a mouth which sucks out the body fluids from other microscopic organisms, plants, algae, etc., they have no bones but do have a tough exo-skeleton like an insect. They tend to be found in damp(ish) places such as within mosses and lichens, though there are marine species as well.

---

[380] As an interesting aside, it is the only mammal to be cold-blooded (like reptiles, etc.) and possibly because their healthy cells have a high resistance to change, the species is almost completely immune to cancer.

Tardigrades claim to fame is their extraordinary durability under many types of high-stress situations. Most of this durability is due to their ability to undergo cryptobiosis – a state so little removed from death as to make no difference. Except that there *is* a difference; when the high stress situation alleviates, the Tardigrades *do* come back to life. Thus, they can:

- Survive dehydration
- Survive extreme accelerations and zero gravity
- Survive for several decades without food or water (by cryptobiosis – their natural life span is a few months to a year or two)
- Survive high levels of high-energy particle radiation, UV, x-rays and γ rays
- Survive in a vacuum
- Survive in extreme pH environments
- Survive in space
- Survive lack of oxygen
- Survive many normally toxic chemicals
- Survive until the end of the Earth? (Some people suggest that, when the Sun starts to expand into a red giant in 5,000,000,000 years, tardigrades will still be around to watch it all happen (or they would be doing so, if they had eyes.)
- Survive very high pressures
- Survive, when dry, temperatures from $-200\,°C$ to $150\,°C$ – although $80\,°C$ can kill them if they are not in a dry state.

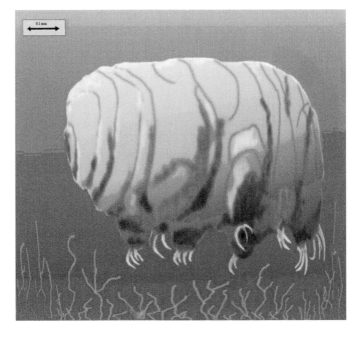

FIGURE 12.3   A sketch of a Tardigrade. (© C. R. Kitchin 2023. Reproduced by permission.)

Terrestrial life is thus surprisingly tough and can occupy an unexpectedly wide range of environments. Perhaps, therefore, we may also presume similar levels of durability for ETs which are based upon similar biochemistries and this is something to bear in mind when we come to look at habitable zones (Chapter 13). It seems unlikely that the First-Living-Molecule would have originated within one of these extreme environments, although the anaerobic chemoautotrophic thermophiles[270] might be a possibility for that role.

TELEPATHY AND TELEKINESIS: THE REAL McCOY

This seems to be a good point at which to include some other, rather controversial, types of extreme capabilities for (perhaps) humankind and (more likely) for ETs and ETIs.

Telepathy and telekinesis are generally classed, along with other speculative phenomena such as precognition and clairvoyance, as Psychic or Paranormal occurrences. There are many fervent advocates for the reality of such happenings and any reader may research a vast number of publications on the internet, if interested in such matters. However, a more rational assessment would be:

"Parapsychology research rarely appears in mainstream scientific journals …"[381]

We shall not therefore venture into the more esoteric *Blithe Spirit*[382] types of speculation, but stick with just examining whether or not we might someday, somewhere or somewhen encounter ETs or ETIs who *do* operate via telepathy and, maybe, also use telekinesis.

Telepathy may be defined as:

"… the communication of impressions of any kind from one mind to another, independently
of the recognised channels of sense."[383]

Stated in that fashion, telepathy sounds far less miraculous.

Indeed, we can easily conceive of circumstances wherein (normal) humankind might appear to be telepathic to another form of life. Imagine that we are visited by some ETIs who are completely deaf and know nothing of the ability to hear or to generate meaningful sounds. Our ability to communicate over long distances via soundwaves would appear to be telepathic to those beings and our enchanted rapturous enjoyment whilst listening to (say) Beethoven's Pastoral symphony or Gershwin's Rhapsody in Blue,[384] would be quite incomprehensible to them.

Thus, the above definition of telepathy should be altered to read:

"… the communication of impressions of any kind from one mind to another, independently
of *those normal* channels of sense *possessed by humankind*."

and now we may see how telepathy could easily be the means of communication between other entities; all it requires is a sense which they possess and which we do not.

So, what might be the sense which we do not possess? We cannot sense *directly* the following phenomena:

radio waves, x-rays, γ rays, high energy particles, neutrinos, gravitational waves,[385] magnetic
fields, electric fields,[386] the weak nuclear force or the strong nuclear force

and there may be more of which we are currently unaware, even in theory. Any of these might therefore form a basis for telepathy.

---

[381] https://en.wikipedia.org/wiki/Parapsychology#:~:text=Parapsychology%20is%20the%20study%20of,synchronicity%2C%20apparitional%20experiences%2C%20etc.

[382] Coward N. A play first performed in London in 1941. It features the spirit of the dead wife, Elvira, who through a complex series of events returns to haunt her still-living husband, Charles.

[383] Venkatasubramaniam G. *et al.* International Journal of Yoga, 66, 1, 2008. (Author's note: I have been looking for a reason to reference this Journal for many years).

[384] Replace with your own choices if you so wish.

[385] Neutrinos and gravitational waves seem unlikely to form the basis for telepathy given that current terrestrial neutrino detectors require cubic kilometres of water or ice and current gravitational wave detectors are several kilometres in length (or millions of kilometres for proposed space-based detectors).

[386] Unless you count your 'hair standing on end' from electrostatic charges.

The phenomenon which does seem to be a distinct possibility as a basis for telepathy and of which we already make extensive pseudo-telepathic use is a combination of electric and magnetic field interactions called induction. These would also provide for telekinesis, which is:

the power to move[387] objects at a distance independently of the normal methods used by humankind.

Induction *is* already in widespread use as one of our tools (Box 12.1), but we do not, ourselves, intrinsically possess the ability to generate inductive signals, nor to receive them. If it were to evolve, then it would seem likely to do so amongst social life-forms, such as the Honey Bees, where large groups of individuals are in very close proximity to each other for much of the time.

BOX 12.1 – ELECTROMAGNETIC INDUCTION

Electromagnetic induction was discovered in 1831 by Michael Faraday. Its essence is that a varying electric current generates a varying magnetic field and a varying magnetic field generates a varying electric current.

We utilise the phenomenon widely – in transformers, electric motors, dynamos and generators, electric induction hobs, induction welding, inductive loops in concert halls (for hearing-aid users) and so on and so on. The principle of an induction system is shown in Figure 12.4.

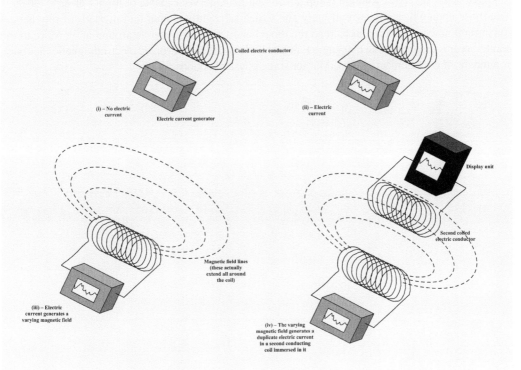

FIGURE 12.4   A simple induction loop system.

The varying electric current input to the coiled conductor on the left sets up a magnetic field which surrounds the coil and which then varies in unison with the electric current.

A second, similarly coiled conductor (on the right) is immersed in the varying magnetic field and a varying electric current is induced within the second coil which also varies in unison with the first current.

In effect, the variation in the first electric current has been replicated, without, as far as an unaided human would be concerned, any detectable link between the two currents. (© C. R. Kitchin 2023. Reproduced by permission.)

From Box 12.1 we may see that if two conductors are close(ish) to each other and one carries a varying electric current, then the other will carry a duplicate varying current. Potentially, therefore, since there are electric currents flowing in our brains, those flowing in one person's brain might be duplicated in the brain of another nearby person and – Hey Presto – telepathy!

The fact that we do not appear to duplicate currents in each other's brains in this way is down to the fact that in such a situation the induction process would be *very* inefficient. A set-up like that shown in Figure 12.4 with two multi-coiled conductors close together is more efficient, but real transformers and electric motors have thousands of coils, and the individual wires are separated by fractions of millimetres.

Nonetheless, looking amongst terrestrial organisms, some can generate very high electric potentials (~600 V – electric eels, Chapter 2) and all animals[388] have nerves along which electric pulses can pass. Thus, had our evolution taken a different turn, nerves capable of higher electric currents than those upon which we currently rely and which are tightly wound into myriads of tiny helical coils might now be enabling us to read the thoughts directly of everyone around us.[389] More to the point, it seems quite possible that some ETs and ETIs *could* have evolved electromagnetic induction as a means of communication and even for some physical actions.

---

[388] Plant cells can also generate voltages called 'action potentials'.

[389] Come to think of it, it is probably just as well that evolution on Earth did not take the route towards telepathy; imagine being in a crowd of 100,000 at some sports event and inducting the thoughts of everyone there (although perhaps evolution would also have provided an 'off' switch for such occasions).

13 ETs and (Perhaps) ETIs

INTRODUCTION

Science fiction and fantasy abound with fantabulous concepts of what shapes and forms ETs and, especially, ETIs might take. Despite the apparently superheated imaginations of the writers/illustrators of such stories, their actual imaginations are mostly very limited. Their heroes and heroines are almost all very thinly disguised human beings (Figure 13.1). Their evil, disintegrator-gun-wielding opponents are slightly less thinly disguised human beings, and the story lines are taken straight from Desperate Dan and Hopalong Cassidy, with the journeys over distances of thousands of light years being regarded as a minor impediment thanks to their warp-factor-42 space drives.

FIGURE 13.1 Left – an artist's concept of a possible ETI/Robot and its ETdog. (Reproduced by kind permission of Pixabay and Stefan Keller.) Right – A typical popular concept of a possible ETI. (Reproduced by kind permission of Pixabay and Mystic Art Design.)

Of course, I am not being fair to a good many SF writers[390] with this analysis, but I *am* going to leave it up to you if you want to watch/read such 'literature' – and, having said that,

I am now going to speculate about ETs and ETIs based largely upon what we know about terrestrial-type life,[391] throughout most of this chapter!

Evolution, as we have taken some pains to argue in the previous chapters, ensures living organisms conform as best as may be possible to their environments. Thus, to speculate about the natures

---

[390] Thus, Adrian Tchaikovsky's *Children of Time* (Pan Books, 2015) envisages an exoplanet which has been terraformed for humankind but which is now populated by convincing spider-types who have evolved to ⁹✖ level (Table 13.4) intelligence and abilities, whilst Larry Niven does concede that three sexes might be involved for some ETIs' reproductive processes ('*Pierson's Puppeteers*' - who appear in many of his books).

[391] That is, life based upon the sort of biochemistry considered earlier. In other words, based around organic molecules (Box 9.1) and amino acids, nucleobases, proteins, etc. which, if not identical to the terrestrial versions, are at least close cousins of them. It does not mean that, even at the micro-organism level, ET life-forms will closely mimic actual terrestrial life-forms, although they may fulfil similar functions in similar ways within their environment when this is similar to some terrestrial environment (see convergent evolution, Chapter 10).

Writing as a human being, my (and everyone else's) viewpoint is inevitably biased by my (our) own experiences and knowledge; a phenomenon given the label of 'Anthropomorphism'. Although we can, perhaps, imagine some life-forms which are very different from terrestrial examples, it will not be until at least one more and preferably 1,000's more examples of independently originating ETs and ETIs, have been studied, that an authoritative version of this chapter can be written. However, what is sometimes called the 'Carter Argument' (after Brandon Carter) which maintains that the existence of life on Earth does not tell us *anything* useful about the probability of the origin, existence or nature of ETs is probably excessive.

DOI: 10.1201/9781003246459-15

169

of ETs we first need to look at their environments. Before *that* however, a check back on the possible incidence of ETs throughout the galaxy/Universe may be advisable (Table 2.1).

In Table 2.1, I have given estimates for the probabilities of existence of life at various levels of complexity. To call these estimates, though, is being generous; they are really just wild guesses. They could be about right, or highly optimistic, I doubt, though, that they are pessimistic. Further discussion of this topic will also be found when we come to the Drake Equation (Chapter 16).

On the plus side (assuming that the existence of ETs and ETIs *is* actually a plus) we have seen (Chapters 9 and 10) that life on Earth originated very quickly, possibly within 200 million years of the moment when the early Earth became potentially inhabitable (about 4,000 million years ago) and that it had evolved sufficiently to leave fossils behind only 300 million years later.[392] Thus, we might deduce (extrapolating from one example again – Figure 2.3) that life begins very easily.

Since the Universe is about 13,700 million years old and the Earth/Solar system is about 4,500 million years old, there has been plenty of time for any exoplanet, potentially capable of supporting terrestrial-type life and which is 1,000 million years old or more, to have become inhabited.

Out of the ~5,580 currently discovered exoplanets (Chapters 4 to 8), we find, though, that there are perhaps a few tens or so which *might* be potentially capable of supporting terrestrial-type life, i.e., a few times 0.1% of all exoplanets. Thus, my guess for row #6 of Table 2.1 does have some possible factual basis. The guesses for the probabilities of more advanced life-forms than micro-organisms are, though, just that, guesses.

What I am reasonably sure about is that, *if* life exists away from the Earth at all, there will be:

1. For every exoplanet inhabited by entities with abilities roughly equivalent to those of humankind, there will be *many* exoplanets inhabited by life-forms which have not evolved much further than multi-cellular organisms (Figure 9.3 (ii)).
2. For every exoplanet inhabited by multi-cellular organisms, there will be *many* exoplanets inhabited by life-forms which have not evolved much further than the stage of single cells or the equivalent (Figure 9.3, (iii), (iv) and (v)).
3. For every exoplanet inhabited by single cells or the equivalent, there will be *many* exoplanets inhabited by life-forms which have not evolved much further than the First-Living-Molecules (Figure 9.3 (vi)).
4. For every exoplanet inhabited by life-forms which have not evolved much further than the First-Living-Molecules, there will be *many* totally uninhabited exoplanets.

I leave it as an exercise for you to decide the numbers appropriate to substitute for each of the '*many*' words appearing in this list.

Whatever the true probability of exoplanets' 'potentially capable of supporting terrestrial-type life' might eventually turn out to be, it is going to be a *small* number.

Although in Chapter 12, we saw that terrestrial-type life can flourish within some quite remarkable environments, from the results which we already have on the numbers and natures of exoplanets, those which come even remotely close to being suitable environments are few and far between; other types of exoplanets from Hot Jupiters to Cold Small Bodies (Table 8.1) outnumber them by far. Nonetheless, as we have seen (Chapter 7), exoplanets may outnumber stars within the Universe and perhaps outnumber them by a large factor. So, there are *lots* of exoplanets to play with and even if the ones suitable for sustaining terrestrial-type life are only a very small proportion of the total, that still gives us plenty to be going on with.

---

[392] Carbon and sulphur isotopes dating back to perhaps 3,900 million to 4,000 million years ago may be the remnants of even earlier life-forms. Such isotopes are sometimes called 'Chemofossils'. Stromatolites, the rocky calcium carbonate remnants of early cyanobacteria colonies, date back to 3,400 million years ago and perhaps to 3,700 million years ago.

HABITABLE ZONES

A common way in which an exoplanet's suitability for sustaining terrestrial-type life is envisaged is via the Habitable Zone of its host star (see also Chapter 2). The habitable zone of a star may be defined as:

> The region around a star wherein a planet could have liquid water[393] present, at least in some places and at some times.

This definition is clearly based upon terrestrial-type life's requirements.[391] Should there be another (or many) quite different chemistry(ies) capable of forming the basis for life (see below), then their habitable zones will be different and will need to be defined differently. A common alternative name, though, for the (humankind) habitable zones is the 'Goldilocks Zone', based upon the children's story *'Goldilocks and the Three Bears'*,[394] because the conditions within the habitable zone are 'Just Right' for terrestrial-type life.

Whilst the whereabouts of the habitable zone of a star might seem easy to ascertain, there are a number of complicating factors:

- Even for terrestrial life-forms, the habitable zone for humankind will differ from that for, say, thermophiles (Chapter 12).
- Marine, land-based and atmosphere-based life-forms could well have differing habitable zones.
- The presence or absence or the varying nature of a planet's atmosphere may change the distance of the habitable zone from the host star. The Earth's equilibrium temperature,[103] for example is ~ 255 K, whilst the greenhouse effect of the atmosphere raises it to ~288 K. Without the greenhouse effect, the Earth's orbit would need to shrink to a radius of ~0.78 AU (32 million km closer to the Sun and be only 12 million km out from Venus' orbit) for the temperature to reach 288 K.

  Whilst it would seem conceivable that terrestrial-type life-forms might be able to evolve to live in a vacuum, if that is not the case, then the absence of an atmosphere for an exoplanet would mean that there would be no habitable zone, whatsoever type of star its host star might be.

- The pressure of an atmosphere will change the temperature range over which water is a liquid (Chapter 9).

---

[393] Planets/exoplanets with liquid water present on them now or in the past, are sometimes divided into four classes; I – Liquid water present freely on the surface now, II – Liquid water present on the surface in the past, III – Sub-surface oceans of liquid water in contact with the planet's core and IV – Sub-surface oceans of liquid water sandwiched between layers of ice.

Water is central to our present concept of a habitable zone, because it is central to terrestrial type life-forms' existences. Darwin's 'warm little pond' is envisaged as a water pond, not a molten iron pond or a liquid methane pond, although these will exist as we have seen in Chapter 8. Many, but not all, organic/pre-biotic molecules are soluble in water, thus facilitating their mixing and reacting together.

But now comes an unexpected twist; water may also facilitate reactions between organic/pre-biotic molecules which are *not* soluble in water. This phenomenon is called 'On-Water Reactions'. It arises because the insoluble molecules will try to avoid the water molecules and so may become concentrated in water-poor regions (see also the discussion on the formation of membranes from phospholipid molecules, later in this chapter). The reaction rates between the water insoluble molecules may thus be increased up to 10,000 times through this concentration process. Even though On-Water Reactions do not use water in their reactions, it is still required to produce their concentrations, so even for this process, water-based definitions for Habitable Zones are appropriate.

[394] An old English folk tale, first written-down in 1837 by Robert Southey. There are several versions, but the one from which this name for the habitable zone is derived has a household with a family of three bears; Father Bear, Mother Bear and Baby Bear. The Bears have gone out for a walk leaving their house ready for their return, with porridge cooling, chairs ready and beds made. The young girl, Goldilocks, finds and enters the empty house and samples the porridge, chairs and beds, in each case finding that the ones for Baby Bear are the only ones which she likes, i.e., which are 'Just Right' for her.

- Significant variability for the host star, such as stellar flares,[395] or large-scale physical variations such as with Cepheid stars (Chapter 7) may mean that the habitable zone moves towards or away from the host star or disappears or is rendered non-habitable at times.

- The Universe was almost entirely formed from just hydrogen, deuterium and helium immediately following the Big Bang. Most other chemical elements have been synthesised inside stars since then. The compositions of stars and exoplanets therefore change with time; more of the heavier elements necessary for life (Box 9.1) are thus present in young (say < 5,000 million years) stars and exoplanets than in the older (say > 10,000 million years) ones. Thus, even if everything else is favourable for life-forms, on very old exoplanets, there simply may not be the required chemical elements available.

- High levels of particle radiation, x-rays or γ rays from the host star may render an otherwise habitable zone uninhabitable.

- A habitable zone might not exist for surface-based life but might exist for life-forms high in the atmosphere (see Venus, Chapter 9) or for life-forms in the (still hypothetical) sub-surface oceans of Enceladus or Europa or deep underground.

  Very deep valleys or high mountains might similarly provide small regions where life might exist on an otherwise inhospitable planet.

- Binary/multiple star systems may have no stable orbits for planets and/or the habitable zone may change its position (Chapter 8) so that the exoplanet is only inside the habitable zone for a part of the time.

- Massive, hot stars evolve and change their sizes, temperatures and luminosities on time scales of a few million years; any habitable zone would move in or out from the star to follow these changes.

- Even if we have managed to determine a reasonably precise position for a habitable zone, the exoplanet has only to be in an elliptical orbit for it to dip in and out of that zone for some of the time. Thus, the red dwarf star, GL 524, has a habitable zone extending between ~0.18 AU and ~0.36 AU, but its possible super-Earth exoplanet may have an orbit with a periastron at ~0.23 AU (well inside the habitable zone) and an apoastron at ~0.61 AU (well outside the habitable zone). Hibernating/Cryptobiotic (Chapter 12) organisms might be able to survive such a regime once they have evolved, but whether or not life could originate under those conditions seems much more doubtful.

- One point not often considered is that of the duration of the habitability zone. One with a lifetime of one second would be unlikely to develop life-forms even if it could. Minimum durations of millions of years, perhaps billions of years, for habitable zones are likely to be needed before there is much hope of life-forms actually developing within them, especially those whom we might hope to be detectable.

There are probably more things which could be added to this list, but these will suffice to show that even for the same type of life, the habitable zone could vary widely from one host star to another and from one exoplanet to another.

THE SUN – HABITABLE ZONE

The various factors listed above, which can affect whether the conditions on an exoplanet are life-friendly or not, mean that estimates for the extent of the habitable zone for a single star will vary markedly depending upon the assumptions made about those factors. Thus, taking five different habitable-zone algorithms and feeding-in the details for the Sun, the outer limit of its habitable zone is given as 1.37, 1.68, 1.77, 1.78 or 1.88 AU, with a simple average of 1.70 AU. Similarly, the inner

---

[395] Some recent work has shown that flares can perhaps heat the upper mantles of exoplanets so that exoplanets nominally too cold for any types of terrestrial-type life-forms could have their equilibrium temperatures raised to more comfortable values – see also below; sub-surface oceans on icy Solar System moons.

radius of the Sun's habitability zone is given by those same algorithms as 0.75, 0.75, 0.95, 0.95 or 0.95 AU, with an average of 0.87 AU (Figure 13.2).

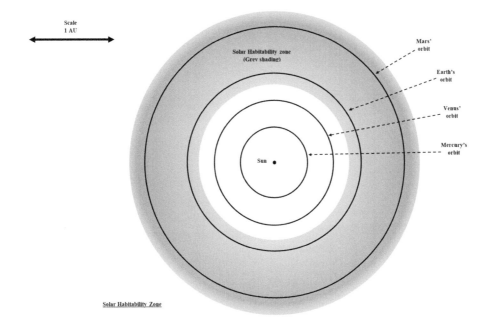

FIGURE 13.2 An averaged habitable zone for the Sun. The positions of the terrestrial planets' orbits are also shown. Both the Earth's and Mars' orbits lie entirely within the zone, but the Earth would be too cold for almost any of the present-day terrestrial life-forms without the atmospheric greenhouse effect, which raises its surface temperature by some 33 °C. Mars is perhaps border-line inhabitable by some of the extremophiles (Chapter 12). Venus is outside the habitable zone, but as discussed earlier (Chapter 9), it might be able to support terrestrial-types of life high in its atmosphere. If some of the natural satellites of the outer planets do have sub-surface oceans and there is any sort of life in those oceans, then that life would be existing a factor of three beyond the outer limit of the Sun's habitable zone. (© C. R. Kitchin 2023. Reproduced by permission.)

Thus, habitable zones should not be regarded as definitive when it comes to estimating whether or not an exoplanetary system could support life; an exoplanet could orbit entirely within the nominal habitable zone and yet be unable to support life, or life could exist within special environments found on an exoplanet well outside the habitable zone. There is little point, therefore, in trying to define the habitable zone for a star precisely, at least until we know some actual details about the planet involved.

THE STARS – HABITABLE ZONES

However, if we do not worry too much about the details and bear the above comments in mind, we can get a useful broad picture of the natures of exoplanetary systems from their calculated humankind-friendly habitable zones (i.e., their Goldilocks zones) by using averaged values for the systems' various properties (similar to those shown in Figure 13.2 for the Sun). These averaged habitable zones are illustrated to scale in their correct relative physical sizes in Figure 13.3. The host stars' temperatures range from ~2,700 K (~2,400 °C; spectral type[215] M8) to 9040 K (~8750 °C; spectral type A2). Stars with lower surface temperatures than 2,700 K, including brown dwarfs, may be expected to act as exoplanet hosts, but their habitability zones will be small. Stars hotter than 9040 K may also host exoplanets, but their main sequence lifetimes are probably too short for

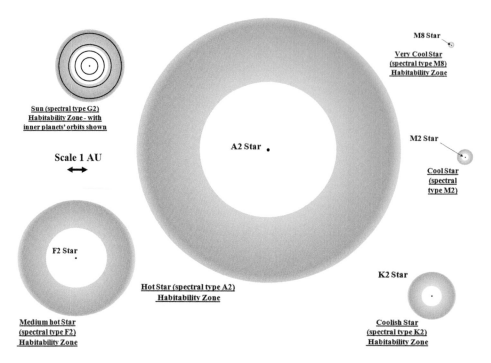

FIGURE 13.3 Habitable zones, drawn to scale, for a range of main sequence star types. (© C. R. Kitchin 2023. Reproduced by permission.)

life to originate on them before the stars leave the main sequence and their increasing sizes and luminosities sterilise the exoplanets.

THE REAL THING #1 – INHABITANTS OF EXOPLANETS WITHIN HABITABLE ZONES

Out of the 5,580 or so currently known exoplanets, the vast majority are orbiting small, cool stars (red dwarfs/M spectral types), and many of those exoplanets are orbiting very near to their host stars (Figure 7.3). The Sun, as a large, massive (amongst the current list of host stars) star with the Earth being a long way out from the Sun, is therefore quite a rare circumstance.

Many exoplanets of small, cool host stars are tidally locked onto their star and that star is often a flare star, so that any life-forms will be having a tough time.[396] On the other hand, evolution will have ensured that those life-forms are suited, in so far as may be possible, to exactly those conditions. Furthermore, small stars have long main sequence life times; for an M8 star it is probably between 1,000,000 million years and 10,000,000 million years (~200 times the Sun's main sequence lifetime/~70 times the current age of the Universe to ~2,000 times the Sun's main sequence lifetime/~700 times the current age of the Universe – see also Box 8.2). So, for small, cool stars' exoplanets, life, if it hasn't originated on them yet, still has plenty of time left to do so.

The Best of the Best

The exoplanetary system of the host star TRAPPIST-1 has been encountered several times earlier (Cover image, Figure 7.2, Chapter 8), and we choose it again here as our first specimen for detailed inspection.

---

[396] One recent estimate suggests that two out of three exoplanets orbiting red dwarfs would be uninhabitable because of tidal heating effects; but that still leaves some 30,000 million red dwarfs in the Milky Way Galaxy alone, where there *might* be inhabitable exoplanets.

TRAPPIST-1 is a very cool (spectral type M8) small star about 40.7 ly away from us. Estimates of the star's current age range from 3,000 million to 8,000 million years. It can be seen visually (if you have a $\geq$ 1.0-m telescope) in the Northern part of Aquarius. Its mass is ~ 0.09 M_{Sun} and its visual luminosity is ~ 0.000 003 7 L_{Sun}. However, TRAPPIST-1 is so cool that most of its energy is emitted at infrared wavelengths, so its actual total energy emission is ~0.000 55 L_{Sun}.

The host star's habitable zone, using the latter figure for its luminosity, extends from 0.022 AU (3.3 million km) to 0.032 AU (4.8 million km) outwards from the star. TRAPPIST-1 is a flare star and the Kepler space telescope detected about 40 flares over a 10-week period. If that rate is typical, any individual planet in the system should be hit by a flare about once every three weeks. Most of the observed flares on TRAPPIST-1 have, though, been of relatively low energy; typically, with a total energy emission of $\leq 10^{25}$ J[397] and, also, most of the systems' exoplanets may have thick to very thick atmospheres (Table 13.2). It is probable, therefore, that life-forms would be able to survive the flares' emissions. Perchance even, they could evolve to take advantage of flares?[201] Furthermore, some recent work has shown that flares can perhaps heat the upper mantles of exoplanets so that the outer exoplanets of the system could have their equilibrium temperatures (Table 13.2) boosted to more clement values.

The main currently known details relevant to this discussion for the seven exoplanets of the system are shown in Tables 13.1 and 13.2.

TABLE 13.1
Orbital Data for TRAPPIST-1 Exoplanets

Exoplanet/Habitable zone	Orbital semi-major axis (AU) ($\approx$ Orbital radius)	Orbital period (days)	Orbital ellipticity (eccentricity) (Earth value; 0.017)	Tidal-lock
TRAPPIST-1 b	0.0115	1.51	0.006	Highly probable
TRAPPIST-1 c	0.0158	2.42	0.007	Highly probable
Inner edge of the habitable zone	0.022			
TRAPPIST-1 d	0.0223	4.05	0.008	Probable
TRAPPIST-1 e	0.0293	6.10	0.005	Probable
Outer edge of the habitable zone	0.032			
TRAPPIST-1 f	0.0385	9.21	0.06 or less	Probable
TRAPPIST-1 g	0.0468	12.42	0.002	Possibly not
TRAPPIST-1 h	0.0619	18.87	0.006	Probably not

[397] TRAPPIST-1, at its normal luminosity, emits 10^{25} J over a two-hour period. Its flares can reach this value but 10^{24} J, or so, is more normal. Since a flare lasts for around 15 minutes, during an average flare, the star is perhaps three times brighter than normal.

Rather than the increase in total energy from the star being the problem, however, it is likely that would be the unusual blue and UV photons which would cause any damage. In London, for example, at a latitude of 51.5° N, the energy reaching the surface in Summer is over nine times that received in winter; so, surviving a mere factor of three energy increase alone for ~20 minutes *should* be a doddle.The most energetic solar flare observed to date was the Carrington event of 1859, whose energy emission is estimated to have been around 10^{25} J. Evidence from the abundances of radioactive carbon-14 trapped within tree rings and ice cores suggests, though, that solar flares up to 10 times the strength of the Carrington event are possible. Such flares are named Miyake events and around ten have been detected as occurring within the last ~15,000 years, the last being in 993 CE. Super flares on other stars, though, have been known to emit up to 10,000 times more energy (10^{29} J). So far, though, these observed super flares have only occurred on solar-type or slightly more massive stars and not on the low mass red dwarfs like TRAPPIST-1. Furthermore, some recent work suggests that these super flares erupt from near the host stars' rotational poles, so that exoplanets with orbits close to the stars' equatorial plane could be safe(ish).

10^{29} J though is a *lot* of energy; it is equivalent to the energy released by the explosion of ~2 x 10^{19} tonnes of TNT, and this is the equivalent of exploding a block of TNT whose volume is ⅔ of the volume of the Earth's Moon ®.

So, you can perhaps see why, if stellar super-flares *do* occur on Red Dwarf stars, they *could* give the life-forms on any nearby exoplanet some really bad hair days.

TABLE 13.2
Physical Data for TRAPPIST-1 Exoplanets

Exoplanet/ Habitable zone	Exoplanet mass (M_{Earth}) (M_{Earth} = 5.97 × 10^{24} kg)	Exoplanet radius (R_{Earth}) (R_{Earth} = 6,371 km)	Exoplanet surface gravity (g) (g = Earth's surface gravity = 9.81 m s^{-2})	Exoplanet equilibrium temperature (K)/ Surface temperature (K)	Exoplanet nature	Exoplanet atmosphere	Probability of ETs?/ ETIs? (Based on terrestrial-type biochemistry)	Notes
TRAPPIST-1 b	1.02	1.12	0.81	390/500 - 750–1,500	Mainly rocky	Yes – Very thick. H_2O? CO_2? No – Recent JWST observations suggest no atmosphere or only a very tenuous one	Zero	Surface may comprise molten rock. Possibly has quite active geology arising from tidal stresses.
TRAPPIST-1 c	1.16	1.10	0.96	310/~335? – ~500 K? – 1,000 K?	Rocky	Yes – Thick. H_2O? CO_2? No – see JWST results discussed above	Very, very low	Possibly has quite an active geology arising from tidal stresses.
Inner edge of the habitable zone								
TRAPPIST-1 d	0.30	0.78	0.49	260–280/?	Rocky, Liquid water? Ice? Oceans?	Probable – H_2O?	Close to 'as good as it gets' (at the moment)	Inside the habitable zone. Possibly has quite active geology arising from tidal stresses. Some models predict a runaway[398] Greenhouse effect raising surface temperatures to near Venus levels (~700 K).

(Continued)

[398] That is, when the temperature increase, due to a greenhouse gas (methane, carbon dioxide, etc.), leads to the production of more of that gas – which raises the temperature further – which then produces yet more of the gas – which raise the temperature again … and so on … until very high temperatures indeed can be reached.

TABLE 13.2 CONTINUED
Physical Data for TRAPPIST-1 Exoplanets

Exoplanet/ Habitable zone	Exoplanet mass (M_{Earth}) (M_{Earth} = 5.97 × 10^{24} kg)	Exoplanet radius (R_{Earth}) (R_{Earth} = 6,371 km)	Exoplanet surface gravity (g) (g = Earth's surface gravity = 9.81 m s^{-2})	Exoplanet equilibrium temperature (K)/ Surface temperature (K)	Exoplanet nature	Exoplanet atmosphere	Probability of ETs?/ ETIs? (Based on terrestrial-type biochemistry)	Notes
TRAPPIST-1 e	0.77	0.910	0.93	225–245/?	Rock/Iron	Not yet detected. Could have an atmosphere comparable in size and density with that of the Earth, but almost certainly without the oxygen.	Reasonable, especially if a greenhouse effect raises its equilibrium temperate somewhat.	Inside the habitable zone. Close to being a twin-Earth in its physical dimensions.
Outer edge of the habitable zone								
TRAPPIST-1 f	0.68	1.05	0.62	220/1,400	Rocky, Liquid water? Oceans?	Yes – Probably very thick. H_2O? CO_2? O_2?	Unlikely, but there are uncertainties in the measurements and there could be errors in the modelling, so that it cannot be ruled out yet.	Just outside the habitable zone, but it could be inside with different assumptions about the conditions.
TRAPPIST-1 g	1.15	1.15	0.87	195/? – Could be much higher due to the greenhouse effect if there is an atmosphere (column 7).	Rocky, Liquid water? Oceans? Ice?	Uncertain. If there is an atmosphere, then it is likely to be very thick. H_2O?	Unlikely, but there are uncertainties in the measurements and there could be errors in the modelling, so that it cannot be ruled out yet.	Outside most of the likely estimates of the extent of the habitable zone.

(Continued)

TABLE 13.2 CONTINUED
Physical Data for TRAPPIST-1 Exoplanets

Exoplanet/ Habitable zone	Exoplanet mass (M_{Earth}) (M_{Earth} = 5.97 x 10^{24} kg)	Exoplanet radius (R_{Earth}) (R_{Earth} = 6,371 km)	Exoplanet surface gravity (g) (g = Earth's surface gravity = 9.81 m s^{-2})	Exoplanet equilibrium temperature (K)/ Surface temperature (K)	Exoplanet nature	Exoplanet atmosphere	Probability of ETs?/ ETIs? (Based on terrestrial-type biochemistry)	Notes
TRAPPIST-1 h	0.33	0.78	0.54	170/?	Rocky? Sub-surface oceans?[399] Ice on surface?	Probably no atmosphere	Unlikely	Well outside any habitable zone.

[399] See later in this chapter for a fuller discussion of possible sub-surface oceans within Solar System bodies. If these exist (and they probably do), then we may additionally expect sub-surface oceans to occur within suitable exoplanets. For red dwarf host stars, like TRAPPIST-1, their early stages last a long time and can be very active and so lead to the loss of any water from the surfaces of their exoplanets. However, water is still likely to be stored below the surfaces, at least for exoplanets with masses similar to that of the Earth. The water could be in liquid form (sub-surface oceans) and/or stored in the upper mantle incorporated into minerals like olivine and its polymorphs wadsleyite and ringwoodite; which can comprise over 3% water by mass. Once the red dwarf host star has settled down, volcanic activity on the exoplanets could produce sub-surface oceans or even restore surface oceans.

An inspection of Tables 13.1 and 13.2 (bearing in mind the above comments about stellar flares and the possible non-reliability of habitable zone estimates) shows that some of TRAPPIST-1's exoplanets look to be quite good prospects for terrestrial-type life-forms. We also need to bear in mind that there are many factors influencing those prospects about which we know nothing at the moment; after all, something as trivial as a small amount of salt in a water body on the Earth renders it lethal to many types of freshwater fish.

Given the very long lifetimes of red dwarf stars (see above), we may justifiably assume that if First-Living-Molecules *can* come into being on their exoplanets, then they very probably *will* do so, *sometime* – although it may not have happened yet. Given, also, the preponderance in numbers, of low mass, cool stars over higher mass stars, within the Universe, it could be that if ETs and ETIs exist, then the overwhelming majority will have evolved[400] red- and IR-sensitive sight and photosynthetic reactions and flare protection/adaptation systems/life styles. The inhabitants of tidally locked exoplanets may migrate around their planets towards or away from their sub-stellar points as the fancy takes them or as they evolve towards higher or lower temperature environment requirements and they may not have circadian rhythms (awake/sleep cycles), or if they do, then these could be linked to the planet's orbital, not rotational, period.[401]

So, let us get to our business of seeing what might be possible for life on some exoplanets.

TRAPPIST-1 d

With an equilibrium temperature of ~260 K to ~280 K (~ -15 °C to ~10 °C) TRAPPIST-1 d, even without an atmosphere (but with spacesuits), this exoplanet would be habitable by many terrestrial life-forms, including Humankind. With a thin atmosphere and a moderate greenhouse effect, the surface temperature could easily rise to Earth-Temperate or Earth-Tropical levels. With a thick atmosphere, though, the surface temperature could well rise to Venusian levels (much too hot; like Father Bear's porridge).

However, let us assume that the Venusian catastrophe has not happened to TRAPPIST-1 d. The First-Living-Molecule on Earth seems to have appeared within some 500 million years of the Proto-Earth becoming habitable. The minimum estimated age for the TRAPPIST-1 system is 3,000 million years. Extrapolating, again from a single example, we might expect life-forms on TRAPPIST-1 d to be around the level of the Earth's Paleoproterozoic/Mesoproterozoic geological eras. At this time (~2,000 million to ~1,000 million years ago for the Earth), terrestrial life-forms were still anaerobic (Box 13.1) since plentiful quantities of oxygen in the atmosphere did not become available until around 800 million years ago. As we have assumed, for the moment, that the geochemistry of the organisms is akin to that of terrestrial life, TRAPPIST-1 d's postulated life-forms could be fairly similar to those in the Mesoproterozoic era on Earth. The terrestrial life-forms, then, were still mostly single cells although multi-cellular organisms and sexual reproduction had probably begun to appear.

---

[400] This may somewhat relieve worries about interstellar invasions of the Earth (Chapter 14) since such ETIs would probably not see the Earth as a desirable place to live.

[401] Of course, for a fully tidally locked exoplanet, the rotational and orbital periods are the same, but there will be partially tidally locked exoplanets as well; like Mercury.

BOX 13.1 – ANAEROBIC LIFE

Anaerobic life exists without the need for oxygen (oxygen-requiring life being called; Aerobic). Anaerobic life may not need oxygen at all; indeed, oxygen may be poisonous to it, or it may use oxygen when it can get it, but the essential criterion is that it *can* do without oxygen completely. Anaerobes, all the same, do still need energy, and amongst the alternative energy sources utilised by them, there are various forms of fermentation or nitrates or sulphates being used as a substitute for oxygen.

Current terrestrial anaerobic organisms are mostly single cells. A multi-cellular parasite of salmon, *Henneguya salminicola* is the currently largest known example of an anaerobe and it is about 10 μm in size. Generally aerobic metabolism is expected to produce more energy, more quickly, which is why, on the Earth, aerobic organisms are dominant. However, this may not always be the case; Zimorski *et al.* suggest[402]:

> We also address the widespread assumption that oxygen improves the overall energetic state of a cell. While it is true that ATP yield from glucose or amino acids is increased in the presence of oxygen, it is also true that the synthesis of biomass costs thirteen times more energy per cell in the presence of oxygen than in anoxic conditions. … The absence of oxygen offers energetic benefits of the same magnitude as the presence of oxygen.

(My underlining)

We do not know how common oxygen-rich atmospheres on exoplanets may be, but the Earth's atmosphere was oxygen-free or low in oxygen for the initial ~80% of the time that life has existed on the Earth. Hence it seems quite likely that many exoplanet atmospheres will remain permanently anoxic, since, unless there are other sources for the oxygen, the presence of oxygen would require the evolution of photosynthesising (or equivalent) life-forms on the exoplanets.

Thus, whether an oxygen atmosphere aids the development of complex/high energy-using life-forms or not, 'needs must when the devil drives' and if an anoxic atmosphere is all that there is, evolution will have to make the best of it.

In later discussions, therefore, I have taken the liberty of assuming that anaerobic life-forms may develop similarly, though perhaps more slowly than terrestrial-type aerobic life-forms and they may also move around more sluggishly as well.

However, the evolution of cellular life is in itself a major and radical step from the previous situation and it may have occurred sooner or later or not at all for the TRAPPIST-1 d inhabitants.

The enclosure of one or more self-replicating organic molecules (perhaps RNA-types – Chapters 9 and 10) within a wrapping of some sort would provide those molecules with a safer and more stable environment, so it would be a probable, but not an inevitable,[403] evolutionary development. Furthermore, in aqueous solutions, phospholipid molecules naturally form such wrapping material (usually called a membrane in this context).

Phospholipids are long chain molecules with one end of the chain having a molecular structure which is compatible with water (hydrophilic – a phosphorous-bearing structure) and the other end incompatible with water (hydrophobic – a fatty-acid-based structure). In aqueous environments, a pair of phospholipid molecules will tend to join their hydrophobic ends together, thus 'protecting'

---

[402] Free Radic. Biol. Med, 279, **140**, 2019

[403] An aqueous solution of some complex molecules, including proteins and amino acids, can form stable microscopic droplets of concentrated solution suspended within the identical, but much more diluted, remainder of the solution. This substance is a type of colloid and the membraneless droplets are called coacervates (Box 9.1).

them from the water, whilst their hydrophilic ends frolic around in *their* favourite environment. Many such pairs will then join side-by-side to form a thin sheet which has its outer surfaces formed from continuous layers of the hydrophilic chain ends and with the hydrophobic ends now being almost entirely protected from the water in the middle of the sheet.

Recently, the hydrophilic component, ethanolamine, of a common phospholipid-forming-terrestrial-life-membrane (phosphatidylethanolamine) has been detected to be present in the molecular cloud, G+0.693-0.027, which is close to the centre of the Milky Way galaxy.

Other membrane-forming structures may be possible, but in the case of terrestrial life, it was a phospholipid membrane that, somehow, somewhere, somewhen, first wrapped itself around some self-replicating organic molecules and formed the First-Living-Cell. Since all terrestrial cells now have phospholipid-based membranes, it would seem that the First-Living-Cell has continued to propagate itself (and evolve) until now.

The interiors of the early cell life-forms on the Earth were largely unstructured and they are called Prokaryotes (Figure 9.3 (v)). Prokaryotes remain to this day to be single-cell organisms.

Later, some cells evolved to contain parts of their internal contents within membranes inside that first outer membrane. In particular, the genetic material of the cell (DNA, RNA, etc.) became enclosed in a membrane and we now call that structure the nucleus of the cell. Cells containing such nuclei are called Eukaryotes. There are still single-celled eukaryotes living today, but they tend to be very much physically larger than any prokaryote (Figures 9.2 (iii) and (iv)). More significantly, it was only the eukaryotes which were able to evolve into multi-cellular organisms.

Once multi-cellular organisms had evolved, then we have essentially come to the end of the story of the evolution of life on Earth – the rest is minor detail.

Of course, if some of those *minor* details were to have changed, we would not have *exactly* today's plants and animals around. If, for example, a certain large meteorite had not, by chance, landed just off what is now the Yucatan peninsula, some 66 million years ago, the dinosaurs might have continued to evolve towards ever-larger sizes and could now weigh some 1,000 tons,[404] or they might have died-out anyway and the apex life-forms might now be ginkgoes and lungfish[405] (or their equivalents), with no hominids in sight anywhere. But whilst this last may be an important detail to us, it is neither here nor there in the greater scheme of things.

Insofar as we can predict, therefore, if TRAPPIST-1 d life reaches the stage of multi-cellular organisms, or its equivalent, then barring catastrophes, it is likely to continue evolving much along the lines followed by terrestrial life-forms. One major exception to this though, may be that, since TRAPPIST-1's surface temperature is less than half that of the Sun, its energy is predominantly emitted into infrared and deep red parts of the spectrum. Now the active component for photosynthesis on Earth, chlorophyll, absorbs energy principally within the blue and orange parts of the spectrum. Light in the green part of the spectrum and in the infrared is reflected (hence the green colours of most plant life). The emitted IR radiation from TRAPPIST-1 would also be reflected by chlorophyll. For plant life, or its equivalent, to evolve on TRAPPIST-1 d, therefore, an alternative to photosynthesis would also need to evolve.[406] Since the early forms of life on the Earth were anaero-

---

[404] This is one estimate of the largest weight possible for a land-based animal on the Earth, given the strength of current bone-forming materials (i.e., if it were to be much bigger, then the animal would break all the bones in its body, every time it turned over in its sleep). With the water to give them buoyancy, marine creatures can easily be much larger. The modern blue whale, at up to 150 tonnes, is often quoted as the largest animal ever to exist on the Earth. However, a recently discovered partial skeleton of a fossil whale from 40 million years ago (*Perucetus colossus*), may have had relatives which topped 300 tons.

[405] 'Living fossils'; modern species which are indistinguishable, or nearly so, from their precursors of tens to hundreds of millions of years ago, although there probably *are* small evolutionary changes between the old and new versions. The modern *Ginkgo biloba* (a.k.a. Maidenhair Tree) is very similar to fossils of its ancestors of 170 million years ago, for example and the modern Queensland lungfish (*Neroceratodus fosteri*) seems to be identical to fossils aged over 100 million years.

[406] Recent laboratory experiments, though, have demonstrated that terrestrial cyanobacteria *can* photosynthesise when illuminated by simulated red dwarf star radiation.

bic, it is quite possible that whatever (unknown) processes replace photosynthesis, oxygen will not be produced as a by-product[256] and all types of life, including animals, will remain anaerobic.

Should cells, or something to take their place, *not* evolve, then it seems likely that the self-replicating molecules would simply continue to self-replicate – perhaps forever. Our descendants, when they visit/colonise ✿✿✿✿✿ TRAPPIST-1 d may then be able to live on a never-ending supply of the nutritious soup covering much of the planet.

We should also re-emphasise at this point, that it is the time-scale for terrestrial life evolution which has been used as the model for the scenario of life evolution on TRAPPIST-1 d discussed above. This could easily, in practice, be much slower or much faster. Also, the youngest estimated age for the formation of the TRAPPIST-1 system (3,000 million years) has been assumed. However, the estimates for that age range from 3,000 million years to 8,000 million years. The evolutionary process could therefore still be at the First-Living-Molecule stage or to have progressed to what it might be like on Earth 3,000 million to 4,000 million years into *our* future.

Now, whilst taking note of the previous paragraph, 'evolving much along the lines followed by terrestrial life-forms' does *not* mean that hominids will emerge in due course, nor even ginkgoes or lungfish, because we still have those 'minor details' to take into account.

The differences between TRAPPIST-1 d and the Earth which comprise those 'minor details' and which are known, or suspected, by us at the moment are:

- Tidal-locking
  - Seasons and related matters
- Flares
- Nature of the atmosphere
- Meteorology/climate
- Nature of the surface
- Low gravity

Tidal-locking is frequently regarded as being a complete bar to even the small possibility of the origin/evolution/presence of life on an exoplanet. That could, indeed, be the case if the tidal-locking is only partial like that of Mercury (Appendix D).

However, with full tidal-locking and a close-to-circular orbit (TRAPPIST-1 d's orbital ellipticity = 0.008 – Table 13.1), so that there is very little libration (Appendix D), the host star when viewed from the surface of TRAPPIST-1 d will be fixed at some position in the sky. It will remain at that particular sky position for each point on the planet's surface, for time intervals which are lengthy even by astronomical standards. Hence this situation (tidal-locking) could actually be quite favourable for the development of life.

With full tidal-locking, various annuli/zones of the exoplanet centred on the line from the middle of the exoplanet to the middle of the host star would receive constant, but differing, levels of energy from the star for long periods of time (Figure 13.4). Assuming atmospheric circulation does not redistribute some of this energy (which it probably will), there should be a simple, smooth temperature gradient around the exoplanet, with the sub-stellar point being hottest and the anti-stellar point being the coldest. The temperature requirements for all forms of life (providing that these lie between the maximum and minimum available temperatures) will thus be found in one or another zone around the exoplanet.

TRAPPIST-1 d could thus be an ideal holiday destination for many types of ETIs, with the thermophiles gathering to star-bathe near the sub-stellar point and the cryophiles ice skating and skiing further round the exoplanet in the cooler regions. Alternatively and/or additionally, if the origin of life can occur over a range of temperatures, a complex situation in inter-species relationships could arise if there are several First-Living-Molecules originating in different temperature zones which duly evolved into species suited to each of those temperature zones and then, when sufficiently advanced in their evolutions, those species try to expand out from those zones.

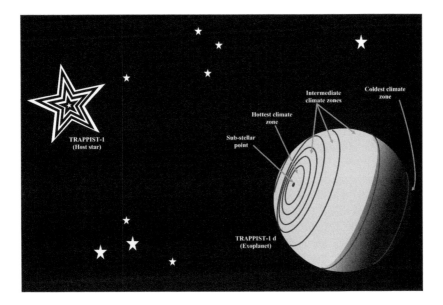

FIGURE 13.4 The constant temperature zones for a fully tidally locked planet such as TRAPPIST-1 d. (© C. R. Kitchin 2023. Reproduced by permission.)

As dwellers on the Earth, we are very familiar with the climate changes arising from the seasons of the year. Those seasons arise because the inclination of the Earth's rotational axis to the plane of its orbit (a.k.a. Axial Tilt or Obliquity)[407] is *not* at a right angle. Many, probably most, exoplanets will also have their rotational axes inclined to their orbital planes at angles other than 90° and so will also experience seasonal variations in their climates. Some may even be as extreme as Uranus (see also retrograde rotation, Chapter 8) whose rotational axis inclination is close to 90°; a Uranian living near one of his/her/its planet's poles will thus have the Sun constantly in its sky for some 21 years and not visible at all for another 21 years. At the moment, though, determining an exoplanet's rotational axis inclination to its orbital plane is beyond our observational capabilities. Exoplanets in *highly* elliptical orbits (c.f. HD 190360 b, below) *will*, though, experience climate changes from this cause.

Although the formation of planets via the collapse of an interstellar gas cloud suggests that the orbital plane(s) of the exoplanet(s) should tend to be more or less aligned, with the host star's equator (spin-orbit alignment), this need not always be the case. Interactions between protoplanets and other nearby stars during the later stages of the star formation process may lead to polar and retrograde orbits. Thus, HAT-P-6 b, a hot Jovian exoplanet, is in an orbit which is inclined at ~170 degrees to its host star's equator (i.e., it is in an almost exactly retrograde orbit). Such anti-grade (opposite to the star's rotation) orbital motions may affect the environment in/on the exoplanet but probably not by very much.

Flares, except perhaps for super flares (as we have seen above), are probably survivable. Early forms of life could simply choose[408] their 'warm, little ponds' to be at the bottoms of valleys, inside crevices, etc. where they would be largely sheltered from the flare's effects. More complex forms of life might evolve and then be able to take active measures to mitigate a flare's effects (see below).

---

[407] *Not*, as is a common misconception, because of the slight ellipticity of the Earth's orbit and the resulting slight change in the Earth-Sun distance. In fact, at the moment, the Earth-Sun distance is at its *maximum* (aphelion) in early July during the Northern hemisphere *summer* and at its minimum (perihelion) in early January during the Northern hemisphere *winter*.

[408] In fact, there would be no 'choosing'; the harsh regime of 'Mother' nature, means that any life developing in 'warm little ponds' which were not sufficiently sheltered from flares would be killed off, leaving only the sheltered ponds to bring forth the life (if it could be done).

It is not clear at the moment whether or not TRAPPIST-1 d has an atmosphere or not (Table 13.2), although the planet's low density (~60% that of the Earth) suggests the presence of largish quantities of volatile compounds, including water and gases. There is evidence for atmospheres on other TRAPPIST-1 exoplanets, so there probably *is* an atmosphere of some sort on TRAPPIST-1 d. The composition of the atmosphere, though, may not include oxygen (see above).

Once you have an atmosphere, though, you also have weather. Since we know almost nothing about the atmosphere, it is difficult to predict any meteorology. However, we do know that with full tidal-locking, the input to meteorological effects will be constant and unchanging, probably over millions of years. The climate patterns are therefore likely to be stable and fixed. Whether those patterns are hurricane-force winds constantly blowing from the sub-stellar point towards the anti-stellar point,[409] or gentle zephyrs, or some sort of cyclic pattern is quite unknown.

Although TRAPPIST-1 d is not rotating with respect to its host star, it is rotating with respect to the rest of the Universe – and with the same period as its orbit; 4.05 days (Table 13.1). There could, therefore, be some version(s) of the Coriolis storms (hurricanes, tornadoes, cyclones, etc.) which occur on Earth. One current suggestion for the meteorology on fully tidally locked exoplanets with atmospheres is that it will develop a circular symmetry of some type, centred on the sub-stellar point. The name 'Eyeball planet' is suggested for such planets, although for all that anyone knows 'Archery-butt planets', 'Sunflower planets', 'Discus planets', etc. might be better descriptors. Another suggestion is that two counter-rotating storms, which have been given the name 'Modons', would form on either side of the exoplanet's equator.

Although TRAPPIST-1 d is tidally locked onto its star, there will still be tide-induced stresses and strains acting upon it and, with its short orbital period, these are likely to keep the planet's interior well stirred up. There is thus likely to be an active geology at the planet's surface. This will probably be different from Earth's geology, but if there is a molten interior, then we may expect volcanoes, mountain building, large-scale lava floods and perhaps some form of plate tectonics.[410] If volatile compounds are present in the atmosphere, then there could on the anti-stellar half of the planet be massive build-ups of ice formations.

All these developments would probably affect the planet's climate and meteorology on both large and small scales. The thermals from an extensive lava flood plain are likely to disturb atmospheric circulation patterns and a First-Living-Molecule is likely to have its whole day ruined if a new volcano erupts under its 'warm, little pond'.

Finally, then, we come to the effects of the surface gravity, which is just less than half that of the Earth (Table 13.2). This will let mountain ranges be built up higher and rivers (whatsoever might be their liquid) will flow more slowly. For life-forms living within a liquid, though, it will have no/little effect since their weight is countered by their buoyancy whatsoever the surface gravity might be.

When/if life-forms evolve to live away from a liquid environment though, the lower gravity is likely to have a significant influence. Given also the frequent stellar flares (see above), the more complex life-forms are likely to evolve active means of reducing the flares' effects. A flare gives little notice of its occurrence; the rise time of the event being just a minute or two. So, one defence mechanism might be to develop the ability to move very fast into the nearest shelter – and the fastest animals on the Earth are the birds. As we have seen, the Peregrine Falcon (Chapter 10) can reach well over 200 mph. Thus, once life-forms on TRAPPIST-1 d evolve to that sort of level and given the reduced gravity, we might expect the apex life on the planet to be flying creatures of some sort.

We have seen above that the brief increase in the total energy which the planet receives from the star would probably not be a serious problem. It is the shortwave radiation, which in an oxygen-free atmosphere could extend well into the dangerous[411] UV-B region, which would be the hazard. Life-

---

[409] Actually, these could not last for very long. The atmosphere at the sub-stellar point would soon disappear completely and/or the pressure gradient between the sub- and anti-stellar points would bring the strong winds to a halt.

[410] The massive tides experienced by TRAPPIST-1 d make it unlikely that its surface is of the static lid variety (Box 9.1).

[411] UV-B radiation is the part of the spectrum where, on Earth, Ozone (a molecule made up from three oxygen atoms) absorbs strongly and so at sea level we are not much exposed to that radiation from the Sun. Exposure to it at high

(Continued)

forms on TRAPPIST-1 d would therefore also be likely to evolve protective mechanisms against that damaging radiation (Box 13.2).

BOX 13.2 – ON HOW TO IMAGINE AN ALIEN

Most peoples' concepts (including mine) of what ET life-forms might *actually* look like are heavily influenced by what we see around us on the Earth and, even so, are pretty ridiculous (Figure 13.1). Furthermore, some products of evolution on Earth *are*, indeed, pretty ridiculous – think; Giraffes, Angler fish, Titan arum and Human beings.

You will find that most of my imagined ETs and ETIs are based upon terrestrial analogues, with appropriate modifications for the differing conditions. This is for several reasons:

  (i) Terrestrial life-forms are generally quite efficiently designed by evolution and thus are workable prototypes for other similar(ish) situations.
 (ii) Terrestrial life-forms are generally of sensible shapes (i.e., there are few, if any, of the function-free appendages, etc. which most artists attach to their impressions of ETs' appearances).
(iii) "All the thoughts of a turtle are turtles, and of a rabbit, rabbits." (Emmerson R.W. *Natural History of Intellect and Other Papers*. P 54, Solar Press, 1893).
      (Please substitute 'terrestrial intelligences' for 'turtles' and 'rabbits').
 (iv) I am no good at drawing imaginary cartoons.

So, recognising this bias as being inevitable anyway, I will generally only suggest Earth-analogues for ETs and ETIs, leaving the details (fire breathing, 10 or 20 tentacles, slobbering multi-fanged mouths, gigantic almond-shaped eyes, even more gigantic brains, green skin, zap guns, flying saucers, etc.) to your own imaginations and preferences.

In a few cases, where the available material seems appropriate, I have also included some artists' impressions to fuel those imaginations (e.g., Figure 13.1).

Obvious defence mechanisms against flares – and evolution could certainly find plenty more – would likely be based upon protection and/or shelter. We have just seen that the lower gravity might lead to fast flying organisms, which could move into shelter very quickly and some land-based life-forms might well choose the same route. Land-based life-forms might also seek shelter by burrowing into the ground, where this is soft and friable. On hard ground, the evolution of 'tortoise-oids' might occur. That is to say the evolution on some animal of an impervious, hard, thick upper shell/carapace, perhaps of keratin or some similar analogue, perhaps of limestone (Figure 13.5 Left). When a flare occurs, our tortoise-oid simply withdraws any of its exposed vulnerable parts under the shell and hunkers down for its duration.

Liquid-based life is probably not in much danger from flares if it lives more than even a short distance below the surface. Surface dwellers, e.g. TRAPPIST-1 d exo-jellyfish, would probably simply evolve an ability to swim downwards for a few metres for the duration of the flare. Micro-organisms might just only live in the permanently well-sheltered spots (remember, the host star is fixed at one place in the sky) or develop their own, more active, molecular-level defences (see *Deinococcus*

---

altitudes or from artificial sources can, however lead to skin aging, skin cancer, snow blindness and sunburn and also to a general lowering of the effectiveness of the body's immune response systems.

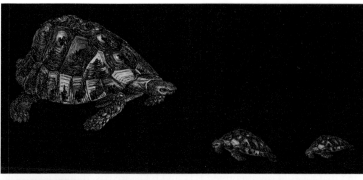

FIGURE 13.5   Possible life-forms for TRAPPIST-1 d. Top – Tortoise-oids – On TRAPPIST-1 d, life-forms resembling terrestrial tortoises could protect themselves from flares by a thick radiation-opaque carapace into which their legs and head could be withdrawn during a flare. The carapace would also be a defence against predators, such as the Peretahs, especially when allied with the tortoise-oids' poisonous-oxygen emissions (see below). (Reproduced by courtesy of Pixabay and Gordon Johnson.) Bottom – A Peretah might resemble this gargoyle (of a winged lion) although my chosen model would be a chimaera of a Peregrine falcon (Figure 10.1) and a Cheetah because of their extreme speeds. In this image, the wings are too small for the body shown (even under TRAPPIST-1 d's lower gravity), but, I guess, stone masons in the 16th century were not given much training in aerodynamics. (Reproduced by courtesy of Pixabay and Dean Moriarty.)

radiodurans, Chapter 12). Plant-life analogues might just tough-it out, or only grow in sheltered spots, or develop molecular-level defences, or develop resistant outer layers (c.f. terrestrial cacti), or develop the ability to move their vulnerable parts into some sort of shelter (c.f., Venus Fly Traps, whose paired leaves can snap shut onto a fly in just 0.1 seconds), or … well, I leave further suggestions to your own, more-than-capable, imaginations.

If we assume that the postulated fast, mobile creature can sense the beginning of a flare within one minute, that would give it another minute or so to find shelter – and, if it were me, I would make sure that I was never more than about a kilometre or two from such shelter. So, a speed requirement of 100 km/h (~ 60 mph) or more is indicated.

If/when TRAPPIST-1 d develops life and it has evolved for some 4,000 million years, then the apex predator could be something like a flying, streamlined cheetah/peregrine falcon chimaera (Figure 13.5 Right) capable of reaching a speed (say) of ~150 km/h (~100 mph) within a couple of seconds of realising that a flare had started. I have exercised the discoverer's naming privilege here and have hence called these chimaeras, 'Peretahs'.[412] Such a level of performance would make Peretahs a truly fearsome creature here on the Earth; except that, here, they would be too heavy to fly.

---

[412] From; PEREgrine falcon and cheeTAH.

A Day in the Lives of Simple TRAPPIST-1 d Country Folk

The Peretah's speed, developed because of the flares, would also come in handy for catching its prey. However, on its home planet, the Peretah's prey will have evolved their own defence mechanisms against the predator.

Suppose that the Tortoise-oids are a favourite food for the Peretahs. The tortoise-oids' defence against the flares is to hunker down inside their shells (see above), and this they can do very quickly, burying the rims of their shells into the soft sandy soil. They are not though, fast movers over the ground.

Despite this, the Peretahs will still need their speed to hunt the tortoise-oids. This arises because the tortoise-oids have evolved a separate defence mechanism against predators and this is the release of the highly toxic gas, oxygen (remember we have assumed life on TRAPPIST-1 d to be anaerobic). The oxygen is released, to be trapped under their bodies, once they are hunkered down into the surface using their flare-defence mechanism. The oxygen, of course, does not affect the tortoise-oids because they go into suspended animation.

A passing hungry Peretah will try to flip over a tortoise-oid, but if the tortoise-oid has completed its defence process, the oxygen will flood out as soon as the rim of the shell is lifted, instantly and fatally poisoning the predator. To catch its dinner, therefore, the Peretah has to flip over the tortoise-oid before it can drop down seal its shell to the ground and release the oxygen – and since the tortoise-oid has heat sensors atop its shell, the flying Peretah has to be able to pounce with extreme rapidity.

Today, a hunting tribe of Peretahs has just spotted a family group of tortoise-oids emerging from the shelter of some rocks after a recent flare. From their altitude of ~500 metres, the stooping Peretahs must reach a speed of 180 km/h, or more, if they are to reach the tortoise-oids in the 15 seconds it takes the latter to seal themselves down and to trigger their oxygen defences. The low gravity of TRAPPIST-1 d is now unhelpful, since it gives much less assistance in accelerating into a stoop than the Peregrine Falcon gets from the Earth's gravity. Despite their speed, the Peretahs fail in their hunts about 75% of the time, perhaps partly because of this slower acceleration and so, on this occasion they are again doomed to go hungry and the tortoise-oid family survives. However, some eggs laid nearby under the rocks, by a burrowing exo-lizard do provide the Peretahs with a little sustenance.

TRAPPIST-1 – The Remainder of the Family

The details of the six other exoplanets belonging to the TRAPPIST-1 system – and perhaps there are more TRAPPIST-1 exoplanets remaining to be discovered – are shown in Tables 13.1 and 13.2.

TRAPPIST-1 b and c are probably Venus-like, possibly with even higher surface temperatures. Unless, as is speculated for Venus (Chapter 9), there are habitable regions high in their atmospheres, they seem unlikely to be inhabited. Even if habitable regions do exist in their atmospheres, the life-forms would be exposed almost directly to the emissions from the star's flares, and these would be ~380% and ~200% more intense than at TRAPPIST-1 d because these two exoplanets are orbiting even closer to their host star.

Some recent IR measurements by the JWST of the surface temperature of TRAPPIST-1 b suggest a value of 500 K (200 °C). This also suggests that the exoplanet has no or little atmosphere and so stellar flares would certainly have long since sterilised its surface. A small pure(ish) carbon dioxide atmosphere, though, could still be fitted to the JWST's observations.

TRAPPIST-1 e is close to the outer edge of its host star's habitable zone. It is quite similar to the Earth in terms of its mass and size, but whether or not it has an atmosphere and what that atmosphere's density and composition are like, if it is present, are not known. The exoplanet's equilibrium temperature, though is 10 to 30 K cooler than that of the Earth, so a thickish atmosphere with a significant greenhouse effect would be needed to allow even the equivalents of Earth's life-forms to survive.

A possible plus factor for life on TRAPPIST-1 e, though, is that it might not need to have that life originate in some 'cold little pond' on its surface. Life could be transferred there via natural panspermia (Chapter 9) from TRAPPIST-1 d; if, indeed, it has succeeded in appearing on the latter. When TRAPPIST-1 d and e pass each other in their orbital motions, their separation is only just over 1 million km – and we know that the Earth receives meteorites from the Moon (0.48 million km away) and Mars (55 million km away). So, we should expect meteorites originating on each planet to travel to the other, carrying life-forms (if there are any to be carried).

Indeed, it is quite possible that all seven exoplanets are merrily enjoying snowball fights and flinging meteorites at each other continuously, since even the innermost and outermost planets approach within 7.5 million km of each other at times; ~14% of the minimum distance between Earth and Mars. So, if life originates on any of the planets of the system, or on any asteroids, satellites, comets, etc., which there may be, there's probably a good chance of it 'panspermia-ing' its way to one or more of the others.

Having said that, the three outer exoplanets, TRAPPIST-1 f, g and h do not seem to be very likely prospects for supporting life. Without atmospheres, they will be cold to very cold and the surfaces would be exposed directly to the flare emissions. With a thick atmosphere, the temperatures could reach Venus-levels. If one, two or all three of these planets have liquid oceans capped with thick ice crusts, then heat from their interiors produced from tidal stressing could render them habitable. The forms of life would, most probably, be microscopic, but larger forms would be constrained by the convergent evolution acting on them to resemble the shapes of terrestrial marine life (Chapter 10, Figure 10.1), thanks to their moving within in a resistant liquid medium.

We will be able to make better assessments/predictions about whether or not life is found there when spacecraft have visited one or more of the (still hypothetical) sub-surface oceans of Europa and her sisters and brothers within the Solar System (see below).

The Best of the Rest

With the TRAPPIST-1 system we have killed many exoplanet birds with one stone, examining a cool red dwarf host star and its exoplanets, ranging in masses from 0.3 M_{Earth} to 1.15 M_{Earth}, in temperatures ranging (possibly) from ~170 K (about -100 °C) to ~1,500 K (~1,200 °C) and with some exoplanets which are tidally locked and some which may be rotating with respect to their host star. In fact, with TRAPPIST-1 we have covered a large fraction of the host star/exo-planet pairings in terms of the numbers which have been discovered to date, i.e., cool red dwarf host stars with small-ish exoplanets in orbits close to the star (Figure 7.3).

Now, these cool red dwarf stars form the majority of all stars (Chapter 7), so they are likely to remain, even when we get less biased means of detecting exoplanets (Chapters 3 and 7), the commonest types of host stars. However, there *are* other types of exoplanets in larger orbits around more massive stars (Table 8.1). There are also the free-floating exoplanets, small bodies and even the Interstellar Medium to consider so, our work is not over yet.

Kepler-452 b

The host star, Kepler-452 is close to being a twin of our own Sun (Table 13.3). It lies about 1,800 ly away from us in Cygnus, 0.85° South of δ Cyg in the sky and it should just be visible, from a reasonable site, through a 150 mm (6-inch) to 200 mm (8-inch) telescope.

The similarity to the Solar System continues when we look at its (single) exoplanet, since this orbits its host star at a distance of 1.05 AU in a period of 385 days (Earth equivalents; 1 AU and 365 days). The exoplanet's equilibrium temperature is ~265 K, depending upon its albedo (Chapter 8), compared with that for the Earth of 255 K.

TABLE 13.3
The Physical Properties of Kepler-452 and Our Own Sun, Compared

	Kepler-452	Sun
Mass (M_{Sun})	1.04	1.0
Radius (R_{Sun})	1.1	1.0
Temperature (K)	5,760	5,780
Luminosity (L_{Sun})	1.2	1.0
Spectral type	G2V	G2V
Age (millions of years)	~6,000	4,500

Thereafter, though, 'it all starts to fall apart', since the exoplanet, Kepler-452 b, has a mass between 3 M_{Earth} and 7 M_{Earth} and a radius about 50% larger than that of the Earth. So, it is a Super Earth (Table 8.1).

Taking average(ish) values for the exoplanet's mass of 5 M_{Earth} (3 x 10^{25} kg) and of 9,600 km for its radius would give the planet a mean density of 8,000 kg m^{-3} compared with the Earth's mean density of 5,500 kg m^{-3}. The surface gravity would then be about 22 m s^{-2}, i.e., about 2.2 g. Such data would suggest a rocky composition like that of the Earth, but the stronger gravitational field would suggest a larger and/or more compressed core.

It is not yet established whether or not Kepler-452 b has an atmosphere. If it does, then the atmosphere's density and composition are still a matter of guesswork.

Now, although we yet lack some essential information about Kepler-452 b, apart from its mass, it has very much in common with the Earth. We might, therefore, justifiably speculate that it does indeed have an atmosphere. The exoplanet's equilibrium temperature (above) is 10 K higher than that of the Earth, so a thin(ish) atmosphere would probably leave its surface conditions very comparable with those of the Earth today. A thicker atmosphere and/or a stronger greenhouse effect would lead to significantly higher surface temperatures, possibly even to Venusian levels. Nonetheless, there is a possibility that surface conditions on Kepler-452 b could be very Earth-like, save for the two significant differences: the much higher surface gravity and the unknown composition of the atmosphere.

So, if Kepler-452 b has had its First-Living-Molecule moment, what might be the natures of life-forms there now? Unless an oxygen-producing version of photosynthesis has evolved, the atmosphere is likely to be some variation on a $CO_2/CH_4/H_2/N_2/NH_3/$ etc. mix and the life-forms will have anaerobic metabolisms. If free oxygen has been produced, then metabolisms similar to Earth's aerobic life may have evolved.

In either case, the principal remaining environmental/evolutionary influence will be the stronger gravity at the surface. Thus, the possibilities would include:

- Life never started.
- Life started, but found it too tough and has now died-out.
- Life started many times, but each time found it too tough and died-out.
- Life has not evolved beyond the replicating big molecule stage.
- Life-forms have evolved to produce unicellular and multi-cellular organisms or their equivalents.
- Liquid-based multi-cellular organisms have evolved to levels comparable with or beyond those of marine life on Earth. As noted several times earlier, convergent evolution is likely to make the body-shapes of any liquid-based life resemble those life-forms living in the Earth's rivers, lakes, seas and oceans.

- Life-forms have evolved to living outside their original liquid environments (always assuming that there is dry land somewhere for them to climb onto) and may bear some comparison with Earth life-forms at some stage between ~3,000 million years ago and now.
- Life-forms and intelligences have evolved way beyond pitiful terrestrial levels, and we can no longer understand what is happening with them at all.

This last possibility will be further subjected to our 'pitiful' speculations in Chapter 17. Here, we may sensibly consider the second-to-last option.

Once a life-form is living out of a liquid environment, it has to support its own weight and so it will have to evolve to the 220% greater weight that it will be bearing compared with similar terrestrial life-forms. Now, 220% heavier is not actually *that* much of a burden. A heavily built adult human could, on Earth, weigh ~120 kg (~20 stone, ~280 lb) whilst one who was lightly built might weigh ~50 kg (~8 stone, ~110 lb) – and that is a 240% difference. So, life-forms akin to (the larger examples of) humankind in size might quite successfully exist on Kepler-452 b's solid surfaces.

Of course, adaptions to the stronger gravity will almost certainly include stumpier body shapes and thicker bones and stronger muscles (Figure 13.6 Left), providing that those particular structures have evolved. Also, the Kepler-452 b high jump world record would probably be less than one metre.

FIGURE 13.6   Some imaginary, but conceivable, life-forms on Kepler-452 b; Left – a Kelamb[413] with a swarm of Bubbleflies. The kelamb is very stocky and has massive legs to support it against the 2.2 g surface gravity. The bubbleflies have evolved an extra wing-pair compared with terrestrial dragonflies and have bladders/bubbles all along their bodies filled with a light gas to provide buoyancy. (Adapted from original images with thanks to Pixabay, Alexa, Gerhard and Robert Balog.) Right – A terrestrial albatross using dynamic soaring within the complex winds above the ocean waves to obtain its energy for flying. (Reproduced by kind permission of Pixabay and jmarti20.)

Flying life-forms will be as hindered from evolving by the high gravity on Kepler-452 b as much as the Peretahs on TRAPPIST-1 d are aided by its low gravity. Possibly different biochemistries and alternative lighter/stronger materials, if they can be evolved, might enable Kepler-452 b birds to fly, but terrestrial birds would struggle. Smaller flying life-forms akin, say, to terrestrial mosquitoes might be much better placed, since the ratio of strength to weight[414] will improve as the size decreases. Larger flying forms might be able to exist if they evolved buoyancy aids (Figure 13.6) – something like a large bladder filled with one of the lighter gases (i.e., a balloon) might thus be

---

[413] KEpler-452 b / LAMB.

[414] If we consider an arm bone (say), then its weight will be proportional to its volume (i.e., to the cube of its size) while its strength will be proportional to its cross-sectional area (i.e., to the square of its size).

   If we have two arm bones, identical in all respects, except that one is twice the size of the other, then the larger bone will be eight times heavier than the smaller bone (2^3) while its strength will be only four times that of the smaller bone (2^2). So, the strength to weight ratio of the smaller bone will be twice that of the larger bone.

developed. Another possibility, provided that the winds are suitable, is dynamic soaring (Figure 13.6 Right), as practiced by, *inter alia*, albatrosses on the Earth. Dynamic soaring requires layers in the atmosphere with differing wind speeds and the albatross can remain in the air for years[415] by moving between the layers and so extracting energy from the winds.

Strength to weight ratio can help in another way; if the life-form is quite small, it is likely to cope with the strong gravity quite well, but then be too small for some of the bigger tasks which it might like to undertake. However, we can look to the Earth again for ideas and find that Ants, Bees, Termites – i.e., the social insects – achieve more as a cooperating group than the same number of individuals working independently would do. Indeed, an ETI visiting the Earth might well put Humankind into the same grouping, since we (sometimes) cooperate in large numbers to undertake major efforts. This is done by choice and by using our intelligence rather than by instinct, but the effect is similar.

There are many other half-way cases of cooperative behaviour; wolves hunt in packs and consequently have more success, and deer gather in herds, the better to detect approaching predators and to give more chances for individuals to escape amongst their many other panicking and stampeding compatriots. Life-forms with the individuals being the size of, say, terrestrial rats and with similar levels of intelligence and ability could probably cope quite well with conditions on Kepler-452 b. If, in addition, they had a similar level of social cooperation akin that of terrestrial army ants, then nothing much could oppose them – and a horde of ten thousand of *those* chasing *you* would be the stuff of real, wide-awake, screaming-out-loud nightmares.

These imaginings and others will depend on many factors and a thicker atmosphere would probably help any flying life-forms. If intelligent life-forms originated and evolved on Kepler-452-b on a similar timescale to that of ourselves on Earth, then they will have had another 1,500 million years of evolution after reaching levels of capability similar to those of current humankind. So, it is literally anyone's guess where they are now.

"Often the hands know how to solve a riddle with which the intellect has wrestled in vain."[416]

At this point we need to pause and consider quite what we mean by 'Intelligence' and how our perception of it interacts with an organism's practical manipulative skills – as Jung points out in the quotation forming the title of this sub-section.

We will then return to whether or not Kepler-452 b could have ETIs in some form.

According to the first published dictionary[417] of the English language, published in 1604 and compiled by Robert Cawdrey, intelligence is:

"Knowledge from others"

whilst a modern source[418] defines it as:

"The capacity for abstraction, logic, understanding, self-awareness, learning, emotional knowledge, reasoning, planning, creativity, critical thinking, and problem-solving. More generally, it can be described as the ability to perceive or infer information, and to retain it as knowledge to be applied towards adaptive behaviours within an environment or context."

---

[415] The birds do touch the water in order to catch their prey, but they only *have* to alight on land for breeding. See also 'Plasma Wings' in Chapter 18.

[416] Jung C.G. *Collected Works*, Vol **8**. '*The Structure and Dynamics of the Psyche*' Para 180, 1972. Princeton University Press.

[417] *A Table Alphabeticall, conteyning and teaching the true writing, and vnderſtanding of hard uſuall Engliſh words, borrowed from the Hebrew, Greeke, Latine, or French, &c. With the interpretation thereof by plaine Engliſh words, gathered for the benefit and helpe of ladies, gentlewomen, or any other vnskilfull persons. Whereby they may the more eaſily and better vnderſtand many hard Engliſh words, vvhich they ſhall heare or read in Scriptures, Sermons, or elſe vvhere, and alſo be made able to vſe the same aptly themſelues,* Edmund Weauer, 1613.

[418] https://en.wikipedia.org/wiki/Intelligence.

Clearly, Cawdrey was thinking of the term as we use it today in relationship to espionage, not mental capacity, whereas Wikipedia's definition is more in line with this book's purposes. Nonetheless, Wikipedia's definition does not seem to encompass quite everything we might look for in ETIs and we need to add 'Sapience' and 'Sentience' to the concept, which Wiki also defines as:

"… possessing wisdom and discernment …"

With sapience, these characteristics are based upon knowledge, whilst with sentience, they are emotionally based.

I'm sure that you can think of your own, undoubtedly better definitions. However, there is a physical aspect to Humankind's 'intelligence' which has not yet been included above, although Wikipedia's definition of 'Intelligence' almost gets there. It is, though, encompassed by the quote in the title of this section:

"Often the hands know how to solve a riddle with which the intellect has wrestled in vain."

The mental development of Humankind alone would still leave us as vulnerable to our environment as is a new-born infant of any species.[419] It is the practical application of the characteristics itemised by Wikipedia which result in our present levels of attainment, i.e.:

HANDs

or as it is more frequently expressed:

BEING A TOOL MAKER AND USER

and this is considered further below and within the section entitled, 'Capability'.

Terrestrial dolphins are widely thought (by humans[420]) to be highly intelligent animals. However, they are not able to *make* tools. Dolphins do, however, *use* natural objects as tools such as Hexactinellid (glass) sponges ripped up from the sea floor to protect their noses when they are foraging through loose, sandy areas and they use conch shells to trap fish. Dolphins, humpback and other whales also communicate with each other by sounds and gestures, but whether or not this amounts to a language is still unclear to us obtuse humans.

Being a tool-user will obviously give a life-form much greater control over its circumstances and environment. But in humankind's experience it often synergises with intelligence to the benefit of both. The quote in the title of this section from Jung sums up this interaction, as does the aphorism attributed to Confucius (but also sometimes credited to other writers):

"I hear and I forget
I see and I remember
I do and I understand"

---

[419] An apposite quotation (which I cannot now trace) also expresses this. It is along the lines of
An Encyclopaedia is Full of Knowledge, but it is Powerless to Do Anything!

[420] This is probably as good an estimate as any. There are various algorithms/numerical ways of 'measuring' intelligence, mostly based upon the properties of brains, but they all have serious deficiencies. If you wish to know more on this topic, then try researching 'Encephalisation Quotient', 'Sentience Quotient', 'Forebrain Neuron Count' and the much over-rated 'IQ test'.

The Encephalisation Quotient ranks the top four species as; Humans (~7.6), Bottlenose Dolphins (5.3), Ravens (2.49) and Chimpanzees (~2.4).

The Forebrain Neuron Count ranks the top four species as Orca (~43,000 million), Long-finned pilot whales (37,000 million), Short-finned pilot whales (35,000 million) and Risso's dolphins (19,000 million). Humans come only fifth at 16,000 million to 21,000 million on this scale, although we do come out better on the scale than natural sponges (which, incidentally, are classed as animals) which rate bottom of the list, at a neuron count of zero.

Between otherwise similar life-forms, we may therefore expect the one which is able to use tools to be more *effectively* intelligent (i.e., Capable) as well as more able to control its environment, than the life-form which has no practical skills. This goes against (and rightly so), the fairly common SF story-line wherein the 'Pure Mind' (whose body has almost withered away, but whose brain is now vast) rules and exercises control over the less intelligent, but physically capable 'lower' orders.

Artificial Intelligence

At this point, we also need to mention the vexed question of 'Artificial Intelligence' and whether or not we could discover ETIs which are non-organic in their constructions and functions.

The basic answer to this AI question is 'Yes – we could come across non-organic intelligences and also find hybrids with both non-organic and organic components'.

Indeed, it could easily be argued that humankind is already of the latter (hybrid) type of intelligence; many types of operations today are performed human-computer-actuator combinations which could not be performed by an un-assisted human. For example, nearly half a century ago, the Grumman X-29 fighter aeroplane was deliberately designed to be so aerodynamically unstable that human responses alone were too slow even to be able to fly it in a straight line; the three on-board digital flight computers had to make their own flight path corrections every 25 milliseconds.

However, to deal with the possibilities inherent in completely artificial, non-organic, ETIs would require another book, which would be outdated before it could be published, so, a brief review of the topic is included in Appendix E and completely-artificial-non-organic-ETIs, otherwise, will not be considered further in this book. Hybrid types of ETIs will be included on the grounds that their 'digital flight computers' or their equivalents are really just tools, albeit very complex ones.

For those readers who never look at appendices, the concluding sentence of Appendix E is also repeated here:

'Quite possibly/probably and, especially for environments inhospitable to organic life, when SETI finds life, it will therefore be of the Non-Organic form.'

SETI

SETI (see also the main discussions in Chapter 14 *et. seq.*) stands for 'Search for Extraterrestrial Intelligence' and its scientists use state-of-the-art observing equipment, mainly within the radio region but recently extending into other parts of the spectrum, to look for signals coming to us from life-forms living on/in/around exoplanets and other environments lying somewhere 'out there'.

That said, let us for a moment imagine dolphin-type life-forms living on, say, Prox Cen b (Chapter 8). Despite the possibility that the dolphin intelligence might equal or perhaps even be above humankind's intelligence (which some people claim, although I always wonder; "if it were true, how would mere humans be able to know it?"), they could still never develop the simplest mechanical devices. Dolphins do not have hands (or arms, claws, feelers, fingers, horns, pincers, tentacles, etc.) with which they could make and then use tools, they thus have no means of manipulating items within their environment to make them more suited for some desirable dolphin-ish purpose.[421] Hence, even though Prox Cen b is only 4¼ ly away from us, *none* of the current SETI searches could detect such dolphin-type's presence,[422] because the dolphin-types have no way of sending out signals.

---

[421] Terrestrial dolphins can, though, use some naturally occurring objects as tools (see above) so their C Factor (Table 13.2, below) may be around ³✶.

[422] The presence of ETs, even ETIs, on an exoplanet may, however, be detectable from passive (on the ETs' side) signals such as the presence in the exoplanet's spectrum, or the spectrum of its atmosphere, of the signatures from chlorophyll, industrial pollution, oxygen, etc.

An additional impediment for any water-based life-form would be that they could not even discover combustion or electricity,[423] until they had devised a means of working within a non-aquatic environment.[424]

Thus, SETI is not actually a

'Search for Extraterrestrial Intelligences',

but it is a

'Search for Extraterrestrial Tool-Making-and-Using-and-Communicating Intelligences'.

However, 'SETTMUCI' does not trip quite so easily off the tongue as does 'SETI', so I doubt if many people will adopt its usage.

Capability

Because, as just discussed above, their intelligence alone will not guarantee that ETIs will be discoverable, for the purposes of this book, which will attempt to consider all possibilities for ET and ETI life-forms, an additional criterion will be defined. This criterion is a gauge of the *Capability* of a *life-form* to manipulate and change its *environment* and to improve it from that *life-form*'s point of view. I have called it called the 'C-factor', and in some ways it is the inverse of evolution, since evolution adapts the *life-form* to be a better fit to its *environment*. The C-factor is based upon the practical abilities[425] of a life-form and uses crossed hammer and spanner symbols; ⚒.[426] Combined, then, with an intelligence measure (see above – footnote [420]), it will give a more realistic gauge of the attainment levels of ETs and ETIs.

The C-Factor definitions[427] and criteria are listed in Table 13.4.

---

[423] Not true of electric eels,[54] but their electric charges are simply discharged through the water anyway.

[424] Larry Niven and Edward Lerner's SF novel '*Fleet of Worlds*', Tor Books, 2007, has a reasonably convincing account of how water-based life-forms (which they call the 'Gw'oth') might discover and apply many of the laws of physics including electricity. Basically, a number of the Gw'oth (which are starfish/octopus-types) can link mentally and form a living super-computer which can conduct physics experiments via simulations until the point is reached where the Gw'oth can devise dry environments for actual experimentation and for manufacturing, etc.

[425] There are already various scales designed to allocate levels of performance in some way to ETIs, such as the Kardashev and Balbi scales (Chapter 17), but they are concerned with much later developments. There seems not to be any assessment rating for the practical abilities of intelligent beings much before they get to the stage of colonising the galaxy.

[426] From the from the Webding font, if you wish to use it.

[427] At the lower ability end of this scale, it may be difficult to distinguish between normal actions and the use of tools. Thus, triggerfish will squirt a jet of water at a sea urchin in order to expose (and eat) their undersides. Is the water being used as a tool in this instance? I would say probably not, but you may well disagree.

There *are* several definitions proposed for when an animal is using a tool, some quite complex. One, although dating back over half a century, which is still useful and is due to Ronald Hall, is: "The use by an animal of an object, or of another living organism as a means of achieving an advantage … . The mediating object is required … to be something extraneous to the bodily equipment of the animal and its use allows the animal to extend the range of its movements or to increase their efficiency." Hall, K. R. L. '*Tool-Using Performances as Indicators of Behavioural Adaptability*'. Current Anthropology, 479, **4**, 1963.

TABLE 13.4
C-Factor Criteria

Symbol	Definition	Examples and comments
$_0 \bigstar$	No ability to use tools in any way at all.	Goldfish, earthworms.
$_1 \bigstar$	Can make use of naturally occurring items to aid normal activities.	Song thrushes, who will often use a flat stone (a Thrush's 'anvil') to smash the shells of the larger snails. Some ant species will use leaves to carry water their nests.
$_2 \bigstar$	Can make use of naturally occurring items to aid unusual or new activities.	The male blanket octopus often carries the stingers from Portuguese man o' war jellyfish; perhaps as a defence, perhaps to catch prey, or both. Some heron species will drop insects and other food items into water as bait to attract fish for them to catch.
$_3 \bigstar$	Can modify naturally occurring items to aid normal, unusual or new activities.	Some octopuses will carry one or more suitable items (such as coconut shells), quite a distance in order to build a shelter. Ravens and other corvids will strip bark from a twig in order to better use it as a spear for probing crevices for food. In the laboratory, at least, they will also fashion hooks out of suitable materials.
$_4 \bigstar$	Can combine two or more items to aid various activities.	Under laboratory conditions, crows can assemble short segments (provided by the researchers), to produce a composite tool long enough to reach the desired food morsels. Cockatoos, again under laboratory conditions, have been able to use two different tools and in the correct order to obtain their rewards.
$_5 \bigstar$	Can design, make and use many complex tools for many differing activities.	*Homo neanderthalensis*, whose implements included shafted tools, separate tools for hunting and food preparation and tools for sharpening other tools.
$_6 \bigstar$	Can plan ahead to ensure that items needed to make/use complex tools are available when needed	Under laboratory conditions, ravens trained to use a tool to obtain a reward will choose that tool from a collection of different tools (when the reward is not on offer) and save it for use later. Orangutans will also save a tool for later use, even selecting the ones which give access to their favourite rewards rather than to the less desirable ones when offered both tools.
$_7 \bigstar$	Use of artificial energy (fires) for heating, cooking, defence, etc. and to smelt ores for metals and similar levels of activity	Black kites (may) start new fires using embers from naturally occurring fires and then feed on the escaping insects, etc. *Homo erectus* seems to have been the first to have had controlled use of fire. *Homo sapiens* was the first to smelt metals (probably copper and probably around 5,000 BCE).
$_8 \bigstar$	Discovery and use of electricity, radio emissions,[428] x-rays, γ rays, advanced medical procedures, manufactured food stuffs, complex mechanical machinery.	*Homo sapiens* This is about the lowest C-factor level to enable discovery/ detection of ETIs by current terrestrial SETI efforts.
$_9 \bigstar$	Nuclear energy, robots, computer control of many processes. Space flight. Ability to control/change the environment on a planetary scale.	*Homo sapiens*

(Continued)

---

[428] It is quite possible that radio and microwave emissions from an advanced civilisation will only be observable for a short while. As technology advances, the wasteful practice of 'broadcasting' may be replaced by 'narrowcasting' whereby the signal is only sent to the intended recipient(s) via optical fibres and the like. This process has already begun on Earth. The signals sought by SETI may, therefore, only be there to be found from an ETI culture for a century or two – see also the discussion of the 'Fermi Paradox' in Chapter 16.

TABLE 13.4 CONTINUED
C-Factor Criteria

Symbol	Definition	Examples and comments
10 ✖	The near future – possibly direct communication between humankind and AI devices via implants/inductive pick-ups in human brains. Nuclear fusion reactors. Generation of new life-forms in the lab. Colonies/industrial facilities in space and on other Solar System planets, asteroids and moons.	*Homo sapiens*
11 ✖	The far future – perhaps Figure 13.7, or some golden wonderland with unlimited energy from fusion reactors, resources from across the Solar System, terraforming of some planets and whatsoever else your wilder dreams may suggest. Remember, also, A.C. Clarke's third law: "*Any sufficiently advanced technology is indistinguishable from magic.*"[429]	*Homo sapiens'* evolved descendants.
12 ✖	Colonisation of the galaxy/Universe. Build new Universes.	*Homo sapiens'* very highly evolved descendants.

Thus, we might classify terrestrial dolphins, who can use conch shells to catch fish (a new activity – see above) as '[EQ 5.3 / ³✖]'[420], clearly eliminating them from Terrestrial SETI levels of detection if they were ETIs, but recognising their basic intelligence

We might also classify an exoplanet as being inhabited by intelligent life-forms from levels somewhere upwards of [EQ 2.5 / ³✖] (crows). But to be detectable by our present-day SETI efforts it would need to be nearer a level of [EQ 6 / ⁸✖] (*Homo neanderthalensis* perhaps – if they had been given time to 'do their own thing', without *Homo sapiens* barging them out the way).

THE REAL THING #2 – INHABITANTS OF EXOPLANETS WITHIN AND NOT WITHIN HABITABLE ZONES

KEPLER-452 b – CONTINUED

We may now return to considering the life-forms which might be possible on Kepler-452 b (see above) and to the possible effects of Kepler-452 b's 1,500-million-year greater age than that of the Solar System.

On billion-year time scales, few species survive for long. Amongst the larger organisms, we have already encountered ginkgos and lungfish (see earlier in this chapter) which may have survived largely unchanged for 0.17 billion years and 0.1 billion years, respectively. Horsetails (*Equisetum arvense* – actually a very invasive relative of the ferns) may date back to the early Jurassic (0.2 billion years ago) with few changes, whilst horseshoe crabs, however *with* some significant evolutionary changes, may trace their origins back to the Ordovician (0.44 billion years). Only micro-organisms can claim histories over a billion years, although with very significant evolutionary changes to them

---

[429] First published, perhaps, in 'Playboy' in 1968, although that source has proved elusive. It is, however, definitely to be found in Science, 255, **159**, 1968.

along the way. Cyanobacteria and dinoflagellates are two such examples, with their possible fossils dating back to 3.4 billion years and to the Mesoproterozoic (0.8 to 1.9 billion years ago) respectively, although the oldest dinoflagellate fossils with a *clear* resemblance to modern species are just 0.4 billion years old.

Thus, if intelligent life (Box 13.2) did evolve on Kepler-452 b on the terrestrial time scale, its most likely current abode is shown in Figure 13.7.

FIGURE 13.7   Where any ETIs which evolved on Kepler-452 b 1,500 million years ago are now likely to be found. (Reproduced by kind permission of Pixabay and Ciker-Free-Vector-Images.)

Suppose, though, Kepler-452 b's ETIs *are* still living today, what might we expect of them?

We can, I think, eliminate the possibility that they have gone much beyond level ^{10}⚒ (Table 13.4) – because they have not taken-over/conquered the Earth. If we see Kepler-452 b as being close to a twin of the Earth, then the 'Kepler-452-ians' would see the Earth, not merely as a twin of their planet in reverse but as an extremely attractive twin; with less than half the gravity of Kepler-452 b to strain against.[430] Kepler-452 b is some 1,800 ly away of course, but a ^{11}⚒ level society ought to be able to achieve speeds of 5% that of light plus having hibernation (or its equivalent ✺✺✺) so that the journey could take them ~36,000 years – peanuts when you have this pristine, gorgeous new home, Earth, as your destination and you are in suspended animation most of the time.

Even if level ^{9}⚒ is the Kepler-452-ians' current status, though, we ought to be able to detect their radio, other emissions and/or other signs of their presence with our current SETI programmes (Chapter 14) or to detect them given SETI's likely developments in the next few decades/century or so.

Humankind's total energy consumption in 1950 was about 10^{20} J (estimates vary somewhat) and by 2020 it had reached about 6×10^{20} J, increasing steadily over that time by $\sim 7 \times 10^{18}$ J per year.

---

[430] Of course, there *is* the possibility that the 'Kepler-452-ians' *did* take over the Earth, say a few tens of millions of years ago and that humankind *is* therefore what they have now evolved into.

Assuming that this is applicable to the Kepler-452-ians and that they are 1,500 million years in advance of us, then they should now be consuming ~10^{28} J per (Earth) year. That translates into the complete burning of the oil in a cubic-shaped oil tank each of whose sides are ~600 km long, every year – and if a 600 km tank of oil is beyond your imagination, then its size is between that of Ceres and Vesta. So, even at a distance of 1,800 ly we *surely* ought to be able to detect *some* effects of the Kepler-452-ians efforts.

The choices left, if the Kepler-452-ians are not much further forward than humankind might well be in a century or so, are, given that we have not detected them so far are, then:

- Whatsoever their intelligence score, their capability score is less than about ^{9}✘; perhaps because they inhabit seas and oceans.
- They do not/did not have convenient coal/oil energy sources (or equivalents) to jump-start their early technological developments.
- They have never got into space because of their much stronger gravitational field (the escape velocity from Kepler-452 lies between 16 km/s and 24 km/s, depending on the planet's actual mass, compared with Earth's 11.2 km/s) and perhaps also because of having a much thicker atmosphere to get through first.
- They got into space but found nothing except small asteroids and the like and so gave space travel up as a bad job.
- Their civilisation is cyclic;[431] i.e.,
    - (i)   reach a certain level of achievement
    - (ii)   use-up resources and/or reach dangerous levels of pollution
    - (iii)   environmental disaster reduces the ETIs back to a very basic survival level
    - (iv)   eventually get back to (i)
    - (v)   repeat.
- Kepler-452 b's First-Living-Molecule moment has yet to occur or only occurred recently.
- They have decided that they are happy with whatever they've got now.
- They have visited us in the past and decided that they didn't like us.
- They are on their way *now* to conquer the Earth – and will be here next month.
- This space is for you to add your own explanation(s) – please use extra sheets of paper if needed.

K2-3 d

Located just inside the edge of Leo's sky area and ~140 ly away from us, K2-3 is a borderline K9/M0[432] red dwarf star with three known exoplanets. It should be visible in a 100 mm (4-inch) telescope from a good site. The host star has just 2% of the Sun's luminosity and its habitable zone extends out from about 0.12 AU to 0.28 AU. Two of its three exoplanets lie within this zone with orbital radii of 0.14 AU (K2-3 c) and 0.20 AU (K2-3 d), and it is the latter which we will examine for possible ETs.

K2-3 d has an equilibrium temperature of ~215 K (-60 °C) which is on the low side for most life-forms. However, its mass is, very approximately 2.7 M_{Earth} and its radius, ~1.7 R_{Earth}, making its density ~3,300 kg m^{-3} (~60% Earth's density). This is a very low density; hence, the planet is classed as a miniature Neptune (Table 8.1). It is likely to have a very deep atmosphere and somewhere within that atmosphere, temperatures to suit many life-forms are possibly to be found; ranging upwards from its surface temperature of ~280 K (~10 °C). Its surface gravity is similar to that of the Earth.

---

[431] Larry Niven's 'Motie' ETI SF novels explore a possible scenario for such a cyclic civilisation; see *'The Mote in God's Eye'*, Weidenfeld and Nicolson, 1975, and *'The Mote Around Murchison's Eye'*, Harper Collins, 1993.

[432] Some recent work speculates that within ~300 ly of the Sun there may be up to ~11,000 temperate terrestrial-type exoplanets with spectral type K host stars and that if microbial life-forms arose as quickly on them as on the Earth (Chapter 9) on just ~1% of them, then such an inhabited (or one-time inhabited) exoplanet might be less then ~60 ly away from us.

The low density probably implies a high abundance of volatile elements and compounds, so the surface may well be covered with liquids, with few, or no, areas of solid ground emerging above those seas and oceans. Life-forms are thus likely to be liquid-dwellers and/or to occupy levels within the atmosphere.

The predicted surface temperature is too high for methane oceans to exist, but ammonia will be a liquid, if the atmospheric pressure is about ten or more times that of the Earth. Most likely, though, it will be water which covers much, if not all, of the exoplanet's surface.

Possible EQ 6 to 7 / ^{6}✖ level ETIs are shown in Figure 13.8 Left, such life-forms may, though, be anaerobic (see earlier discussions). The escape velocity from the surface of K2-3 d is ~14 km/s; only about 25% more than for the Earth. So, if level ^{9}✖ ETIs develop, either on land or able to overcome the problems of operating within a liquid, then spacecraft should be possible. If the ETIs still require a liquid environment when in space, though, the mass required for that will probably be prohibitive and they will be limited to sending out robotic missions only.[433]

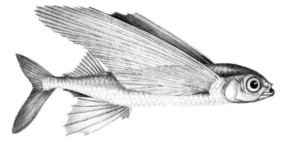

FIGURE 13.8   Some imaginary, but conceivable, life-forms on K2-3 d. Top – A three-eyed Quadropus[434] fleeing from its main predator, a large, well-armed Mer-chimaera, in the shallows of K2-3 d's second largest ocean. (Adapted from original images with thanks to Pixabay, Sergei Tokmakov. Esq, Parker West, Thomas and Harmony Lawrence.) Bottom – A sketch of a terrestrial flying (actually 'Gliding') fish; *Exocoetus* sp. Although unable to flap their 'wings', by using thermals and winds deflected from waves, terrestrial flying fish can fly for up to 400 metres. Perhaps similar evolutionary paths might lead to 'proper' flying life-forms for K2-3 d? (Reproduced by kind permission of Pixabay and WikiImages.)

---

[433] If they achieve level ^{10}✖ (nuclear fusion), then they will probably be able to go into space themselves.

[434] Based on the terrestrial octopus.

 The octopus' intelligence is difficult to relate to an EQ score, which is intended for vertebrates, but their C-factors are at least ^{3}✖ and they can solve mazes and remember individual humans over long periods. Their tentacles are also strong (they can remove lids from glass jars) and versatile and make a good substitute for arms and hands. Their main drawback from further advances is living in a liquid environment and short life spans (~3 years).

 I have therefore given these Quadropuses natural lifespans of 50 (terrestrial) years and the ability to survive and move around on dry land for short periods of time.

 The Mer-chimaeras cannot leave the liquid environment completely, but they can bring the top halves of their bodies above the surface and breath the atmosphere. The specimen shown here has therefore been able to develop tools (including its trident) by working with its upper half on land on ledges of dry land which are on the margins of seas/lakes, etc., of sufficient depth. Their level is around EQ 6 to 7 / ^{6}✖ to ^{7}✖ and, although the quadropuses are an important part of their diet, they have also trained some to obey commands (such as bringing fuel materials to within their reach), for extended land operations.

The suggested surface conditions seem ideal for tropical rain-forest-type plant life, so, if there is land above the liquid surface, then it may be densely carpeted with vegetation and, perhaps, also, with some land life-forms. Given the thick, dense atmosphere, avian life-forms would seem to be a possibility. However, if there is no land, evolving straight from being a liquid-dweller into a gas-dweller might be difficult, though terrestrial flying-fish (Figure 13.8 Right) show that it *is* possible. Lighter-than-'air' (balloons, feathers, microbes, etc. – Figure 13.9) atmospheric life-forms might evolve more readily, though.

HD 190360 b[435]

This exoplanet orbits a star which is fairly similar to the Sun, but some 2,000 million years older. The star, to be found in Cygnus and just about visible to the naked eye, has therefore started its evolution off the main sequence.[436] Thus although slightly less massive than the Sun (0.82 M_{Sun}), HD 190360 is both brighter and larger than the Sun (1.13 L_{Sun}, 1.06 R_{Sun}).

It has two exoplanets. The innermost one, HD 190360 c, is very close to the star and has an equilibrium temperature of ~730 K (~455 °C). With a mass $\geq$ 19 M_{Earth}, it is likely to have a thick atmosphere and so even higher temperatures may be expected below the top of its atmosphere. It is therefore unlikely to host any life-forms which we can imagine.

The second exoplanet, HD 190360 b is in a highly elliptical orbit which ranges from 2.6 AU to 5.2 AU out from the star in a 7.8-year-long orbit. Its equilibrium temperature therefore ranges from ~160 K to ~115 K (-110 °C to -160 °C). These temperatures are thus also unlikely to enable the existence of any life-forms which we can imagine.

Nonetheless, it is HD 190360 b which we are going to look at in more detail. This is because HD 190360 b's mass is ~570 M_{Earth} (1.8 $M_{Jupiter}$), and it is thus a Jovian class exoplanet (Table 8.1), which has, almost certainly, an extremely deep atmosphere of hydrogen and helium together with smaller quantities of heavier and/or more complex gases. Although older than Jupiter, it is also nearly twice its mass, and so, like Jupiter, it will still retain some primaeval heat at its centre.[437] So, whether or not there is greenhouse effect in operation, the temperature will increase towards the centre of the planet. For Jupiter (whose equilibrium temperature is ~110 K (-160 °C)) its atmosphere reaches a temperature of ~300 K (27 °C) some ~100 km below the tops of the visible clouds. The pressure there is similar to that at sea level on the Earth (remember Jupiter's atmosphere is mostly hydrogen, not nitrogen). Thus, providing that we could breathe hydrogen, humankind could comfortably live,[438] supported by balloons, just below Jupiter's top cloud level – and the same is very probably true of HD 190360 b and possibly a few/some/many/most other exoplanets with very thick atmospheres.

It seems unlikely that a nearly pure hydrogen atmosphere *can* support respiration (Chapter 12), but that still leaves plant-type life-forms, micro-organisms and so on. Thus, the HD 190360 b-ians are likely to be mono-cellular photosynthesising (or the equivalent) organisms free floating in the atmospheric winds, or, if membranes can be evolved, hot 'air', blimp-types also floating high in the atmosphere. In the latter case, at least, the exoplanet's varying distance from its host star would not be a problem, since they would float at the same atmospheric temperature all the time and so move up and down within the atmosphere automatically as the atmospheric temperature profile changes.

Whilst HD 190360 may have life-forms dwelling somewhere upon it, if those life-forms are intelligent and wish to become space travellers, then they are out of luck until they reach level [10]✦ and nuclear fusion energy generation; the escape velocity from the exoplanet is ~80 km/s (~50 miles per second, ~7 times that from the Earth).

---

[435] a.k.a. Gliese 777.
[436] Spectral class G6 IV.
[437] One current estimate of the temperature at the centre of Jupiter is ~25,000 K (~25,000 °C)
[438] There is no evidence that humankind, or any other life-form, is actually doing this at the moment. C.f. also, life in Venus' atmosphere – Chapter 9.

FIGURE 13.9 Balloons filled with a lighter gas and/or a warmer gas floating high in the atmosphere of HD 190360. The balloons are typically around 1 km in size and their surfaces are colonised by photosynthesising micro-organisms and/or small, plant-type life-forms. (© C. R. Kitchin 2023. Reproduced by permission.)

TOI- 4603 b

TOI-4603 b is an exoplanet of a star considerably hotter than the Sun[439] and it orbits very close-in to its star. Furthermore, with a mass of ~4,100 M_{Earth} (12.9 $M_{Jupiter}$) it is, itself, near to being a brown-dwarf star. The equilibrium temperature is variously estimated at between ~1,000 K (~700 °C) and 1,700 K (~1,400 °C). For a Super-Jovian planet (Table 8.1), its mean density is very high at ~ 14,000 kg m^{-3} (~11 times Jupiter's density and ~ 2.5 times the Earth's density), so that its surface gravity would be ~ 32 g and its escape velocity ~210 km/s (130 miles/s). The exoplanet is also in a strongly elliptical orbit (c.f., HD 190360 b).

This sounds like a pretty unlikely home for any type of life-forms,[440] does it not? – and that is exactly why it is in this list; an example of an exoplanet where we come as near to being as certain as possible that it cannot be a host to any ETs or ETIs. Even any natural satellites, if it has them, would have temperatures of 1,000 K or more.

So why do such high temperatures rule out life? Until we have lots of examples of ETs and ETIs, we cannot be absolutely certain that it does; there is at least one SF short story which speculates about life-forms on the *Sun*! However, when we look at even the simplest known terrestrial

---

[439] It is a spectral class F sub-giant with a temperature around 6,300 K (~6,000 °C); ~500 K higher than that of the Sun.
[440] But see Appendix E regarding non-organic life-forms.

life-forms and the organic molecules which form them (Figure 9.3), they are very complex and highly structured.

Of the 94 elements whose boiling points are currently known, 29 are gaseous at 1,000 K and 39 are gaseous at 1,700 K. This includes the important (for terrestrial life) elements: calcium, magnesium, phosphorous, potassium, sodium and sulphur.[441] At 1,700 K only 18 elements are still solid and the only one of those which has much significance for life is carbon. Also at 1,700 K, the hydrogen atoms will be moving around at some 6 km/s and any free electrons at 250 km/s.

In such a maelstrom it is difficult to see structures as complex as tRNA (Figure 9.1 (vi)) surviving for more than fractions of a second, even if they could be formed from the available 18 solid elements. Almost by definition, complex structures cannot last in such hot gases or liquids.[442] We have, of course, though, already seen that complex organic molecules *can* be formed *from* those gases and liquids at much lower temperatures (Chapter 9).

The substance with the highest melting point currently known is Tantalum Hafnium Carbide alloy which melts at about 4,200 K to 4,500 K (3,900 °C to 4,200 °C). Very high pressures, though, will solidify substances at temperatures well above those we find for normal atmospheric pressures. Iron forming the Earth's inner core has temperatures estimated around 5,000 to 6,000 K, but still behaves as a solid, for example.

If, however, life-forms can be built from Ta_4HfC_5, or enjoy living at the centre of the Earth, then we can probably never know about them and I, for one, do not relish shaking hands with a 5,000 K Iron life-form who has just tunnelled its way 6,371 km from the centre of the Earth to its surface.

Thus, when it comes to searching for ETs and ETIs (Chapter 15) we probably have nothing much to lose if we ignore all environments above 1,700 K and very little more to lose if we extend that limit down to 700 K (which temperature allows most of the terrestrial organically important elements to be solids except for those which, at room temperature, are already liquids or gases).

Extremely high pressures are probably less life-limiting, because, providing there are no gas-filled or empty spaces within the organism, it will compress very little, until it gets to centre-of-the-Earth-type pressures.

If, against all the above evidence, there is life somewhere on TOI-4603 b, then it is probably formed from red-hot diamonds[443] (Figure 13.10).

---

[441] Plus, of course, all the elements which are normally gaseous anyway at terrestrial temperatures.

[442] Buckyballs (a.k.a. Buckminsterfullerene, Fullerene, C_{60}, C_{70}, C_{80}, etc.) are molecules formed purely from 60 carbon atoms or more and which have a spherical physical structure. They and other related pure carbon structures, called nanotubes, have been detected in the interstellar medium and GMCs. They would be unlikely to form in hydrogen-rich environments and it is suggested that they build-up on the surfaces of silicon carbide dust grains in the relatively cool outer atmospheres/ejected nebulae of stars collapsing towards becoming white dwarfs. Since silicon carbide and Buckminsterfullerene have melting points of ~3,000 K and ~1,000 K (2,700 °C and ~600 °C), respectively this is a possibility. How such structures might contribute to the formation of organic molecules or even to life-forms is much less clear, though.

[443] At very high pressures (10 million x Earth's atmosphere) diamond does not melt until around 9,000 K.

FIGURE 13.10 A completely imaginary and quite implausible life-form on TOI-4603 b (see text). The Heddimond[444] family is enjoying an early spring outing on the planet's non-surface, now that the temperature has risen again to 1,350 K and the pressure is a safe 10,000 times that of the Earth's atmosphere. (© C. R. Kitchin 2023. Reproduced by permission. The 'Diamonds' are adapted from an image, by courtesy of Pixabay and Biju Toha.)

PSR B1257+12 b, c, AND d (AND LOTS MORE)

We have looked in some detail in Chapter 5 at these first exoplanets of any type to be discovered. Scenario #1 in that chapter looks at what is most likely their present states; very cold, very dark, little or no internal energy sources and without any atmospheres.

This description will also characterise extremely well a very large number of other 'normal' exoplanets, i.e.:

- pretty well all low-mass free-floating planets,
- most low mass bodies in orbits well distant from their host star

and

- all small bodies (asteroids, comets, small natural satellites, etc.).

Though if the last of these are close-in to their host star, they will at least be somewhere in the temperature range very chilly to quite extraordinarily hot, rather than 'very cold'.

Although, therefore, we shall base this case study around the PSR B1257+12 exoplanets,[445] we shall really be discussing the life-form possibilities for any very cold, atmosphere-less exoplanets (termed 'Exo-Popsicles' from here onwards).

---

[444] HEnry Draper DIaMOND
[445] So, it would probably be a good idea to refresh your memory from Chapter 5 at this point.

So, exactly what environmental conditions are we now considering? The main ones are temperatures < ~50 K (-220 °C) and no atmosphere. The lack of an atmosphere suggests that ETs (ETIs would seem to be most unlikely, unless they are space travellers from elsewhere) will seek environments well below the exoplanet's surface. The low temperatures and the probable lack of much internal heat will restrict the organisms to being extremely frugal in their energy requirements. One plus side, their lack of doing very much, may give them extremely long lifetimes.[446]

Even if life can exist under the conditions just described, it seems difficult to see how a First-Living-Molecule moment could ever occur. Perhaps, if there is life on these exo-popsicles, then it has been bred for those conditions by ETIs elsewhere and seeded onto them[447] (for reasons quite unknown and probably unknowable to us).

On the whole, my feeling would go for this group of exoplanets *not* having life-forms on them. But, if they do, then primitive unicellular-type or non-cellular organisms occupying caves and cavities well below the planets' surfaces and existing on any residual primaeval energy leaking out from the planets' centres would seem the main possibility. If multi-cellular life can exist, then something akin to terrestrial lichens[448] would probably form the apex of any exo-popsicle's ambitions (Figure 13.11).

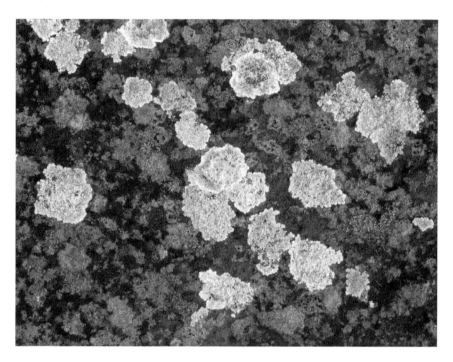

FIGURE 13.11   A terrestrial lichen – probably far too advanced for exo-popsicles – but who knows? – the Universe is *very* big. (© C. R. Kitchin 2023. Reproduced by permission.)

Kepler-16 b

We have seen earlier (Chapter 8) that a binary star system can be the host of an exoplanet. Most triple systems have unstable and/or chaotic orbits. But when two of the components are much more massive than the third, then they simply get on with orbiting each other, and it is the low mass third

---

[446] I'm not sure, though, that I would *want* to live for (say) 100,000 terrestrial years with an exo-popsicle's life-style.

[447] C.f., Directed panspermia (Chapter 9).

[448] Lichens are not really that primitive but are a composite of early-type life-forms such as terrestrial cyanobacteria or algae with fungal species – and so they could evolve soon after unicellular organisms appeared.

object that gets flung around willy-nilly. Mostly, the third object will be flung out of the system, or it will collide with and be absorbed by one of the stars.

However, two situations (Chapter 8) provide the third object (the exoplanet) with a reasonably long-term stable orbit. These are when the exoplanet is close to one of the stars and so that star's gravitational field dominates the exoplanet's movements. The second is when the two stars are close to each other and the exoplanet in a much larger orbit so that it is almost like orbiting a single star (Figure 13.12).

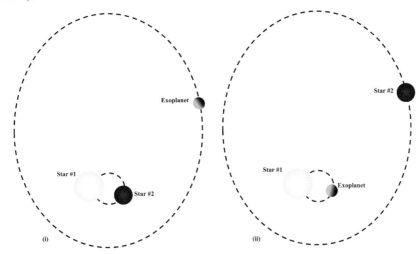

FIGURE 13.12   The two main stable(ish) orbital arrangements for triple systems. (i) – The exoplanet orbiting two close stars (a.k.a.; Tatooine system/Circumbinary exoplanet) (ii) – The exoplanet in a close orbit around one star with the second star in a larger orbit. The exoplanet is shown orbiting the larger star, but it could equally well orbit the smaller star. (Circumstellar exoplanet in a binary star system). (iii) – There is also the Trojan configuration as with the Trojan and Greek asteroids of Jupiter. In this, the three bodies have high, medium and low masses. The medium mass body orbits the high mass body. The low mass body (or bodies) has the same orbit around the high mass body as that of the medium mass body, but the low mass body is 60° ahead or behind the medium mass body.[449] This is a moderately stable situation, although the low mass body can drift quite some distance away from its 'proper' position and for an exoplanet this would result in a very variable and largely unpredictable climate. (© C. R. Kitchin 2023. Reproduced by permission.)

Kepler-16 is a triple system of the type (i) in Figure 13.12 (Tatooine – see also Chapter 8). Both stars are cooler and of lower mass than the Sun. The primary star is ~4,500 K (4,200 °C, spectral type K) and ~0.69 M_{Sun}, whilst the secondary star is ~3,300 K (3,000 °C, spectral type M) and ~0.20 M_{Sun}. The two stars are in a moderately elliptical orbit with a semi-major axis of 0.22 AU and an orbital period of 41.1 days.

The exoplanet has a near-circular orbit around the barycentre of the two stars. It is ~0.705 AU out and has a period of 229 days. From the exoplanet, the primary star would be about 30% as bright as our Sun appears from the Earth and the secondary would have about 1% of the Sun's brightness. These figures suggest an equilibrium temperature of ~190 K (-80 °C). The exoplanet also has a mass

---

[449] When two bodies are orbiting each other, there are five positions in space, called the Lagrange points, where gravitational and centripetal forces within the system are in balance. There, a third, much lower mass body, may be able to find a moderately stable orbit.

The Lagrange points are labelled L1 to L5.

L1 to L3 lie on the line joining the two massive bodies. L1 is between the bodies, whilst L2 and L3 lie on the extended line outside the system. L4 and L5 lie on a line perpendicular to that of L1 to L3 and are 60° ahead and behind the lower mass body of the two main bodies.

In the Sun-Jupiter system the Trojan asteroids oscillate around the L4 position, whilst the Greek asteroids occupy the L5 position and (as a complete aside) the JWST is positioned at the Sun-Earth's L2 position.

of around 100 M_{Earth} (~ 0.33 $M_{Jupiter}$) and so is quite unlike the roughly Earth-mass Tatooine envisaged in the *Star-Wars*[232] stories.[450] It is classed as a Jovian-type exoplanet (Table 8.1), and, although its equilibrium temperature is low, higher temperatures may be expected lower down in its atmosphere (c.f., K2-3 d and HD 190360 b for related comments and suggestions of possible life-forms).

An example of the type (ii) configuration is Kepler-13A b. The binary star system comprises two large hot stars with masses around 1.7 M_{Sun}, separated by ~700 AU, with a mutual orbital period of ~9,500 years. The system is ~2,000 ly away from us in Lyra. The exoplanet orbits the slightly more massive star, Kepler-13A, at a distance of 0.36 AU in 1.76 days. It is named Kepler-13 b or Kepler-13A b. Kepler-13A b is a super Jupiter with a mass of ~9.3 $M_{Jupiter}$. Kepler-13 is actually a triple system since Kepler-13B has a smaller companion star in a highly elliptical 66.8-day orbit.

PSO J318.5-22

Low(ish) mass free-floating planets have been mentioned above (section entitled PSR B1257+12 b, c, and d). However, free-floating planets come in all sizes and we looked briefly in Chapter 7 at PSO J318.5-22.

There, it was estimated that the surface temperature might reach ~1,100 K (~800 °C), but this would not be due to the energy from any star. Instead, this exoplanet may be only around 12 million years old and so its energy source is the potential energy released during its collapse from an interstellar gas cloud. That energy is now stored in the form of an extremely hot core and is slowly leaking out from the core to the surface. With no external energy source (i.e., no host star), temperatures in the atmosphere above the surface are likely to be much lower than 1,100 K.

PSO J318.5-22 has a mass of around 2,000 M_{Earth} (~6 $M_{Jupiter}$), so it is probably similar in many ways to TOI-4603 b (above). However, since it is 'living on its capital', over the next few millions, billions or trillions of years, it will cool down, eventually nearing absolute zero.

During this cooling process it will obviously spend long periods of time at all the various temperatures that any/all ETs and ETIs could possibly want. Thus, unless prevented by the lack of shorter wave radiation from a host star (the principal emissions from objects at ~1,100 K are in the near-infrared spectral region), PSO J318.5-22 could be populated by the whole gamut of ETs and ETIs in sequence. The very slowly reducing temperature could allow one type of ET to originate, evolve as it may and then die out/emigrate as the temperature became too cold. Then, after a little more cooling, a different ET suited to the lower temperatures repeats the pattern. Then ..., – well you've probably got the picture by now.

Of course, if PSO J318.5-22 should have ETIs on board, who favour terrestrial temperatures and find right now that it is getting a bit too chilly for them; well, here is the beautiful, tempting Earth just 80 ly away from them (Chapter 14).

The Interstellar Medium and Interstellar Gas Clouds

We have seen in Chapter 9 that the interstellar medium and giant molecular clouds are replete with a multitude of organic molecules and that these may form more complex pre-biotic molecules on the surfaces of ices, etc. as those clouds condense into proto-stars and proto-planets.

So why should First-Living-Molecule moments not occur a little earlier, within the still-condensing nebulae? There are only two reasons against such an event, and the first is just our prejudices and fondness for 'warm little ponds'.

The second reason is more serious, and it is the low density of atoms and molecules which means that the First-Living-Molecule moments become *much* less likely to occur. On the other hand, we have also seen in Chapter 9, the incredibly vast quantities of material that is gathered into the denser

---

[450] The fact of an exoplanet being a Tatooine or a circumstellar exoplanet in a binary star system imposes no restrictions on the nature of that exoplanet; it could be of any of the types listed in Table 8.1. If it is a Trojan-type system and the two stars are both of low mass, then more massive exoplanets, however, might not find even moderately stable orbits.

GMCs. So, if it is possible for life to originate in GMCs, etc., then it probably will do so. Moreover and probably more likely is the possibility that life-forms originating within more conventional environments are carried out into space by atmospheric winds and stellar radiation pressure and once there evolve into true space-life-forms, i.e., a variation on panspermia which stops halfway.

If there are, then, space-dwelling life-forms what might they be like? Well, numerous SF authors have given this some thought and, *inter alia*, Arthur C. Clarke and (Prof.) Fred Hoyle have gone for broke and suggested that smallish interstellar clouds/the vacuum of space itself might be living entities. They furthermore suggest that these entities will be sentient ETIs.

Clarke's *The City and the Stars*[451] has two space entities, though these have gone beyond needing material support and are pure intellects. Both, in some unspecified way, were brought into being by more conventional ETIs from throughout the galaxy and over billions of years ago. The first, called the 'Mad Mind', was a disaster; it was/is insane and it did tremendous damage to large parts of the galaxy before (again in an unspecified manner) the remaining ETIs managed to imprison it. The second attempt was benign, friendly and named 'Vanamond', and contact with it was made by a human youth 'Alvin' – both Alvin and Vanamond, of course, could travel at many times the speed of light with very little effort. Vanamond's 'job', at the end of time, would be to destroy the Mad Mind and be destroyed itself in its turn.

Hoyle's *The Black Cloud*[452] at least retains the speed of light limit,[453] but it involves a small (on an astronomical scale), dense, sentient and intelligent interstellar gas cloud, which decides that the Sun would be a good place to take a rest for a while. Unfortunately, this shields the Earth from much of the Sun's radiation, causing an instant ice-age and leading to the ruination and death of most terrestrial life-forms. Terrestrial scientists eventually realise that the cloud is sentient and that, actually, it is far *more* intelligent than Humankind. They then manage to communicate with it. Once the cloud realises that there is life on Earth (it is amazed that life-forms *can* develop on such an unlikely environment as a planet's surface, instead of occupying the nice empty spaces between them), it thins part of its structure to allow normal sunlight through a tunnel and so reach the Earth again; somewhat belatedly therefore, the 'US Cavalry' comes to the rescue again.

Unfortunately for both these ideas, the *actual* inability to travel faster than the speed of light is a major problem for their realisation in practice.

Clarke's scenario simply falls to pieces because FTL travel 🌀🌀🌀🌀🌀 is intrinsic to his plot and also, whilst there may be some very strange life-forms 'out there', a pure[454] intellect is literally "inconceivable".[455]

Hoyle's plot runs into rather more subtle objections. As humans, we are accustomed to 'Thinking Fast', and the electric pulses which this involves travel along nerves at a speed of about 50 m s⁻¹ (180 km/h, 110 mph). For most of humankind's purposes this is adequate.

But – have you ever tried to catch a fly with your hands as it goes past? I would bet that nine times out of ten, you failed. This is not because the fly's nerves transmit their electrical impulses any faster than your's do but because their pulses have a much shorter distance to travel. If the distance from a fly's eyes, via its brain to its legs and wings, is, say, 3 mm, then it can detect an incoming pair of

---

[451] 1975, Corgi. This is a re-write by Clarke of an earlier book, called *Against the Fall of Night*.

[452] 1970, Penguin.

[453] As, indeed, any respectable top astrophysicist should do. Hoyle (who died in 2001) was Plumian Professor of Astronomy and Experimental Philosophy at Cambridge (UK) and had earlier, together with Margaret and Geoffrey Burbidge and William Fowler, deciphered the nuclear fusion reactions which power the Sun and other stars. As an after-thought, he, Tommy Gold and Herman Bondi also came up with the Steady-State theory of the origin of the Universe.

[454] Clarke may have been thinking that it comprised only energy, or perhaps only forces, but in today's physics, energy and particles are just different aspects of the same basic entity and forces are mediated by the exchange of particles. For example, the interaction between atomic nuclei is called the Weak Force and involves the exchange between nuclei of the W and Z high mass intermediate vector bosons ®. Readers interested (or incredulous) about this topic should research 'Exchange Particles'.

[455] 'Inconceivable' literally means 'cannot be conceived', i.e., that the very *first* requirement of a new life-form *cannot* take place.

hands and take action to avoid them in about 100 µs. For your eyes to send an impulse indicating that they have seen a fly to your brain and then back to the muscles controlling your arms and hands is probably a distance of about 1 m, giving you a (minimum) reaction time of about 2 ms. No wonder that the fly wins most of the time.

Thus the 'speed of thought' depends upon both the speed of signals along nerves (or their equivalent) *and* on the distance which the signals have to travel. Many of you will doubtless remember in the infant school being told that large dinosaurs had a subsidiary brain at their rear end to control their tails, etc., because the signals from the brain in their head took too long to get there. This is actually a myth, but it illustrates well the problems which could be caused by the slow(ish) speed of signals along nerve fibres.

Now, Hoyle's Black Cloud presumably used e-m radiation beams in place of signals along nerve fibres for its internal 'thinking' signals, and most of the time,[456] any type of e-m radiation travels at the speed of light; 300,000 km/s. That sounds fast enough for anybody does it not? – but it is still *too slow* at times.

If dinosaurs, at a length of up to ~30 m, have now turned out not to need second brains after all, then there still must come a size at which nerves' slow transmission speeds *do* cause problems.

To be generous, let us make that problem size to be a length of ~1 km and call the resulting monstrosity a 'super-hyper-brachiosaurus'. A conventional nerve signal would take ~40 seconds to travel that distance there and back, so a hungry T Rex would have probably eaten several big chunks from the rear end of the super-hyper-brachiosaurus before a signal that it was happening and the order to move the tail could be effected.

What then, if the nerve signals were to travel at the speed of light? The super-hyper-brachiosaurus would be able to flick the annoying T Rex off into the far distance in just 6.7 µs.

But the speed of light is 'only' 6 million times faster than the signal speed along nerves. So, a 6 million km long super-super-hyper-brachiosaurus would still be left with a 40-second response to an attacking super-super-hyper-T Rex.

In practice, 'thinking' at human levels of intelligence and above probably requires maximum response times around 1 ms (i.e., the distance for a signal-along-a-nerve to travel about 50 mm). Thus, the Black Cloud, in order to have a similar thinking speed to ourselves and using speed-of-light signalling, must thus have a maximum brain size of just ~300 km and a maximum physical size (for adequate response times to threats) of ~3,000 km. Hoyle's Black Cloud, though, was at least large enough to surround the, i.e., Sun (i.e., $\geq$ ~1,400,000 km diameter), so its response speed can have been no faster than ~2 seconds and probably very much slower, i.e., hardly up to humankind's thinking speeds.

There are lots of 'ifs and buts' and other considerations not mentioned here, but in general the answer to the question 'are GMCs and/or larger or smaller interstellar gas clouds able to be individual ETs or ETIs?' is 'No'. In terms of thinking speeds, nebulous ETIs of around human levels might be possible if their sizes were less than about that of our Moon, but as a nebula, their C-factors would seem likely to be in the region of ^{0}✗ or worse – so, pretty dismal prospects all round for interstellar gas clouds actually being living entities.

The answer, however, to the question 'are GMCs and/or larger or smaller interstellar gas clouds able to *host* individual ETs or ETIs?' seems quite likely to be 'Possibly', 'Probably' or even 'Yes'. In other words, life-forms not too dissimilar to some of those on Earth might be able to live as permanent space-dwellers.

In fact, we have already considered the essentials of such life-forms under the headings of 'Panspermia' (Chapter 9) and PSR B1257+12 b, c, and d (above). The entities discussed therein survive (perhaps) in conditions not dissimilar to those which would be encountered by free-living

---

[456] In a refracting medium, light, etc., can travel more slowly, but *never* faster. Inside ordinary glass, for example, light moves at about 200,000 km/s (~2/3 c).

space-life. The main difference would be in the lack of protection from x-rays, γ rays and particle radiations.

Either, therefore, the space-life needs to evolve an immunity to x-rays, γ rays and particle radiations (don't ask me how) or they need to evolve shielding as a part of their structures.

In the former case, space-life might be very plentiful indeed throughout the whole Universe and be in the form of complex, self-reproducing molecules and/or small aggregates akin to unicellular terrestrial organisms. We have already seen (*Deinococcus radiodurans*, Chapter 12) that some significant resistance to radiation *can* be evolved. Perhaps a living panspermia colony on board a small interstellar body and so exposed to radiations for millions of years whilst travelling between the stars, might develop sufficient resistance to leave home and journey on alone. But – don't ask what it would do for food.

Evolving shielding might be much easier; see the tortoise-oids of TRAPPIST-1 d (Figure 13.5). However, the entity would then face the problem of getting into space and how it would then live in space. Perhaps some kind-hearted ETIs might help them with their evolution and then give them a hand to get into space?

When it comes to free-space-living-entities, my verdict would be 'I'll believe in them when I see (feel/hear/taste/smell) them'.

LIFE ON STARS

There isn't any.

For certain.

No life today, no life yesterday, no life tomorrow, no life, ever.[457]

On the other hand, once a star finishes generating its own energy, then just like PSO J318.5-22 (above) it will cool down and most will then eventually end up as Black Holes, Neutron Stars or White Dwarfs.

So, could there be life on/inside any of these candidates?

- Black holes – We will never know. One of those fascinating 'Wow!!, Gee Whizz!!, Oh I Say!!', aspects of black holes is that at their surfaces (their event horizons) time does not pass[458] as seen by an outside observer ®. So, even if we could see the surface (which we cannot), nothing would be seen to *do* anything and life and non-life would be quite indistinguishable.
- Neutron stars and white dwarfs. These can be considered together since, so far as habitability is concerned, neutron stars are just much more intense versions of white dwarfs. There are two major considerations with these objects: temperature and surface gravity.
  - Temperature simplifies the question of life on these objects; a white dwarf formed soon after the Big Bang will *still*, today, not have cooled sufficiently for any conceivable life-forms to exist on it; it will be around the Sun's surface temperature (5,700 K, 5,400 °C) or thereabouts. Estimates of how long it might take for a neutron star to cool down to similar temperatures range from a million times to a trillion times the current age of the Universe.[39] So, we do not really need to concern ourselves (yet) with the White-Dwarf-ians, or Neutron Star-ians.
  - However, if/when white dwarfs or neutron stars are cool enough for life-forms potentially to originate on them, then those life-forms will have to contend with those objects' surface gravities. These will not be the piffling few times the Earth's surface gravity which we have considered several times already, but ~100,000 g for a white dwarf and ~100,000,000,000 g for a neutron star ®.

---

[457] See the discussion under the Section on TOI- 4603 b above (but see also Box 9.4).

[458] There's insufficient room here to pursue this subject. Readers will find a more detailed treatment of it in the Author's *'Understanding Gravitational Waves'*, 2021, Springer, and via an internet search for 'Relativistic Time Dilation'.

An amoeba (Figure 9.3 (iii)) weighing about 0.000 25 μN (2.5 x 10⁻¹¹ kgf) on Earth would thus weigh some 25 N (2.5 kgf) on the surface of a neutron star and a 70 kg human on the neutron star would weigh the same as ~7,000 cubic kilometres (1,700 cubic miles) of water on the Earth ®.

Perhaps a more graphic comparison for this latter case, though, would be that the human on the surface of a neutron star would weigh one and a half times as much as *all* the water in Lake Michigan (Northern USA) does on Earth.

- Clearly neither the amoeba nor the human would be able to move at all if they lived on some neutron star in the very far distant future. So, not even the First-Living-Molecule moment could occur. Perhaps they might be seeded as panspermia by some antigravity-capable ETIs? – though I doubt it.
- The problems for life on a white dwarf (although it would be a MIR to FIR dwarf when cool enough for life) are a thousand times less than for neutron star life. But they are still pretty well insuperable. At least, though, some white dwarfs are largely formed from carbon. However, my vote still goes for them all being quite life-less.

THE SOLAR SYSTEM

By now, we have considered quite a range of possible and impossible places where ETs and ETIs might dwell, but all are outside the Solar System. Also, although a lot of the ETs and ETIs imagined above are inevitably based on terrestrial-life-form models, we have not considered whether/where actual terrestrial life-forms might find comfortable habitats away from the Earth.

We shall look at the possibilities of terrestrial life (in practice, that is to say, mainly humankind) developing colonies beyond the Solar System in Chapter 18. Here, then, we consider environments beyond the Earth, but within the Solar System, as possible hosts for terrestrial life and/or ETs and/or ETIs.

Homes for Terrestrials

We already know that given spacesuits and spacecraft, humans and a few other terrestrial life-forms (Chapter 9) can survive in space for periods up to a year or so and on the Moon for a few days. At the time of writing, there is a considerable amount of discussion about sending people to Mars, although that is probably still a decade away. Even on Mars, though, spacesuits will need to be worn the whole time when outside the landing spacecraft. However, let us be optimistic and follow Stuart Clark's opinion[459];

"… The first person to walk on Mars is almost certainly alive now …"

Apart from the Earth, is there anywhere else within the Solar System where any type of terrestrial life could exist *now* without spacesuits or equivalent levels of protection?

No.

Well … actually it is *really* only almost … 'No'.

In recent high-altitude balloon-borne experiments, samples of various algae, bacteria and fungi were exposed above the ozone layer directly to the Earth's stratosphere, where the environment closely resembles that on Mars. In particular, the UV radiation levels at that altitude are some 1,000 times higher than at sea level. Whilst many of the samples failed to survive, some algae did and spores of *Aspergillus niger* (black mould) were still viable and were revived later. The experimenters concluded that *Aspergillus niger* spores would be able to survive (temporarily at least) on the present Martian surface. Several experiments on board the ISS,[460] which exposed various micro-organisms to various levels of exposure to 'naked' space conditions, have showed that, if sheltered from the

---

[459] Private communication.
[460] International Space Station.

solar UV radiation, they would be likely to survive for at least a journey to Mars and probably on Mars itself. Laboratory simulations of Martian conditions have shown that several bacteria which are infectious to humans can survive those conditions for several weeks, with one, *Pseudomonas aeruginosa* (an anaerobic organism which can produce pneumonia and sepsis) perhaps even being able to multiply and thrive under those conditions.

Also, as we have seen above, there may be regions high in Venus' atmosphere, where at least the temperature conditions are survivable. However, with Venus' atmosphere being 96.5% carbon dioxide, only anaerobes could consider colonising that planet.

Thus, if humankind and most other terrestrial life-forms are to visit, explore, live-on, colonise, etc. objects within the Solar System, spacesuits and other protective measures are a must everywhere. Permanently occupied outposts may be possible within the domes so beloved of SF writers (Figure 13.1 Left). Inflatable habitats (think very large 'Bouncy Castles' or 'Zeppelins', etc.) are also popular with SF writers, who park their FTL, but cramped, spacecraft at a convenient spot in space and blow up their inflatable kitchens, bedrooms, showers, or whatever in order to take a comfortable break and stretch their legs for a few weeks, before resuming their FTL journey.

More realistically for the Moon and Mars,[461] at least, much larger habitats may be constructable, based upon existing caves, tunnels, lava tubes and the like to be found on/near the lunar and martian surfaces (Figure 13.13), by sealing them off to make pressurised living spaces.[462] A recent idea is that frozen water would need to be mined to support such habitats and that this could contain samples of material from all over the planet/moon in the form of dust particles. Analyses of these particles might then quickly reveal if there are any organic molecules to be found anywhere and perhaps whether or not they have been produced organically.

FIGURE 13.13   An example of a terrestrial lava tube. (Reproduced by kind permission of Pixabay and David Intersimone.)

---

[461] The Mars Reconnaissance Orbiter spacecraft has recently imaged a collapsed lava tunnel within Mars' Hephaestus Fossae region.

[462] A recent estimate places the number of currently identified caves found on Solar System bodies, other than the Earth, at ~3,500. Another recent estimate, based upon computer modelling of thermal observations of a large lunar pit (small deep holes, tens of metres deep, many of which are partially collapsed lava tubes), suggests that shaded areas of lunar pits may maintain almost constant temperatures throughout the whole lunar day and night of around 17 °C. Other researchers, though, suggest temperatures will be much lower; ~150 K (-150 °C). One of several concept-level space missions to explore lunar caves is given the name 'Daedalus' and might comprise a spherical probe which could move through the cave by rolling along the cave's floor.

More conventional-type houses might even be possible. Recent work has shown that simulated Moon dust and Mars dust when mixed with potato starch and salt produces a concrete (called 'Starcrete') having up to three times the strength of the ordinary concrete with which we are familiar – and just 1 kg of potato starch will produce 20 kg of Starcrete. Other investigations have shown that simulated Martian soil can, with the aid of a bacterium which converts urea $(CO(NH_2)_2)$[463] into calcium carbonate, plus a few other additives, be made into a slurry which is castable into complex shapes and which then sets hard over a few days.

So, bricks, mortar and the materials for making pre-cast concrete structures might be available locally with only a minimal number of imports having to come from the Earth. Perhaps even without imports; once greenhouses are going, potatoes can be grown on site and astronauts will always have to answer their urea-providing calls of nature.

On icy worlds, building construction could, perhaps, be even simpler; Igloos.

All the icy worlds have significantly lower gravities than does the Earth, and so quite big buildings (Figure 13.14) could be produced using blocks of water ice from the surface and by applying a coating of a sealant to pressure-proof them. Of course, they might be a bit cold inside, but reportedly, the inside of an igloo can reach a temperature of +16 °C – warmer than many houses once were in the days before central heating. Compacted snow is a very good insulator, so when the inside temperature rises above freezing point, only a very thin layer of the snow melts and then quickly refreezes into ice when it contacts the still very cold snow a little further inside the snow blocks. It then provides a skin on the inside to seal the structure even further.

FIGURE 13.14   Is this what the First Explorers of Mars or Europa will call home? (Reproduced by kind permission of Pixabay and diapicard.)

---

[463] Doubtless donated F.O.C. by the Mars colonists (think about; it especially next time you feel a call of nature).

Ice, as a building material, can be made much stronger and longer-lasting via the addition of quite small quantities of a suitable filler/binder. Experiments during the Second World War mixed water ice and sawdust in a ratio of 6:1 to produce 'Pykrete'[464] with a tensile strength three times that of concrete. Despite looking promising, Pykrete has never been used seriously on Earth because temperatures are not low enough. It may yet, though, come into its own on Mars (average temperature 210 K, -60 °C), Europa (average temperature ~80 K, -190 °C) and Titan (average temperature 180 K, -90 °C). The lunar samples obtained by the Chang'e-5 lunar contain natural glass fibres (probably produced during meteorite impacts) and these could be an ideal filler for Pykrete-type construction materials.

Pressurised and UV-proof greenhouses on the surface could provide some, if not all, of the food requirements for such outposts, and solar panels would provide for the inhabitants' energy supplies. Over most of the lunar surface, though, the ~340-hour long nights would require significant amounts of battery storage to be built. However, mountains close to the South pole of the Moon, called 'The Peaks of Eternal Light', are illuminated by the Sun almost permanently, and so outposts sited in that region would be able to have continuous energy supplies. Quite a number of experiments have now been conducted into farming/horticulture/gardening in space environments (mostly on board the International Space Station) and/or using simulated soil derived from materials available on the Moon or Mars within artificial habitats on the Earth. On the very small scales of these experiments, some successes have been achieved, but they are still a very long way short of being able to provide the long-term food requirements for even two or three people, never mind outposts of ten or twenty or colonies of a million. The Chang'e-4 lunar lander in 2019, for example, managed to germinate cotton and rape seeds, amongst other plants, on the lunar surface during the lunar daytime, but all these were killed-off during the lunar night when the temperature fell to 80 K (-190 °C).

Outposts, such as those just considered, will be exactly that, not colonies. They will be temporary settlements of a few to perhaps a few tens of humans for some specific purpose(s); research, industrial processes which cannot take place or are too dangerous to take place on Earth, perhaps also for some medical procedures and so on. Although they may exist for years, even tens of years and be partially self-sufficient, outposts will still depend on supplies from Earth for many of their essential requirements.

Mining, of various sorts, is much promoted as being an early pay-off for all the investments in space to date.[465] Although the word 'early' here seems inappropriate for something that is still likely to be decades away. For outposts, settlements and colonies on the Moon, Mars and other cold objects though, if they have water present at all, it will be in the form of ice and so will need to be mined in some way. Depending on the nature of the ice deposits and the amount of available energy, this may be possible by melting the ice with hot jets of gases/liquids. 'Real' mining, probably using AI-controlled machines, is generally envisaged for small asteroids where the weak gravitational field will make for easier working and for lower energy launches from the asteroid surface into space.

Getting the mined material down to the Earth, though, is likely still to require a lot of energy (= a lot of money). So, it is likely that mining operations will be linked to manufacturing operations in space. That way, only the high-value end products would actually be sent down to Earth. For example, a tonne of iron ore would cost the same to send to the Earth as a tonne of high-end mobile phones, the end value of the latter, though, is probably some 10,000 to 20,000 times the value of the former. Alternatively, some industrially essential and valuable elements such as Rhodium (market price per tonne at the time of writing; ~$140,000,000), Gold (~$70,000,000), Palladium

---

[464] Named for Geoffrey Pyke; spy, inventor and journalist.

[465] It is also *criticised* on the grounds that we should learn from our mistakes in wreaking havoc with the Earth's natural resources as though there were to be no tomorrow. It is by no means clear that the pitiful efforts currently underway to avert runaway climate change, even if (ever) fully implemented, will actually stabilise the situation adequately (but *PLEASE* do not stop).

Thus, with this possible new start, we should plan to utilise the resources in space entirely in a 100% efficient, recyclable way – *we* may need them still to be out there, when the Earth gets too hot for us.

(~\$30,000,000) and Platinum (~\$30,000,000) are likely to be worth shipping down to Earth in their fairly raw states.[466]

Colonies are another matter; they will have to be self-sufficient in almost all their needs. At what point an outpost becomes a colony is a debatable point, but for the purposes of this book we will take it to occur at the point when, if the Earth were to become uninhabitable, then the colony could continue to exist, expand and thrive, i.e., it could become a back-up for human and other terrestrial organisms' continuing survival.

Taking note that any colony surviving under the harsh and unforgiving conditions available away from the Earth will need people with some very advanced industrial, scientific and technological skills, the number of colonists needed for it to be viable must surely be at least a million and perhaps as many as ten million people.

At a very rough guess, therefore, we are looking a century or more into the future for outposts to become widely established and several centuries[467] for colonies to get going.

A much-touted alternative to transporting or building little bits of the Earth's environment out to/on other bodies within the Solar System or even within free-floating habitats in space is to convert the whole of an inconvenient body into an appropriate Earth-like living space (Chapter 18). This process is called 'Terraforming', and SF authors love it, without, usually, specifying how it can be done.

In fact, we are already in the process of terraforming the Earth, with climate change now well underway. Unfortunately, we've managed to get it wrong. The usual idea behind terraforming would be to make an environment *more* suitable for Humankind, not *less*.

Also, purpose-designed terraforming is not easy. There have been recent suggestions, for example, of intensifying the magnetic field of Mars so that it can develop a magnetosphere and so obtain some protection from the Solar charged particle radiations which currently bombard its surface ༄༄༄.

If Mars has a liquid interior (which we do not know) and if the circulating currents in that liquid move in the right sort of ways (which we do not know) and if we could get down to those liquids (which we cannot) and generate static electric charges on them (which we cannot), then this would certainly produce magnetic fields. But would it be worth the effort? – I doubt it greatly; just get the colonists to dig a few metres deeper down for their living quarters.

The alternatives of manufactured electric currents/magnetic fields on Mars' surface or in its atmosphere or in orbit around the planet would be just as impossible, if not more so, to accomplish, as setting up the magnetic fields internally.

Even if we managed to produce the right magnetic fields around Mars to provide protection from high energy charged particles, there would still be the solar UV, x-ray and γ ray radiations and Mars' carbon dioxide atmosphere and freezing temperatures to deal with. The latter might be addressed by positioning huge mirrors in orbit around Mars in order to reflect additional solar energy onto its surface (40 or so mirrors each with 2,000 km diameters, might do the job, they could raise Mars' equilibrium temperature to around 290 K (~20 °C[468])). The carbon dioxide might then have oxygen added to it, once the planet has warmed up a bit, by seeding it with photosynthesising plant life and waiting a million years.

A suggestion for 'terraforming' Venus, although it is really just a case of finding the best bit of the unmodified planet upon which to plant our colony, would be of 'land masses' (= balloons) floating high in Venus' atmosphere. The idea would be for nitrogen to be the lifting gas, and for the

---

[466] Of course, *finding* an asteroid containing a tonne of rhodium is probably the most difficult part of the whole process by far. Although to help you make a start, there is already an online catalogue of promising near-Earth exploitable asteroids available – see https://www.sciencedirect.com/science/article/abs/pii/S0032063322000496. It is called ECOCEL (Exploitation des Ressources des Corps Célestes).

[467] Generally, such futurological predictions are wildly wrong; either things happen *much* more quickly than expected, or take *much* longer, or things develop in quite a different manner from that expected. So, you have been warned.

[468] This would cause the frozen carbon dioxide at Mars poles to sublime into gas completely. However, it would only increase Mars atmospheric pressure by about 60% (to ~1% of the Earth's sea-level atmospheric pressure) and increase the current Martian greenhouse effect by ~ 15%.

'balloons' to be of some solid honeycomb construction. Presumably then linkable into larger units. The construction time is estimated at ~200 years.

No wonder the SF authors do not go into much detail on how their terraformed exoplanets have been terraformed.

When/if nuclear fusion reactions ⚛become a viable source of energy, then, with a plentiful source of energy, a good many presently impossible tasks may become possible. But current approaches to nuclear fusion need, not the common or garden (and plentiful) ordinary hydrogen nuclei but the much rarer heavy-hydrogen (deuterium) nuclei (Box 4.2) and the *very* rare second isotope of hydrogen with two neutrons, called tritium. So, whilst viable nuclear fusion power stations may alleviate Humankind's current energy supply problems somewhat, they are not going to give us the power to move planets yet (or even mountains).

If, fusion of ordinary hydrogen nuclei ⚛⚛⚛ can be achieved, though, *then* a lot of possibilities *will* open up. When four ordinary hydrogen nuclei fuse to form one ordinary helium nucleus, 26.72 MeV of energy is released.

Now that number will probably not mean much to many people, but to translate it a little; it means that when one kg of ordinary hydrogen fuses into slightly less than one kg of ordinary helium, 180,000,000 kWh is produced; sufficient to launch a 10,000-tonne spacecraft from the Earth's surface right out into the space beyond the Moon ®. *Then* we *will* be able to start thinking seriously of terraforming, although, still within some limits. Thus, moving Ceres to be conveniently near to the Earth, for example, would require the conversion of some 440,000 million tonnes of hydrogen into helium ®.[469]

Homes for ETs and ETIs

Homes for ETs and ETIs within the Solar System are now the last permutation which we need to consider, and you might immediately think that there's obviously none around because "we would know about them already". It is certain (Box 9.4) though that there is nothing anywhere, other than on the Earth, with large-scale agriculture, pollution-producing industries, significant heat sources, radio transmitters, spacecraft able to visit the Earth and the like anywhere within the Solar System.

That, however, is not quite the whole story. There are several possible scenarios to consider:

1. Micro-organisms and small-scale life-forms could easily have been missed so far and they could take the forms:
   1.1. Terrestrial-type life-forms which have
       1.1.1. Spread from the Earth (panspermia).
       1.1.2. Spread from elsewhere inside or outside the Solar System (panspermia).
       1.1.3. Evolved independently from terrestrial life.
   1.2. Non-terrestrial-type life-forms which have
       1.2.1. Evolved independently within the Solar System.
       1.2.2. Spread from outside the Solar System (panspermia).
2. Artefacts which remain to be found (e.g., remnants of spacecraft, deliberately placed constructs like the monoliths in *Sentinel from Eternity/2001: A Space Odyssey*[470] or traces left such as footprints, marks of landings by spacecraft, etc.).
3. ETIs are here now, but we cannot detect them for some reason and/or they are hiding from us for some reason.

Taking these possibilities in turn:

1.1 and 1.2 scenarios (above) – Many types of potential or non-potential Solar System habitats, such as cold and airless (Pluto)/very hot or cold, very large and with very thick atmosphere (Jupiter)

---

[469] The Sun, though, does this conversion every 12 minutes; so, … if we can start to harness the whole power of the Sun … the sky will not be the limit … we could then really start to organise the Solar System and even change nearby parts of the galaxy to suit our own convenience (see *Ringworld* – Chapter 18).

[470] Clarke A.C. 1951, *10 Story Fantasy* / Clarke A.C., Kubrick S. 1968, Metro-Goldwyn-Mayer.

and so on have already been discussed earlier amidst the discussions for exoplanets. There remain Mars (which has already had some intensive direct biological investigations by several space-craft landers – Table 13.5) and the postulated sub-surface oceans on Callisto, Enceladus, Europa, Ganymede, Pluto and Titan.

1.1 and 1.2 Scenarios – Mars

Mars is the one place in the Solar System where there is a high expectation amongst many people, scientists and the general public alike, of the possibility of life being there to be found. We can prob-ably mostly blame Schiaparelli's Martian 'Canali'[471] and Wells' fictional *The War of the Worlds*[472] for this (see also Box 15.1), although Edgar Rice Burroughs, Ray Bradbury and many other SF authors have contributed to the mythologies as well.

Clearly, Martian life-forms are not going to take the form of intelligent octupoids, mounted in tripod fighting machines and armed with heat rays, as in *The War of the Worlds*. If there is any native form of life on Mars, then we know enough already from over five decades of Martian space probes that it must be at the microbe level.

Mars does, though, have several valid scientific reasons for being the first place to look for life away from the Earth:

- There is clear evidence that in the past ($\geq$ 3,500 million years ago), Mars has had lakes and riv-ers of liquid water (Figure 13.15 Left). Around that same time, the climate may also have been of the current terrestrial cyclical wet/dry nature and so conducive to supporting life-forms.
- Mars still has water, in the form of ice on its surface and in the form of a salty liquid in several large sub-surface lakes near the South Pole.
- There is evidence (sulphate salt deposits) of the existence of glaciers even in the Martian equatorial regions until at least ~3,000 million years ago.
- Mars was warmer and had a thicker atmosphere (but still mostly carbon dioxide) in the past. Hydrogen may have provided a greenhouse effect.
- Various space missions to Mars (Table 13.5) with both static (lander) and mobile (rover) components have detected some basic organic compounds (Box 9.1) such as benzene, butene and propane. Organic carbon, as a whole, comprises ~200 to ~300 ppm of the sam-ples from the Curiosity rover (Table 13.5); comparable with the levels in some terrestrial deserts. Furthermore, nanopore DNA sequencing techniques can now operate using just 2 picograms (2×10^{-15} kg) of DNA so that terrestrial contamination may soon be distinguish-able from native Martian biotic material (if any).
- Volcanoes (Figure 13.15 Right) abound, suggesting the existence of heat sources below Mars' surface and a supply of biologically useful elements and compounds. They were active until about 500 million years ago, so there might be remnants of hot-spring-type and other micro-organisms (Figure 12.1) still to be found. Recently, a fault region on Mars, named Cerberus Fossae, has been imaged and may show signs of explosive volcanism hav-ing occurred within the last 50,000 years.
- It is much easier to send spacecraft to Mars than to any of the other possibilities (see below).

---

[471] Giovanni Schiaparelli observed Mars in 1877 and thought that he had observed numerous gently curving or linear narrow features. He called these, in Italian 'Canali', meaning 'channels' or 'gullies', i.e., natural features. The mis-translation of this into English as 'canals' brought in the unintended implication that these were artificial constructs (i.e., built by Martians).

 This mistake was greatly aggravated by Percival Lowell who spent decades and a great deal of time and money map-ping thousands of claimed 'canals' on Mars at around the turn of the 19th to the 20th centuries. Schiaparelli's, Lowell's and many other's efforts were completely wasted however, because what they were actually seeing were illusions aris-ing from imperfections in human vision; a few barely resolved point features frequently being seen as an apparently linear feature by the eye-brain combination.

[472] Wells H.G. 1897. Pearson's Magazine.

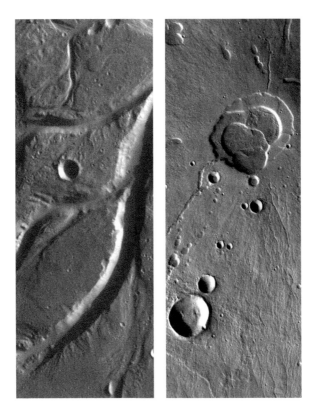

FIGURE 13.15 Left – Dried-up Martian River Valley. Called Osuga Valles, it has a total length of 164 km and is up to 20 km wide and 0.9 km deep in places (for terrestrial comparisons, the Amazon has a maximum width of 11 km, the Congo River has a maximum depth of 0.2 km, the Nile has a length of 6,650 km). (Reproduced by kind permission of NASA/JPL-Caltech/ASU.) Right – The Hecates Tholus Martian Volcano. The caldera (the volcanic crater) is the complex of nested craters in the top half of the image. Calderas are formed when the eruption ceases, leaving a space into which the remaining surface layers collapse. The caldera's appearance is therefore quite different from that of a meteorite impact crater, many of which are also visible within the image. Several (at least five) eruption-collapse sequences must have occurred in order to have formed the nested set of craters forming the caldera. Martian craters are formed from a very liquid basaltic lava and therefore form low mounds rather than the popular sharp peaks concept such as Mount Fujiyama. (Public Domain – NASA.)

On the other hand, Mars' surface is exposed to direct solar ionising radiation, toxic perchlorate compounds[473] are abundant there, the temperature averages 210 K (-65 °C) and the current atmosphere is mostly carbon dioxide.

Any current Martian life-forms are thus likely to be subterranean, living metres to kilometres below the surface; some terrestrial chemoautotrophic microbes (Chapter 7) could probably survive in this manner even on Mars in its current state. The best which *we* can hope for, though, seems to be that space probes might detect the gaseous or other by-products of such life-forms, which filter up to the surface and/or that remnants of ancient life might still be there to be found (Figure 13.6).

---

[473] Compounds based upon a chlorine atom and four oxygen atoms. Perchlorates can affect the thyroid gland in humans and they become bactericides in the presence of the types of UV radiation reaching the Martian surface. Some microorganisms, though, can utilise them in their metabolisms by chemically reducing them to chloride, producing oxygen in the process.

Under terrestrial conditions, mutations of rice have been found which will still grow in soils with perchlorate concentrations of 0.1% by mass, however many Martian soils have concentrations which are ten times higher than this level.

But, who knows? – with a bit more genetic manipulation, future Martian explorers might soon be working in Martian rice fields (if only they can get enough water).

TABLE 13.5

Significant[474] Martian Space Probes

Mission status	Active time on Mars	Instruments carried relevant to searching for life	Results/Notes	Result (Life/No Life)
CONCLUDED				
Viking 1 (Lander) (Figure 13.16 Left) plus Viking 2 (Lander)	20th July 1976 to 11th November 1982. 3rd September 1976 to 11th April 1980	Biological Laboratory Gas Chromatograph/ Mass Spectrometer[302] (GCMS) Imaging System (2 facsimile cameras) Remote Sampler Arm	The imaging systems failed to photograph any macroscopic Martians. This comment applies to all missions to Mars to date and so will not be repeated each time. The remote sampler arm served to feed samples to the more active instruments. Again, this comment again applies to all missions to Mars to date and so will not be repeated each time. The GCMS can identify many chemicals when they are in a gaseous form. It found no evidence of organic molecules. It did detect chloromethane and dichloromethane probably derived from perchlorate (see above and Phoenix, below) when the samples were heated to provide gases for the GCMS. Biological laboratory – Gas exchange; Martian soil samples had various nutrients added to them, sometimes also with water and evidence sought for the emission of, *inter alia*, hydrogen, oxygen, methane, etc. resulting from microbiological activity. Biological laboratory – Labelled release; Nutrients labelled with radioactive carbon were added to Martian soil samples and evidence sought that metabolisation had occurred. Biological laboratory – Pyrolytic release; Radioactive-carbon-labelled carbon monoxide and carbon dioxide used to seek evidence of photosynthesis via a series of exposures to the gases and then sterilisation at 650 °C. No positive result.	No Life
Spirit (Rover) plus Opportunity	4th January 2004 to 22nd March 2010 25th January 2004 to 10th June 2018	Microscopic Imager	Neither rover had, as a primary aim, the detection of life, although the study of evidence for the past presence of water was included. They did carry microscopes which could have imaged small life-forms/ micro-fossils, etc. but none of these were found.	No Life
Phoenix (Lander)	25th May 2008 to 2nd November 2008	Mass spectrometer and furnace Wet chemistry laboratory	One of the scientific objectives was to assess the habitability of the surface layers. During this analysis, Phoenix discovered the presence of perchlorates (see above) in the Martian soil. The wet chemistry laboratory assessed the compatibility of the soil for possibly Martian life and for terrestrial life and gave the soil a pH of ~7.7 (slightly alkaline; asparagus and peas should grow well on Mars, when the temperature has been raised to about 20 °C).	No Life

(Continued)

[474] In the respect that they were searching in some way for evidence of life now or in the past on Mars.

TABLE 13.5 CONTINUED
Significanta Martian Space Probes

Mission status	Active time on Mars	Instruments carried relevant to searching for life	Results/Notes	Result (Life/No Life)
Zhurong (Rover)	14th May 2021 to 20th May 2022	Sub-surface radar	The radar has a penetration depth of ~100 m, but no evidence of water or ice was found.	No Life
ON-GOING				
Curiosity (Rover).	6th August 2012 to present day	Gas chromatograph Laser spectrometer[475] Mass spectrometer	Biological investigations are one of the three primary missions for Curiosity, although it is not able to detect life directly. It has shown that the conditions on the early Mars might well have been suitable for microbial life, and it has searched for the preserved chemical signatures/decay products of such life, although so far without success. By drilling into the mudstone (a sedimentary rock) near its landing site in the Gale crater, the organic molecules, benzene, butene, propane, thiophene and toluene have been detected plus chlorine-bearing organic molecules and atmospheric methane.[476] The atmospheric methane has shown a seasonal variation in its abundance and the possibility that this could have a biological origin cannot yet be completely eliminated.	No Life so far.
Perseverance (Rover) (Figure 13.16 Right)	18th February 2021 to present day	Helicopter Mars Oxygen ISRU Experiment (MOXIE) Sampling Sub-surface radar UV laser spectrometer	Several of Perseverance's mission objectives are life/habitability oriented. The helicopter, 'Ingenuity', in addition to testing flying techniques and route planning for the rover was able to look for life-related artefacts and even macroscopic life over much wider areas than can be covered by the rover (no success though). After exceeding its original mission objectives many times over, Ingenuity has now ceased to operate following some damage to its rotor blades during a crash landing. MOXIE is testing for the feasibility of generating oxygen from carbon dioxide (successful so far) for future manned missions. Samples of apparently promising areas of Mars' surface are being obtained and stashed for future retrieval and return to Earth by a sample retrieval mission (in 2029?), comprising a Mars lander, an ascent vehicle to take the samples to a spacecraft in Martian orbit which will then bring the samples to an Earth orbit and then a re-entry vehicle to bring the samples down, through the atmosphere, to the Earth's surface[477] for *really* detailed bio-chemical analyses (some samples obtained and stashed, so far). The sub-surface radar looks down to 10 metres depth for, *inter alia*, water ice and salty water (results, so far, inconclusive). The UV laser spectrometer can detect organic and mineral substances at sample sizes under a millimetre (results, so far, inconclusive).	No Life so far.

(Continued)

[475] Based upon Raman spectroscopy. See the author's books; *Astrophysical Techniques* 7th Edn., CRC Press, 2020. *Optical Astronomical Spectroscopy*, Institute of Physics, 1995 and *Remote and Robotic Investigations of the Solar System*, CRC Press, 2018 or use an internet search for 'Raman Spectroscopy'.

[476] Also detected by various Mars orbiter spacecraft

[477] Precautions are already being planned against contamination of the samples by terrestrial sources *and* contamination of the Earth by the Martian material (see also Chapters 9 and 11). A sea landing for the returning capsule has therefore been ruled-out (in case it should sink). However, aerobraking only (i.e., no parachutes) *is* planned for the landing in Utah, which will mean that the landing impact will be much harder than normal. Still; I suppose they know what they are doing?

TABLE 13.5 CONTINUED
Significanta Martian Space Probes

Mission status	Active time on Mars	Instruments carried relevant to searching for life	Results/Notes	Result (Life/No Life)
PLANNED				
Martian Moons Exploration (Lander and Sample return to Earth)	2024		A mission to the Martian Moons; Phobos and Deimos. Not intended for the investigation of life-forms (but ... who knows).	Not yet applicable
Rosalind Franklin – ExoMars (Rover)	2028? (originally planned for 2020)	Gas Chromatograph/Mass Spectrometer for identifying organic molecules; Laser spectrometer; Neutron spectrometer[478] for the detection of ice and water-containing minerals; Sub-surface radar	The mission's aim is to search for signs of life having been on Mars in the past.	Not yet applicable
Tianwen-3 (Lander and Sample return to Earth)	2028	No details currently available	The mission's aim would be to return samples of the Martian soil to Earth for detailed bio-chemical analysis	Not yet applicable
Sample return mission for Perseverance's stashed samples	2030?	Will probably employ two upgraded versions of the Ingenuity helicopter for the collections	To collect Perseverance's Martian samples and return them to the Earth	
OTHER				
Mars-Grunt (Lander and Sample return to	?	?	?	Not yet applicable
Manned mission(s)	2030 onwards?	?	Various, very speculative concept proposals abound at the time of writing	Not yet applicable

[478] See the author's book: *Remote and Robotic Investigations of the Solar System*, CRC Press, 2018, or use an internet search for 'Neutron spectroscopy'.

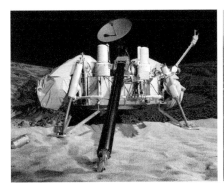

FIGURE 13.16 Left – A replica of the (twin) Viking landers which was used on Earth as a test bed and for fault evaluation whilst the actual landers were operating on Mars. (Reproduced by kind permission of NASA/ JPL-Caltech/University of Arizona.) Right – Perseverance (the rover) and Ingenuity (the helicopter) operating on the surface of Mars. Part of Perseverance is visible to the right and Ingenuity is on the Martian surface just left of centre. (Reproduced by kind permission of NASA/JPL-Caltech/MSSS.)

As we may see from the above discussions and from Table 13.5, a great deal of effort has already been expended on the search for life on Mars without any positive results to date.[479] In the next decade such investigations will continue and be intensified and Mars will remain for at least the next two decades the primary target for such searches (see below).

However, if life is on Mars, it may have been missed and it may continue to be missed by the present and near future Martian landers and rovers. Some recent work has shown that the types of instruments so far sent to Mars, or which are planned to be sent soon, were unable to detect remnants of micro-organisms in samples from a Mars analogue site in the Atacama Desert on the Earth. Those remnants could be detected, though, by state-of-the-art equipment in terrestrial laboratories. So, the detection of Martian life remnants may have to await the return of samples to the Earth, but then, with very few samples to work on the life evidence may *literally* be missed, i.e., the samples being taken from the wrong spots.

For our first Martians, whether originating from inside or outside the Solar System, we will therefore just have to be very patient and 'Wait and See'.

1.1 and 1.2 Scenarios – The Icy Moons and Other Small Bodies

The outer radius of the habitable zone for the Sun (see above) is ~ 1.7 AU, so Mars, with a semi-major axis of 1.5 AU lies well within it. The bodies which we are now about to consider, Enceladus (Figure 13.17 Left), Europa (Figure 13.17 Right), Pluto (Figure 9.6), Titan, Triton and several more, all lie well outside the Sun's habitable zone, so why do we bother to include them here at all?

The reason for the bothering is they may all possess volumes of liquids (water or methane, puddle-size to ocean-size) sheltered below a thick, icy surface layer and warmed by internal energy sources.

There are several lines of evidence for the presence of sub-surface liquids on various Solar System bodies and the main one is volcanic activity. On the Earth, volcanos of various types and other volcanic-related activities such as lava flows, hot springs, geysers and the like are found in many regions (Figure 13.18 Left). Evidence for such activity in the past, but which is now extinct, is also widespread (Figure 13.18 Right). Such terrestrial volcanic activity is due to molten rock (lava) and most of the interior of the Earth is still in a liquid state from the primaeval energy left over from its formation.

---

[479] We have, though, learned a very great deal about how remotely to operate very complex machinery and systems – which has a huge pay-back via terrestrial and commercial space applications.

 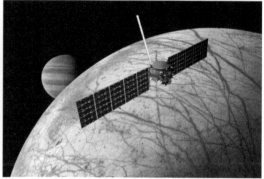

FIGURE 13.17 Left – A plume of water vapour emitted from Enceladus (at the top, the very narrow illumi-nated crescent of the rest of the moon can also be seen). Enceladus was imaged many times during the Cassini mission to Saturn (Chapter 9) and was found to be emitting plumes of gases, mainly comprising water vapour, but also including hydrogen and other volatiles together with small ice particles. (Reproduced by kind per-mission of NASA/JPL/Space Science Institute.) Right – Europa. A photo-montage/artist's impression of the Europa Clipper spacecraft (due for launch in 2024) close to Europa with Jupiter in the distance. (Reproduced by kind permission of NASA/JPL-Caltech.)

Evidence of extinct lava volcanism (Figure 13.15 Right) or of volcanism which has now been dormant for a *very* long time may also be seen on Mars, Mercury and our Moon. Active lava vol-canism still occurs on Io and Venus, and there is at least one claim that Mars' Elysium Planitia had a major lava eruption only a million years ago, whilst Martian seismic evidence shows that there is still sub-surface activity occurring today.

Io shows 'normal' (i.e., molten rock) volcanism, although water and ice may be mixed into it. The reason for this is that the tidal heating of Io is intense due to it being close to Jupiter and having strong orbital resonances with Europa and Ganymede which keep it in an elliptical orbit. Io is thus thought to have a molten core of iron and/or iron sulphide and has some 400 active volcanoes emit-ting sulphur dioxide and sulphur.

Venus, like the Earth, has a molten interior due to stored primaeval heat. Radar-equipped space-craft such as Magellan (1990 to 1994) have mapped its entire solid surface (which is totally hid-den to visible wavelength observation beneath thick clouds) and shown that volcanism on Venus is extremely extensive. Originally it was thought that Venus' volcanoes were all extinct, but recently some 37 which are still active have been identified. For example, in 1991, during the Magellan spacecraft mission, the volcanic vent on the 8-km high Maat Mons was seen to double its size within eight months and perhaps to have become filled with liquid lava.

Lava volcanism is not intrinsically favourable to life, unless such life can exist at ~1,000 K (~700 °C). However, it may make elements and compounds available which are essential to life and its origins (Chapter 9). Volcanism, though, is not restricted to molten rock. Any liquid from the interior of a solid body, which makes its way to that body's surface where the temperature is below that liquid's melting point, will lead to similar phenomena and surface features. In particular, water may behave in this way when it erupts or otherwise appears on the surface of cold bodies. Since the study of materials at very low temperatures is a science known as Cryogenics, such water-to-ice eruptions are termed 'Cryovolcanoes' and the phenomenon, 'Cryovolcanism'.[480]

We have already seen (Chapter 9) that Titan has lakes of methane and that it may have larger bodies of the liquid trapped below its solid surface. Its surface temperature is around 94 K (-180 °C).

---

[480] At the opposite extreme, exoplanets with a high proportion of iron, nickel and other metals might exhibit ferrovolca-nism; volcanism based upon molten metal. No serious candidates for this process have been found, though (yet).

FIGURE 13.18 Left – Mount St Helens, Washington state, USA. It is shown here with evidence (a steam plume) of mild activity. But in 1980 it exploded with an energy of ~24 megatons of TNT, removing 400-m from the top of the old mountain. (Reproduced by kind permission of Pixabay and WikiImages.) Right – Arthur's seat, Edinburgh, Scotland. The much-eroded core of a volcano which was last active about 340 million years ago. The 'Arthur' involved was King Arthur of Round Table fame. (Reproduced by kind permission of Pixabay and James Glen.)

The Cassini radar data has shown that there is at least one area on Titan, named Sotra patera,[481] where the features resemble terrestrial volcanic features quite closely (Figure 13.19). In one respect though it is very different from terrestrial volcanoes; its molten 'rock' was actually liquid water, though probably highly contaminated with hydrocarbons, i.e., it is a (probable) cryovolcano.

FIGURE 13.19   A false-colour radar image of Sotra patera on Titan. Sotra patera itself is the deep depression left of centre. The heights are magnified in this image by a factor of 10. In a normal view though, the depression resembles a volcanic caldera (Figure 13.5). It is about 1,700 m deep and some 30 km in diameter. To the right of Sora patera is the 1,450-metre high and possibly volcanic mountain, Doom Mons.[482] Linear features, which may be up to 300 m thick, may result from solidified flowing liquids ('lava' flows; perhaps, though, we should be calling them 'Glaciers') can be seen on its sides. (Reproduced by kind permission of NASA/JPL-Caltech/USGS/University of Arizona.)

Assuming that the Sotra patera is a cryovolcano (and that is not quite certain yet), then it implies a large supply of relatively warm water under considerable pressure somewhere below the surface, i.e., an ocean. Supporting evidence for this comes from the observed larger-than-expected tidal bulges on Titan and an apparent shift in the positions by as much as 30 km of some of Titan's surface features throughout the period of Cassini's observations. This latter observation suggests that the ocean might cover much, perhaps all, of a layer below Titan's visible surface so that the surface layer can move around relatively easily with respect to the interior of the moon.

The energy to warm such a huge volume of water may come from Titan's core whose temperature some put as high as 1,400 K (1,100 °C), together with a contribution from the aforementioned tides and perhaps a bit of help from an atmospheric greenhouse effect.

Thus, we have some strong circumstantial evidence for a sub-surface ocean on Titan.

Then we come to Enceladus (Figure 13.17 Left) where Cassini both observed and sampled jets (plumes/geysers) of material being emitted from the moon's surface. The jets comprise water vapour together with other gases and as can be seen from the height which was reached above Enceladus' surface (Figure 13.17 Left), that it, too, must be under some pressure. Enceladus also has cryovolcanoes which are still emitting gases near its South pole and both ammonia and phosphorous have been detected in Enceladus' emissions. The heating mechanism for the interior of the moon seems likely to be tidal, since it is in an elliptical orbit and has an orbital period resonance with Dione which maintains that ellipticity. A sub-surface ocean up to 10 km in thickness is thought to exist

---

[481] Sotra after a large Norwegian island just West of Bergen and patera from the Latin for 'bowl'.
[482] Named for 'Mount Doom' in J.R.R. Tolkien's *Lord of the Rings'*.

near the South Pole. Conditions within/on some parts of Enceladus are such that some terrestrial extremophiles (Chapter 12) might be able to exist. However, suggestions that the methane found in the plumes might originate from the ET analogues of such organisms are a long way over-the-top for our present levels of knowledge about the moon.

Extinct, dormant or active cryovolcanism and/or other evidence suggests that sub-surface oceans might also be found on:

(i)   Ariel (Slight indications of cryovolcanic activity in the distant past)
(ii)   Callisto (Magnetometer results suggest the possibility of an electrically conducting sub-surface ocean).
(iii) Ceres (A possible cryovolcano, the 4.1-km high 'Ahuna Mons')
(iv)   Charon (Cryovolcanic processes *may* explain why much of Charon's ice is crystalline, instead of amorphous as conditions there would normally suggest)
(v)   Dione (Slight indications of cryovolcanic activity in the distant past)
(vi)   Europa (Probably currently active emission of plumes of water vapour and possibly icy 'slush' just below the surface in some places. Water vapour atmosphere. Magnetometer results suggest the possibility of an electrically conducting sub-surface ocean)
(vii) Ganymede (Magnetometer results suggest the possibility of an electrically conducting sub-surface ocean. Slight indications of cryovolcanic activity in the distant past. Water vapour atmosphere.)
(viii) Miranda (Slight indications of cryovolcanic activity in the distant past)
(ix)   Mimas (Computer modelling of libration)
(x)   Oberon (Slight indications of cryovolcanic activity in the distant past)
(xi)   Pluto (Possible cryovolcanoes; 'Wright mons' and 'Picard mons' which are ~6 km high and ~150 km in diameter. A sub-surface ocean is possible due to internal heating by radioactive elements)
(xii) Quaoar (Cryovolcanic processes *may* explain why some of Quaoar's ice is crystalline, instead of amorphous as conditions there would normally suggest)
(xiii) Tethys (Slight indications of cryovolcanic activity in the distant past)
(xiv) Titania (Slight indications of cryovolcanic activity in the distant past)
(xv)   Triton (Nitrogen gas plumes. 'Leviathan Patera' a very large, possibly cryovolcanic caldera. Possible flood plains of frozen water. 'Cantaloupe terrain' (intersecting ridges and valleys, etc.) which could be due to a wholesale break-up of Triton's icy surface at some time).
(xvi) Umbriel (Recent computer modelling).

Some of these listed objects may be eliminated or firmed-up as sub-surface-ocean possibles, when more spacecraft visit more Solar System objects[483] and/or as old data archives are re-analysed in the light of new understandings. Over time, other bodies will also doubtless be added to the list.

OK, so what's so great about a sub-surface ocean anyway? They're probably full of nasty chemicals and colder than charity.

You are probably right about the nasty chemicals, except, their true name is 'pre-biotic molecules', i.e., the feed-stock for First-Living-Molecules or the nourishment for life-forms which have passed their First-Living-Molecule moments. Also, they cannot be *too* cold[484] if the water is liquid. So, they are the best substitutes for 'warm little ponds' which we are likely to encounter anywhere within the outer reaches of the Solar System.

---

[483] One proposed mission planned to launch in 2028 is the MBR (after Mohammed bin Rashid Al Maktoum). It is hoped to visit seven asteroids with the aim *inter alia* of looking for organic compounds and water. A lander is included in the mission which will examine the last of the asteroids to be visited; Justitia, in 2034.

[484] 251 K (-22 °C) is the lowest temperature which pure water can reach and remain liquid, though lower temperatures are possible for liquid water with other chemicals dissolved in it. See also 'cryophiles', Chapter 12.

It is the hope that life-forms could exist on/in/within some of these outer bodies that is currently causing excitement amongst a good many exobiologists and why a lot of brain power is now being devoted to ways in which we might discover it, if it's there.

In the immediate future two spacecraft are scheduled to investigate, *inter alia*, the life possibilities on some of the Jovian satellites.

JUICE (Jupiter Icy Moons Explorer) is an ESA mission which aims to study Callisto, Europa and Ganymede. Launched in 2023, it will arrive at Jupiter in 2031 and spend three years remotely investigating the moons. The mission aims include investigation of the sub-surface oceans, detailed mapping and studies of the surface layers' features, physical properties and compositions and an ice-penetrating radar to study the layers immediately below the surface.

NASA's Europa Clipper (Figure 13.17 Right), currently hoping for a 2024 launch, is expected to reach Jupiter in 2030 and to perform 45 flybys of Europa. Its aims are to study the icy surface, the sub-surface ocean, surface features especially those which are or have been recently active and chemical compositions. It is hoped that its ice-penetrating radar will reach down to the ocean itself.

Further into the future are the Dragonfly and Tianwen-4 missions. Dragonfly may launch in 2027 carrying a mass spectrometer, to reach Titan in 2034. The lander will be a helicopter (the Dragonfly) capable of 8-km-long flights and with the aim of assessing Titan's habitability for microscope-level life-forms and of searching for pre-biotic compounds. Tianwen-4 is another Jupiter mission with a launch date around 2029 and possibly intended to be in orbit around Callisto by 2038. Further details of the programme for the spacecraft are still uncertain.

Following on from Tianwen-4, at the moment, there is nothing beyond possible concept ideas – but there are plenty of those. Most include some attempt to sample/penetrate down to an ocean. Most ideas for *that* procedure involve melting a pathway down through the ice. The energy requirements are thus likely to be large by the present-day standards for outer Solar System probes.

One proposal suggests using the recently invented lattice confinement fusion[485] process to provide sufficient power. Alternatively, the vents emitting gas jets (geysers) may provide easier routes to the bodies' interiors and perhaps even down to their oceans. The double ridge lines found on Europa and on some other icy bodies may, by analogy with similar features on Greenland, also be indicators of easier routes to the sub-surfaced oceans and/or of liquid water bodies much nearer to the surface.

A prototype of a worm-like robot, EELS,[486] capable of operating independently of real-time human control and driven by rotating screws is currently being tested for these and other difficult exploration missions. A related concept, SWIM,[487] would use a small narrow probe to melt through the ice, which would then release a horde of small (~0.1 m) submarine robots to explore the ocean itself.

For any/all of these ideas, it is not just the details which remain to be decided, it is entire groups of projects. One thing that is certain, though, is that for any of these missions, where the prospect of finding life which has not come from Earth is not quite of zero probability, the sterilisations of every cubic micron of the equipment must be orders of magnitude better than anything achieved to date.

2 – Artefacts

This line of evidence for ETIs really does not need much said about it except that the artefacts may be quite difficult to identify as not being of natural origin or even to be noticed at all.[488]

---

[485] Yes – the word there *is* 'Fusion'; and No, we are not about to enter the era of free unlimited energy for everyone. Lattice confinement fusion (see; *Nuclear fusion reactions in deuterated metals*, Pines V. *et al*. Phys. Rev. C., 044609, **101**, 2020) traps deuterium (heavy hydrogen) nuclei inside crystals of (for example) erbium. Irradiating those crystals with γ rays then leads to fusion of the deuterium nuclei.

[486] Exobiology Extant Life Surveyor

[487] Sensing With Independent Micro-swimmers

[488] The recently announced Galileo Project (https://projects.iq.harvard.edu/galileo/home) has the aim of monitoring the Earth's atmosphere and beyond for ETI artefacts ranging from UFOs to objects entering the Solar System from interstellar space (like 'Oumuamua' – Chapter 8).

If the artefact has been left deliberately (c.f., Clarke's Sentinel – see earlier[470]), then it is likely to be obvious, as also will be crashed/abandoned spacecraft, machines/equipment, buildings and so on. Though one wonders if even these types of artefacts will still be recognisable after a billion years. On Solar System bodies with atmospheres and/or active geology, the survival times could be very much less. It is doubtful if any of Humankind's Venus landers will exist in a century, for example. Furthermore, artefacts, such as old spacecraft, could well have been abandoned in space and then their detection would even more chancy.

Artefacts such as foot-prints (or tentacle-prints), vehicle tracks, marks left by landing spacecraft, etc., will have lifetimes which depend upon their depths, the firmness of the ground and the planetary environment. Thus, Neil Armstrong's and Buzz Aldrin's footprints on the Moon (Figure 13.20) will probably be imperceptible in 100,000 to 1,000,000 years from now due to disturbance[489] of the Lunar regolith caused by the solar wind, cosmic rays and micrometeorite impacts.

FIGURE 13.20 Armstrong and Aldrin's footprints – Gone in less than a million years? (Reproduced by kind permission of NASA.)

---

[489] Often called 'Regolith Gardening' or 'Lunar Gardening'. Estimates of effectiveness of regolith gardening vary considerably and also, the rate of disturbance decreases rapidly with depth within the regolith. Thus, some recent estimates suggest that complete turn-over/re-mixing/stirring-up will take about; 10,000 years to a depth of 0.5 mm and between 80,000 and 10,000,000 years to a depth of 10 mm.

Recognition lifetimes for artefacts such as pollutants, unusual elements and/or chemical compounds, radioactivity and the like could range from minutes to billions of years. Whilst modifications to a Solar System object by, say, mining would probably be detectable on Venus for less than a century, but for in excess of many millions of years if they were on somewhere like Jupiter's small, atmosphere-less and geologically inactive moon, Amalthea.

All-in-all, the detection of ETIs from their artefacts seems about as probable as winning a National Lottery ten times in a row.

3 – ETIs Are Here Now

There is less to say about this possibility than about ETI artefacts. If ETIs are here now and hiding, then we probably have no means of finding them until they do reveal themselves.

The exception to this would be for microscopic life-forms which have evolved within the Solar System elsewhere than on the Earth and/or for similar organisms arriving here via panspermia (Chapter 9). Such organisms would not be hiding deliberately and so would be detectable by the methods and approaches discussed earlier.

Theme 3

Suppose There Really Are Aliens Somewhere 'Out There'
What Do We Do About It?

PAUSE FOR THOUGHT

Themes 1 (Chapters 3–8) and 2 (Chapters 9–13) have taken us on a long and sometimes tortuous, but I hope also an informative and interesting exploration of some well-known and some less-well-known aspects of our local 'bit' of the Universe.

Our journey started by looking at what we currently know about the thousands of far-flung planets beyond the edge of the Solar System and the environments which they might provide. It then continued by examining how terrestrial-type life-forms might have originated and by speculating how similar processes might operate for those exoplanetary and other possible environments. Finally (and very speculatively) some simple-minded ideas for the forms that some of those ExtraTerrestrials and ExtraTerrestrial Intelligences might take and how they might live on their Exoplanets have been fancifully suggested.

You should by now be able to evaluate news items, scientific discoveries, wildly unlikely SF books and films and highly optimistic 'scientific concept proposals' for further work in this area for what they are – and, when (not 'if') a 'Real Result' comes through, you should be able to recognise that for what it is and to separate the megatonnes of dross which such an event will generate from the nuggets of true gold forming the actual 'Real Result'.

The real result will happen when someone somewhere and somewhen finds some convincing line of evidence that life unrelated to terrestrial life *does* exist somewhere within the Universe.

We have, in the earlier chapters, looked at many aspects of where, when and how ETs might potentially develop and exist. So far, though, we have been essentially looking for actual ETs or bits of them or traces of them as fossils or even 'making our own' and so forth.

A major scientific effort though, often somewhat outside 'mainstream' science (especially as regards funding), is currently underway and has been for some decades. This effort is to find ETs, or more probably ETIs from the evidence of *their* own activities. In particular, this has meant

DOI: 10.1201/9781003246459-16

searching for the ETIs' radio and/or other electromagnetic radiation emissions coming to us from the nearby or the far reaches of the Universe.

That brings us to this book's third theme, the

Search for ExtraTerrestrial Intelligences

or

SETI[490]

and in the next few chapters we will look into that and its possibilities.

We will also look at other possibilities related (closely or more distantly) to SETI such as

- the possibility of humankind expanding out into the universe via space missions to observe/ explore some of the nearer potentially ET and ETI-supporting exoplanetary systems,
- whether humankind might be able to establish colonies in the remote reaches of the Universe (or at least on the closer exoplanets),
- the potential advantages and dangers involved in these studies and missions
- and
- discuss some of the difficult and complex ethics which may be involved.

These discussions will generally be based upon the premiss that there *are* ETIs somewhere to be found; otherwise, there would be little point in your continuing further with this book anyway.

So, now, forward into the future with Chapters 14 to 18.

---

[490] See also Chapter 13 (SETTMUCI).

14 The Current Searches for Extraterrestrial Intelligences (SETI); Should We Be Doing Them?

Extraterrestrial beings in the modern sense could not, of course, be envisaged until after the Earth itself was known to be just one planet amongst many. Before that though and even afterwards, some parts of the Earth itself were so remote that they were labelled by ancient and mediaeval map makers with "*hic sunt dracones*" or "here be dragons" (Figure 14.1) in what seems a very close analogue of widespread modern perception (at least amongst movie makers) of extra-terrestrials as inevitably being hostile towards humankind.

FIGURE 14.1 Dragons – the mediaeval equivalent of ETs? Glance at some of the current computer/internet games and you might suspect that we have not advanced very far during the intervening centuries. The dragon myth, though, may have originated via garbled accounts of volcanic eruptions. (© C. R. Kitchin 2021. Reproduced by permission.)

DOI: 10.1201/9781003246459-17

WHAT IF THE ETIS ARE HOSTILE?

So, if even only a few ETIs are inimical to humankind,[491] should we be searching for them at all? Might not that process call attention to ourselves? Especially those messages which we[492] have already deliberately sent out towards a few known nearby exoplanets (a process sometimes called CETI: Communication with Extraterrestrial Intelligences or METI: Messaging Extraterrestrial Intelligences – Chapter 15)? As Jared Diamond expresses this thought:

> "Think again of those astronomers who beamed radio signals into space from Arecibo, describing Earth's location and its inhabitants. In its suicidal folly that act rivalled the folly of the last Inca emperor, Atahualpa, who described to his gold-crazy Spanish captors the wealth of his capital and provided them with guides for the journey. If there really are any radio civilizations within listening distance of us, then for heaven's sake let's turn off our own transmitters and try to escape detection, or we are doomed."[493]

Similar opinions have been expressed many times, including by Stephen Hawking and by the SF author, Exobiologist and Astronomer, David Brin, who wrote in his review[494] of Milan Ćirković's book *The Great Silence: The Science and Philosophy of Fermi's Paradox* (2018, Oxford University Press):

> "… the controversy over METI, … is primarily about grownup behavior (or its lack) down here on Earth… whether small groups of zealots should bypass all institutions, peer critique, risk appraisal or public opinion, to shout 'yoohoo' into a potentially hazardous cosmos. Ćirković's book offers plenty of grist for discussion and consensus-seeking, before rushing to force a fait accompli on our children."

Finally, although he was not thinking at the time about SETI or exoplanets, since such topics would hardly even be imagined for another 250 years, The Sun King (Louis XIV of France) gave the following very sage advice to his eldest son (also called Louis, but who died before he could become king):

> "Little harm has ever come from not having said enough …. but, infinite misfortunes have resulted from having said too much …"[495]

The 'Bury-Our-Heads-In-The-Sand' advocates will thus say "No; it will be safer to keep very quiet", whilst the 'Better-The-Devil-You-Know' advocates will say "Yes; we'll be OK, after all, we're quite dangerous ourselves".

In practice and since around 1880 (when radio broadcasting started – Chapter 1), even perhaps, since ~1,000,000 BCE (when early hominids started to use fires – Chapter 9), there has been no choice in this matter anyway. ETIs with our level of abilities, or a bit higher (say ^{10}✶ - Table 13.4), will already have been able to detect *us*,[496] whether or not *we* are looking for or signalling to *them*.

---

[491] See also the discussion in Chapter 13 on cool star exoplanets' life-forms; footnote.[400]

[492] Actually, these messages were sent out without the knowledge or consent of some 99.999 9% of all currently living humans, assuming (generously) that ~10,000 people *were* asked about it.

[493] *The Third Chimpanzee; The Evolution and Future of the Human Animal*, 1992, Harper Collins.

[494] American Journal of Physics, 878, **86**, 2018.

[495] *Mémoires for the Instruction of the Dauphin*, Louis XIV, 1661 to 1668. Translated by P. Sonnino, 1970, The Free Press. In modern phraseology: "Least said Soonest mended."

[496] Carl Sagan and collaborators, in 1990, used the fly-by of the Earth by the Galileo space craft (then on its way to Jupiter) to observe the Earth and find if life could be detected upon it – and succeeded. The spacecraft's instruments *were* able to detect modulated radio emissions and its spectroscopic observations were indicative of the presence of vegetation. Of course, Galileo had a closest approach to the Earth of less than 1,000 km – and that is far removed from being able to make those same detections at light year distances, but, we probably *can* do that now.

Thus, whilst it might have been better if ETIs had never known about us, the 'keep-very-quiet' option is no longer available. We might therefore just as well go with the 'Better-The-Devil-You-Know' advocates and find out as much as possible about any ETs and ETIs as we can, as quickly as we can. Then, to use another common expression, we will be fore-warned and so, fore-armed.[497]

Even if the 'keep-very-quiet' option *were* still available, though, it might, in any case, *not* be the best choice. If we were able to keep quiet *enough*, even perfectly friendly ETIs might not know we were there and so stomp on us by accident. As Rabbie (Robert) Burns expressed this type of danger in his poem *To a Mouse*,[498] after he had accidently ploughed-up a mouse's nest:

> "… Thy wee bit housie, too, in ruin!
> It's silly wa's the win's are strewin!
> An' naething, now, to big a new ane,
> O' foggage green!
> An' bleak December's wins ensuin,
> Baith snell an' keen!
>
> Thou saw the fields laid bare an' waste,
> An' weary Winter comin fast,
> An' cozie here, beneath the blast,
> Thou thought to dwell,
> Till crash! The cruel coulter past
> Out thro' thy cell …"

Thus, although Burns was deeply sorry for the mouse, this would not have helped it very much in practice. No more in fact, than if the Friendly ETIs were to be deeply sorry that they had passed through the Earth on their way to the centre of the Galaxy and were then to write a poem about *us*, would help us very much.

Too Late Now

Be all that as it may, as already noted; it probably is *too late now*. So, if ETIs *are* inimical, can they *do* anything about it? – and could we *do* anything about them if they did?

The answer(s) to this/these question(s) is/are essentially indeterminable at present; there are so many factors to take into account, some of which are undoubtedly unknown to (and many not even suspected by) us at present. Nonetheless a few back-of-the-envelope calculations of a few of those factors which we do know about, will perhaps give some idea of some of the obstacles and lucky breaks which may face any hostile ETIs. Those same obstacles and lucky breaks will naturally also be ours should we try to send out colonists to exoplanets (Chapter 18).

---

[497] Although what *we* could do against inimical ETIs at Kardashev Level V (i.e., ETIs having the ability to manipulate universes – Chapter 18) is a quite different question.

[498] Full title; *To a Mouse, on turning her up in her Nest, with the Plough, November, 1785. Poems chiefly in the Scottish Dialect*, 1786, Kilmarnock volume. Verses 4 and 5. Translation below (by Wikipedia, 'To a Mouse' 2023);

Your small house, too, in ruin!	You saw the fields laid bare and empty
Its feeble walls the winds are scattering!	And weary Winter coming fast,
And nothing now, to build a new one'	And cozy here, beneath the blast,
Of coarse green foliage!	You thought to dwell,
And bleak December's winds ensuing	Till crash! The cruel coulter passed
Both bitter and piercing!	Out through your cell.

Interstellar Invasions #1

Science fiction stories abound with ideas for interstellar invasions of the Earth, mostly along much the same lines as those home-grown invasions which have occurred far too often in our own history, i.e., vast numbers of star troopers armed with whatever is the latest in zap-guns, physically crossing space, landing forcibly on the Earth and taking it over.[499]

So, would such a physical invasion be possible? One immediate obstacle, unless the ETIs have lifetimes measured in thousands of years, would be the length of time that the journey would require.

Which brings us to Animals in Zoos. If you have a lion and a lamb in adjacent cages, then neither can do anything to the other, provided that the bars of their cages are close enough and strong enough to stop them physically meeting each other.

For us and these potentially inimical ETIs, those close and strong cage bars are constructed from the Speed of Light, since that is an upper limit to how fast *anything* can move through space.[500] Combined with the immense distances between stars and galaxies, it simply takes *so* long to get *anywhere* within the Universe, that the 'pluses' of doing so will almost always be swamped by the 'minuses'.[501]

Thus, with distances of light years, or more likely, hundreds or thousands of light years from the nearest ETIs to us and with the speed of light limitation, travel times from them to us or vice versa are going to be measured in tens of years to hundreds of thousands of years. So, that looks like a pretty strong cage for both us and the ETIs; we can never attack or be attacked simply because of the length of time which it would take for the journey between our two necks-of-the-woods within the galaxy.

Interstellar Invasions #2

Despite the previous sub-section, there is, however, just a possibility that those speed-of-light cage bars could be very slightly flexible, allowing, in a round-about way, for the lion (= inimical ETIs) to slip through them and for those impossible journeys to be made after all. That flexibility is called 'Relativistic Time Dilation' (Box 14.1).

BOX 14.1 – RELATIVISTIC TIME DILATION

Many people hear the words 'Einstein' and/or 'Relativity' and their minds go blank, even though they probably *do* know some of the important results from the theories. In particular, they may have heard that when two spacecraft (say) are moving past each other, the clocks on the spacecraft will be keeping different times.

This latter phenomenon is *all* there is to Relativistic Time Dilation; clocks *gaining or losing time* with respect to each other, whenever they are *moving* with respect to each other.[502]

---

[499] Usually, of course the hero(ine) is working feverishly in his/her laboratory, aided by his nubile/her handsome young assistant to produce the *Deus ex Machina* which will save the Earth – and naturally succeeding at the last nano-second.

[500] Apologies to Star Trek fans, Star Wars fans and others with similar interests, but the Speed of Light really *is* a limit. Not only do our own experiments and theories tell us that this is the case, but the entire Universe does as well. Travelling faster than light would lead to many observable phenomena which are simply not there to be seen. Causes and their effects could become disconnected so that (as an example) we might see a supernova explode and a more distant exoplanet be affected by the explosion before the radiation from that supernova could possibly have reached it – but we do not do so.

More generally, such a tremendous deviation from our current physical laws would lead to many other massive anomalies. Even a small straying from Standard Physics would invalidate the predictions of many events of all types in both the short and long terms and breaking the speed of light barrier would *not* be a small deviation.

For those of you who have read that the Big Bang's inflationary period involved 'faster-than-light' motions, these motions were *not* of particles/radiation *through* space, but the expansion of space *itself.*

[501] We hope.

[502] An internet search for 'Relativistic time dilation' and/or 'Special Relativity' will provide background to this statement; there just is not enough room to explain everything here. You have already encountered a similar time dilation effect though, for black holes, in Chapter 13 – see the Section entitled 'Life on Stars'.

Although few realise it, many of us will already be affected by Relativistic time dilation in our every-day lives. If you are driving along using your Sat Nav, then that Sat Nav works out your position on the Earth's surface by comparing extremely accurate clocks carried by several spacecraft orbiting in space above you. Since those spacecraft are in orbits, they are moving with respect to you and with respect to each other. The times given by each clock differ then, because their relative speeds differ and each has therefore to be corrected for its own Relativistic time dilation effects before your position (e.g., 'Your destination is 300 metres/yards ahead') can be passed on to you.

So, if an ETI sends a spacecraft towards us, then (if we could see through the hull) we would notice that their clocks were going more slowly than the non-moving clocks close to us. For example, if a year has passed according to our clocks, then we might see that only (say) 11 months had elapsed for the passengers inside the spacecraft[503] and that 11 months would be *their* perceived travel time. Thus, a journey as experienced by the passengers of a spacecraft *can* take a shorter time than would seem to be necessary to an outside observer. Furthermore, the time-shortening effect *increases* as the speed *increases*. Thus, those long journey times might not be so bad after all. To quote Nigel Calder[504]:

> "In a 1g[505] spaceship, you can for example set out at age 20, and travel right out of our Galaxy to the Andromeda Galaxy, which is 2 million light-years away. By starting in good time to slow down (still at 1g) you can land on a planet in that galaxy and celebrate your 50th birthday there. Have a look around before setting off for home, and you can still be back for your 80th birthday. But who knows what state you'll find the Earth to be in, millions of years from now? If stopping is not an objective, nor returning home, you can traverse the entire known Universe during a human lifetime, in your 1g spaceship. Never mind that it is technologically far-fetched[506]. The fact that Uncle Albert's theory says it's permissible by the laws of physics should make the Universe feel a little cosier for us all."

At first sight therefore, Interstellar, even Intergalactic, travel *is* feasible within human time scales and so, almost certainly, also feasible within some ETIs' lifetimes. All that the travellers have to do is to move fast enough and then *their* time will flash by. Nevertheless, to us, outside the spacecraft, it will still take the ETIs at least 1,000 years to travel 1,000 ly.

Calder's 20-year-old traveller (Box 14.1) who does the round trip to the Andromeda galaxy in just ~60 of his/her perceived years would actually arrive back on Earth after a minimum elapsed time of 5 million years and he/she would have had to be moving at around 0.999 999 999 999 699 5 c[507] for much of that time (Box 14.2).

So, say our inimical ETIs are 1,000 ly away, all they have to do to get to us in a short journey time for themselves is to boost their spacecraft up to a speed of 0.999 999 999 999 699 5 c and we will be able to expect their arrival here in 1,000 years plus a fraction of a second (because 0.999 999 999

---

[503] Actually, there are *some* complications – see the Author's book; *'Understanding Gravitational Waves',* 2021, Springer or use an internet search for 'Frames of Reference in Relativity'.

[504] *Magic Universe. A Grand Tour of Modern Science*, 2002, Oxford University Press.

[505] Here, this means an acceleration equal to the force of gravity at the surface of the Earth (~10 m s⁻²). It does *not* mean one gram.

[506] If you are limited to the resources of just one Universe, similar to the one which we occupy, then Calder's traveller's journey is actually *impossible* (for energy reasons), not just ' ... technologically far-fetched ...' – see Box 14.2.

[507] Reminder; 'c' is the symbol representing the speed of light and that has the value of 299,792.458 km/s exactly in a vacuum.

999 699 5 c is not *quite* 1.0 c). The ETIs, though, will have experienced a perceived journey time of just 7 hours and their star troopers will be raring to go.

Interstellar Invasions #3

This last discussed scenario looks entirely feasible and perhaps we should be quite worried about an invasion by multi-headed spiders and/or bright green alligator-types, at least if they live within about 1,000 ly of us.

However – and it is a BIG 'however' – there are more relativistic effects associated with high speeds than just time dilation and one of these may come to our rescue. That rescue will arise because we have yet to consider the *energy* requirements for speeds such as 0.999 999 999 999 699 5 c (Box 14.2).

BOX 14.2 – KINETIC ENERGY, FUEL AND OTHER MATTERS

The energy of a moving object, called its kinetic energy, increases quickly as its speed increases. In fact, if you double the speed, the kinetic energy goes up by a factor of four, triple the speed and the energy goes up by a factor of 9. Thus, to accelerate something up towards near to the speed of light requires a *lot* of energy. On top of that, Einstein's relativity makes the problem far worse as we get to speeds close to that of light.

Thus, if we take a spacecraft travelling at a 0.000 1 c (~30 km/s) and increase its speed by 1% (i.e., to 0.000 101 c), then its kinetic energy will need to increase by 2.01%.

However, if we take a spacecraft travelling at 0.95 c (~285,000 km/s) and increase *its* speed by 1% (i.e., to 0.9595c), then its kinetic energy will need to increase by 15.76% – and that extra 13.66% on top of the earlier figure of 2.01% is the effect of relativity.

The energy requirements get worse and worse as the spacecraft gets closer and closer to light speed; thus, to go 1% faster from a speed already of 0.99c (i.e., to 0.999 9 c) requires a total energy increase of 1,044%.

In Calder's example (Box 14.1), the traveller to the Andromeda galaxy (M31) and back would have had to have speeds peaking at ~0.999 999 999 999 699 5 c in order to get the required time dilations and that is only about 0.1 mm/s less than the actual speed of light.[508] Now at that speed a 1 kg block of something (anything) would have a kinetic energy of rather more than 10^{23} J, whilst typical currently available rocket fuel[509] (on Earth) has a calorific value of around 10 to 100 MJ/kg. Being in a generous mood, we'll take the higher value for the fuel and also assume 100% efficiencies everywhere in the system. We would *still* need around 1,000,000,000,000 tonnes of rocket fuel in order to accelerate 1 kg up to Calder's traveller's maximum speed.

However, Calder's traveller, even with that maximum speed is spending some 60 years of perceived time on board her/his spacecraft, so even if something like hibernation plus re-cycling is used, hundreds of tonnes of supplies will be needed. Add to that the spacecraft itself, say something the size of a small cargo ship and we are ending-up (optimistically) around the 1,000 tonnes (1,000,000 kg) needing to be launched. Thus, the total amount of fuel needed, just to launch the spacecraft plus traveller onto his/her first stage will be in the region of 10^{18} tonnes (10^{21} kg), equivalent to ~10^{29} J.

---

[508] A garden snail can move at over 10 mm/s, when it wants to.

[509] There may, of course, be much better rocket fuels waiting just around the corner, but they would have to be better than our present ones by a factor of some 1,000,000,000,000 before Calder's traveller's journey might become actually possible – and rocket engineers count a 5% increase in rocket fuels' efficiencies as an *enormous* advance in their techniques.

To put this last figure in some perspective, in the year 2020, the entire human race generated and used about 6 x 10^{20} J (~10^{21} J, say, for ease of calculation). The energy requirements for Calder's traveller thus represent the energy usage at humankind's 2020 rate over 100 million years.

Perhaps you are now beginning to see why the time dilation effect is not that practicable. But, '*You ain't seen nuthin' yet*'[510] and that is because Calder's traveller goes to the Andromeda galaxy, stops there and then comes back. So, there are four stages of acceleration needed (slowing down to a halt from 0.999 999 999 999 699 5 c requires the same amount energy as accelerating up to it did).

Thus, working in reverse from when the traveller and her/his spacecraft have returned to Earth:

(i) Halfway back from M31, the traveller will be moving at 0.999 999 999 999 699 5 c and needs to slow down to a halt. From the above discussion, this will mean that there will need to be **~ 10^{21} kg** of fuel on board the spacecraft at that instant.

(ii) When leaving M31, the total mass needing to be accelerated up to 0.999 999 999 999 699 5 c again is now almost entirely the fuel for the last stage, i.e., 10^{21} kg. The acceleration to 0.999 999 999 999 699 5 c though requires ~10^{18} kg of fuel for *each kg* of payload (see above). So, upon leaving M31 the fuel load will need to be **~10^{38} kg.**

(iii) Halfway outbound to M31, the traveller will have reached a speed of 0.999 999 999 999 699 5 c and needs to slow down. The remaining fuel mass when she/he gets to M31 needs to be ~10^{38} kg, so halfway there he/she will need to have **~10^{56} kg** of fuel on board.

(iv) By now it should come as little surprise that when we calculate the launch mass from Earth, it will need to be **~10^{74} kg** – and almost all of that will be fuel.

Our own, visible Universe though is estimated to contain about 10^{53} kg of normal matter. 10^{74} kg is thus the amount of normal matter in ~1,000,000,000,000,000,000,000 Universes similar to our own ®.

After all, then, Calder's traveller's jaunt to the Andromeda galaxy and back would work out to be rather expensive, and, if there were only 999,999,999,999,999,999,999 Universes similar to our own available, it would be impossible

To return to our 1,000-ly-away-from-us-and-hostile-ETI's, the energy requirements outlined in Box 14.2 do not provide us with a *total* reassurance against the possibility of those ETIs' depredations. That is because Calder's scenario is rather extreme. The distance to M31 and back is ~5 million ly, whilst that to the ETIs is just 1,000 ly, the speed can thus be lowered considerably without returning to immense perceived travel times, and, also, the ETIs will probably think that they will not need a return journey. Hence, repeating the calculations in Box 14.2 for a 1,000 ly trip involving:

- a constant acceleration of 0.2 g
- a slow-down to a halt at the destination
- no return home afterwards

and

- a perceived travel time of ~ fifty years

---

[510] The title of a 1974 song by the Canadian rock band, BTO (Bachman-Turner Overdrive).

will require:

- a maximum speed of 0.999 953 974 c; which is still enormous, but less so than Calder's 0.999 999 999 999 699 5 c.

However, these changes will reduce the fuel requirement for the journey to:

- 43.5 tonnes per one kilogram launch load; a reduction from the Calder scenario's requirements by a factor of ~23,000 million.

The downside of this reduction in the fuel requirements is the greatly increased perceived travel time (from 7 hours to ~50 years). So, far more supplies and consumables will be needed and those of the star troopers who have survived after their 50-year journey will be hardly 'raring to go' any longer, based upon reasonable parallels with humankind.

Assuming, therefore, a launch mass of 10,000 tonnes, the fuel requirement will be ~500 million tonnes. This is equivalent to about 5×10^{19} J. The invaders would have a travel time across space of ~1,010 years. These figures probably seem to us likely to prohibit any invasion. The energy requirements though are 'just' a month's energy usage for humankind at the 2020 rate. If the ETIs *really* hate us, then they might still decide that the invasion would be worthwhile.

So, perhaps we *should* still 'be quite worried about an invasion by multi-headed spiders and/or bright green alligator-types, at least if they live within about 1,000 ly of us'; especially if the ETIs first noticed us 1,009 years ago.

Interstellar Invasions #4

Those invading ETIs will have another, not-quite-so-obvious problem which may protect us if they *do* decide to go ahead with an invasion – and that is Cosmic Rays and Dust Particles (Box 14.3).

BOX 14.3 – COSMIC RAYS AND DUST PARTICLES

COSMIC RAYS

Somewhat illogically 'proper' cosmic *rays* are actually sub-atomic *particles*: hydrogen nuclei (protons), other atomic nuclei and electrons. The particles are moving at speeds close to that of light. They originate from most types of stars and from massive explosions such as novae, super-novae, kilo-novae and the like and the particles pervade all of space. The cosmic 'rays' are also accompanied by 'real' high energy rays, i.e., x-rays and γ rays. For our purposes, though, both types of entity can be lumped together and just called Cosmic Rays.

Now, as is widely known, cosmic rays are a hazard to astronauts. When the latter are in space, three days in orbit around the Earth causes as much radiation damage to a human as would a year on the Earth's surface – and vastly more than that during a solar flare.

Cosmic ray energies are typically within a factor of ten of 10^{-11} J per particle but can be very much higher. The Earth's atmosphere protects us from most cosmic rays, but when a really high energy cosmic ray hits the top of the atmosphere, it produces a burst of radiation and high energy particles (known as a Cosmic Ray Shower) and these particles can reach ground level.

In October 1991, the highest energy cosmic ray detected to date was observed via its shower by the Fly's Eye cosmic ray detector in Utah. It had an energy of ~50 J[511] and has become

---

[511] 50 J would be emitted by a 1 kW electric fire in just 20 milliseconds, so this may not seem to be *that* impressive. However, 50 J is about the same energy as that of a 150 km/h tennis ball – but it has all been packed into a particle (if it was a proton or anti-proton) whose mass is about 3×10^{-25} that of the tennis ball. Written out that is:

mass of 1 tennis ball = 30,000,000,000,000,000,000,000,000 'Oh My God' particle masses.

NB: For those readers conversant with Relativity, it is *rest* masses that are being spoken of here. If you are not conversant

(Continued)

known as the 'Oh-My-God' particle (although it is not actually known to be a particle). Not until 30 years later was another comparable cosmic ray to be detected (May 2021 and ~40 J).

Even 'ordinary' light can be dangerous, though, if you are traveling very fast. At a speed of 0.999 953 974 c (Section Interstellar Invasions #3), a visible-light blue-green photon (wavelength 550 nm) is Doppler shifted (Chapter 6) to become a 2.6 nm wavelength x-ray.

Furthermore, when looking outwards from a moving object, the positions of stars and other objects in the sky are shifted towards the point directly ahead by small angles. This effect is called 'Astronomical Aberration' and was discovered by James Bradley around 1726 because he (and we) see the stars from the moving Earth.

The Earth's speed though is only about 0.000 003 c (orbital speed around the Sun) and about 0.000 7 c (orbital speed around the galaxy). Were the Earth to be moving at 0.999 953 974 c, then not only would the star-light illuminating us be mostly x-rays, but the vast majority of the stars normally seen over the whole sky would now be seen to be almost directly ahead of us. A star 'normally' to be seen at 90° to the velocity direction, would be shifted to being just 0.55° away from being straight ahead. It would be almost like being bathed by an intense x-ray laser powered by all the stars in the Universe.

DUST PARTICLES

We know that in some parts of the galaxy there are huge clouds of denser-than-normal interstellar gas and dust particles (Figure 9.5). Doubtless these would be avoided by any ETIs intent upon sending a near-speed-of-light spacecraft to us. Nonetheless, as seen below (main text), the normal interstellar medium has enough atoms and ions (i.e., gas) to ensure that, at least humankind-travellers would die from radiation effects very soon along their journey.

The dust particles in the interstellar medium are far less frequent than the atoms and molecules, but they are there. One estimate suggests that the ISM is ~1% dust by mass. Now, the actual sizes of interstellar-medium dust particles are essentially unknown but probably range upwards from ~10^{-25} kg (~100 atoms) to salt-grain sizes (~0.1 mm, ~10^{-9} kg, ~10^{18} atoms) and upwards. One set of estimates of the number densities in the ISM of the lower mass dust particles, when it is extrapolated linearly to higher masses, suggests a number density for those enormous, salt-grain-sized and larger particles of ~3 x 10^{-12} particles per cubic metre. That is about one salt-grain-sized dust particle per 300 cubic kilometres of the ISM.

The ETIs throughout their 50-year journey to Earth (see previous section) will be being bombarded directly by 'normal' cosmic ray particles and that could be a problem if they have a humankind-level of sensitivity to such radiation. But it is not their *real* problem.

The interstellar medium has around a million atoms and molecules per cubic metre, mostly in the form of molecular hydrogen. If we take the ETI's spacecraft to be a long thin cylinder in shape, say, 10 metres in diameter and perhaps 100 metres long, then it will hit some 10^{27} atoms or molecules directly along its journey to us. Now 10^{27} hydrogen atoms amount to less than a couple of kilograms of matter, so that should not, in itself, be a problem for a 10,000-tonne spacecraft.

These interstellar particles are moving quite slowly; only a few km/s, *but*, the ETI's spacecraft will be moving through the interstellar medium at speeds of up to 0.999 953 974 c. From the spacecraft's view-point,[512] it will be undergoing a continual bombardment by hydrogen and other nuclei, etc. travelling at speeds of up to 0.999 953 974 c throughout its journey.

---

with Relativity, then just take the figures as shown.In terms of speed, the 'Oh My God' particle was travelling at around 0.999 999 999 999 999 999 999 995 089 c. The perceived time (Box 14.1) for the particle to travel 1,000 ly would thus be about 1 milllisecond, although an external observer would still see it taking 1,000 years for the journey. It was also travelling at over 60,000 million times nearer to the actual speed of light than was Calder's traveller (Box 14.2).

[512] This is the essence of Relativity; the spacecraft's point of view is just as valid as the hydrogen atom's point of view.

A hydrogen nucleus (proton) with a speed of 0.999 953 974 c will have an energy of ~10^{-8} J, i.e., a thousand times that of the typical cosmic ray particle. The impact between that hydrogen nucleus and the spaceship atoms at its outermost surface will produce a cosmic ray shower of hundreds to thousands of cosmic rays penetrating into and throughout the whole of the interior of the spacecraft; in other words – directly into the ETIs' living quarters.

Thus, for any form of life which is endangered by high energy particles and radiation, the interior of their spacecraft will be quite unhealthy. When we add-in the fact that all those interstellar atoms along the 1,000-ly journey will hit the spacecraft in just the 50 years of perceived travel time, we get a bombardment rate of ~10^{18} impacts per second.

Now, assuming that the 'cosmic ray shower' particles mostly pass into and are absorbed within the spacecraft (including within the ETIs), which, considering the relative velocity directions, is quite likely, this will give us an absorption rate for the particles and radiation of 1,000 joules per kilogram per second. In SI units, that is a dosage rate is ~1,000 sieverts per second, or ~10^{12} sieverts over the whole journey.

To put this last figure into context, the lethal radiation dose for a human is about 5 sieverts. Furthermore, and to 'gild the lily', just the energy being absorbed from the radiation would raise the recipient's body temperature to fatal humankind levels (say, greater than 45 °C) within the first half-minute or so of the journey.

Now (as if the cosmic ray problem were not enough) let us consider the ISM dust particles (Box 14.3) as well. Taking our examples from those in Box 14.2, at a spacecraft speed of 0.999 953 974 c, an impact with a 100-atom dust particle would cause a ~0.000 002 J 'explosion' on the surface of the spacecraft. Not something to worry about much, you might think. But those 'explosions' will be occurring at a rate of ~2,000 million per second, as we see the situation, or 40,000 million per second in the travellers' perceived time. The bonds holding atoms together in order to form molecules typically require an input of ~10^{-18} J in order to break them. So even just 0.000 002 J could erode ~1,000,000 million atoms from the spacecraft's structure. That is, a loss of ~ 2 x 10^{21} atoms per second from the spacecraft's structure (as seen in our time frame), or about 100 tonnes in total for the journey. For a 10,000-tonne spacecraft such a mass-loss, though, would probably not be catastrophic. The ETIs would *surely* have installed a ~500 tonne shield anyway, over the spacecraft's bow, in readiness for *that* problem.

Fewer encounters with the salt-grain-sized particles are to be expected, but still about 2,000 million for the entire journey, given their possible number density in the ISM (Box 14.3). Each one of those impacts, though, will be equivalent to the explosion of ~4 tonnes of TNT. Over the whole journey this would be equivalent to a total of about 20,000 million naval torpedoes hitting the spacecraft (and just *one* of those can sink an aircraft carrier with ten times the mass of this spacecraft).

Undoubtedly, though, the abundance of salt-grain-sized ISM dust particles will differ from the estimate used here, but even one hit from such a particle is likely to ruin the travellers' entire day.

Thus, the ETI's chances of surviving the journey seem to be quite questionable, unless the ETIs are quite incredibly resistant to high-energy radiation resistant, can survive body temperatures in the hundreds of degrees Centigrade and can easily cope with large explosions. More likely they will just be glowing (from their radioactivity) burnt cinders when they get here.

So, perhaps we should *not* be quite so worried about an invasion by multi-headed spiders and/or bright green alligator-types after all, at least if they live about 1,000 ly distant from us.

On the other hand,

- What have we forgotten?
- What do we not know about?
- What can the ETIs do when their Science and Engineering is 10,000 years (say) in advance of ours?
- See also Chapter 18.

Alternatives to Interstellar Invasions

If the ETIs are seeing the Earth as a new site for a colony, then an interstellar invasion will probably be essential.

If they just feel that the Universe would be better-off without humankind occupying it, then there are many alternative solutions. Just three of those alternatives, which seem to be reasonably possible, will be mentioned here – you can think your own up whilst trying to get to sleep tonight.

- The 'Dinosaur-killer' meteorite (Chicxulub event) released an energy variously estimated to be within a factor of ten of 10^{24} J (1,000,000,000 megatons of TNT). Travelling at 0.999 953 974 c (see Interstellar Invasion #3, above) a mass of 500 tonnes has a kinetic energy of 1,100,000,000 megatons.

  Thus, whilst just using ~5% of the mass of spacecraft discussed in 'Interstellar Invasion #3' as a kinetic energy weapon, the ETIs could reproduce a dinosaur killer (and *it* would have no problems with cosmic rays, etc.). If the ETIs felt that one dinosaur killer might be insufficient, then they could send ten (or whatever) and still be using much less than half the resources required for 'Interstellar Invasion #3'.

  Their gain would actually be even greater, because there would be no need to slow the missile down. Quite the opposite, in fact, if they were to continue the 0.2 g acceleration for the whole journey, they would reach a speed of 0.999 988 376 c and a 250-tonne missile would suffice.

- Only rather stupid ETIs might try getting rid of us by the previous method. Cleverer ones might try a scam.

  Imagine we have been receiving their radio messages[513] for some time and that they have given us lots of very helpful tips in all sorts of areas so that we think that they are wonderful friends. Then there comes a section which reads:

  > "Recipe to increase your life expectancy by a factor of 1,000.
  > Take 1,000,000,000 tonnes of uranium and separate out the U^{238} and U^{235}.
  > Make a big pile of the latter ..."

  We, today, would know better than to do this – but suppose there's some similar process which we do *not* know about?

- The ETIs have lifetimes of 1,000,000 years or do not die at all, so spending a few centuries traveling slowly to the Solar System in order to put us properly in our place would be no sweat at all to them.

Of course, these few simple examples of imagined interstellar travel are exactly that, i.e., *simple* examples. There are many, many other factors/possibilities/problems/variations/things-that-we-do-not-know-about which will change the viability/non-viability of any such mission.

Moreover, this discussion could go on to include many more factors influencing the viability of high-speed interstellar flight, but there is a limit to the space available within the book. Sufficient has been covered, though, to show that there are going to be many problems, some of which we know about now and some which have yet to be discovered. Some of those problems leave the invading ETIs on the verge of a successful invasion, others leave them failing so thoroughly that we would probably never know that there had even been a threat. How all of these factors would interact with each other is probably currently unknowable, at least to humankind.

So, at this time, we really have no answer as to whether the inimical ETIs could or could not mount a successful invasion of the Earth. What we do know is that if we want to send out colonists

---

[513] This would be via a continuous broadcast, not a conversation – see Chapter 15.

to nice looking exoplanets ourselves (Chapter 18), we will face those same problems and so will have to solve them ourselves eventually.

WHAT IF WE ARE THE HOSTILE ONES?

See previous 3,500 words (or so).

In case you think it impossible that we (humankind) could be hostile, then two quotes, attributed to two well-known terrestrial conquerors who were widely separated in space and time, may be evidence that given the opportunity, we (or at least some of us) *would* be out there grabbing whatsoever we could find.

Alexander the Great[514]:

> "Is it not worthy of tears, that when the number of worlds is infinite, we have not yet become the lords of a single one."

Cecil Rhodes[515]:

> "... think of these stars that you see overhead ... , these vast worlds which we can never reach. I would annex the planets if I could ..."

Even without a Rhodes or an Alexander 'helping' on our side, there is still, unfortunately, the 'Better Safe than Sorry'[516] argument.

In the case of an actual visitation from ETIs, this could become a decision which really has to be made.[517] For long-distance, messages-only contacts, we should be safe, so we can then afford to be friendly.

Still and all, another common expression, often attributed to Theodore Roosevelt, may hold the solution to the problem of ETI first contacts:

> 'Talk soft, but carry a big stick.'

However, some rather more in-depth considerations of the situation may be found via an internet search for 'Interstellar Diplomacy', wherein the articles are mostly written by diplomats and other similarly more appropriate specialists than myself and I therefore leave you to explore those further possibilities for yourself.

WHAT IF THE ETIS ARE FRIENDLY?

That will be nice.

---

[514] 356–323 BCE. Alexander conquered a large area lying from Greece to Pakistan. The quote is to be found in Plutarch's *De Tranquillitate Animi* (On the Tranquillity of Mind) which was possibly written within a few years of 55 AD.

[515] 1853–1902. Rhodes funded the conquering of a large area of South-Eastern Africa later to be called 'Rhodesia' and known today as the two countries: Zambia and Zimbabwe. The quote is given in W.T. Stead's 1902 publication entitled *The Last Will and Testament of Cecil John Rhodes with Elucidatory Notes* as a part of Stead's commentary when reporting a verbal statement by Rhodes.

As an aside, Stead, himself, was a controversial figure, but achieved his main, and probably unwanted, fame by being amongst the victims of the Titanic sinking.

[516] A.k.a., the pre-emptive strike, Thomas Hobbes' impasse, between the devil and the deep blue sea, getting your revenge in first, the fear spiral and many other similar common expressions. All have the same meaning; when a potentially life-threatening choice is available, you go for the safer option rather than for the better option.

People not making the safer choice become known as Heroes and Heroines and are called Brave and Superbly Courageous (if they survive to tell the tale).

[517] We may, of course, not be able to decide anything except when to say "We Surrender".

However, friendly ETIs will no-more be able to do anything *for* us than the inimical ETIs could do *against* us.

Thus, if the friendly ETIs are 1,000 ly away and we send them a message saying '*Hello*' (or whatever). It will be 1,000 years before they receive that message and at least another 1,000 years before we get their '*Hi Folks – Nice to hear from you. Please let us have a photo.*' reply.[518]

Not unexpectedly, friendly ETIs who are closer to us than 1,000 ly will be contactable more quickly. For very close exoplanets, we might even be able to exchange physical artefacts within a human lifetime (see Prox Cen B, Chapter 2, footnote[61]).

WHAT IF THE ETIS ARE ALTRUISTIC?

This is frequently likely to be the situation for any ETI discoveries arising from their deliberate broadcasts. That is to say, we detect the ETIs from modulated signals which are deliberately being sent out to attract the attention of other ETIs (humankind included) and which contain information about some of those aspects of 'Life, the Universe and Everything'[519] which the broadcasting ETIs think we would like to know about.

In other words, they are making some or all of their knowledge available to anyone capable of decoding their signals and understanding the information therein, without (provided that they have read *this* book, *this* far) any expectation of a return for it.

WHAT IF THE ETIS ARE UNINTERESTED AND/OR DISINTERESTED?

We don't care about *them* either.

Actually, we would probably regard them in the same way as an exoplanet inhabited by ETs, i.e., we would learn as much about them and from them as we could just by using our own efforts.

IS IT REALLY IMPOSSIBLE *FOR US TO VISIT ETIs?*

The answer to the question posed by this section title, as you will have inferred from the rest of this chapter, is a most definite 'Yes', i.e., It *is* Really Impossible for *Us* to Visit ETIs.

However, you will notice that the 'Us' has now been highlighted and that is because we *might* be able to send 'things' to ETIs or receive 'things' from ETIs, even if personal visits are impossible.

The sending and receiving of e-m signals is covered in Chapter 15 and the viability of physically travelling to the homes of ETIs has just been discussed above. The latter discussion, though, is premised upon the assumption that the journeys must only require reasonable lengths of time (whether this is 'normal' time or 'perceived' time). However, the 'reasonable journey time requirements' are what has been behind the enormous speeds and vast energies which we found would be needed earlier in this chapter. If, somehow, the viable journey times could be significantly increased, then a great deal could be accomplished which currently is judged to be impossible.

For you (or me) to travel in person to visit the distant home of some ETIs remains impossible, unless some form of hibernation/suspended animation capable of operating for thousands to tens of thousands of years can be found. *Humankind* though might make such trips, probably in order to found colonies, via huge self-contained environments which can achieve speeds of a few hundreds of kilometres per second and which generally are called 'Generation ships' or 'Space Arks' by SF writers (who *love* the idea). This possibility is discussed further in Chapter 18, although you can easily see that at a speed of 300 km/s (0.001 c) a 100 ly journey would take 100,000 years.

---

[518] Of course, we would actually send a continuous message (Chapter 15), but it would *still* be 2,000 years before we would hear from them.

[519] The title of one of the books in the 'Hitch-Hiker's Guide to the Galaxy' series. D. Adams, 1982, Pan Books.

Two other ways of travelling throughout the galaxy, both with long journey times, have already been mentioned:

- Simply have a very long-life expectancy (see above) and sit out the journey whilst solving difficult crosswords

and

- Panspermia (Chapter 9).

Another conceivable future possibility, not so far encountered, which comes down to being a very advanced and sophisticated form of panspermia, is to use von Neumann machines. The von Neumann machines concept essentially requires a self-replicating machine (Chapter 9, Appendix E). Such ideas date back centuries and in principle, for the Earth, go back to the 'First-Living-Molecule' (Chapter 9).

John von Neumann expanded on the 'First-Living-Molecule' idea, to dream of populating the entire galaxy, maybe the entire Universe, within a comparatively short period of time. A von Neumann machine, then, is a self-replicating machine which can make all of its own parts from scratch and then assemble them into a duplicate of itself. That is to say, if *one* such machine *were* to be landed on a suitable exoplanet, then it would be able to:

  (i) Prospect the exoplanet for the needed chemical elements
 (ii) Develop suitable energy supplies from local sources
(iii) Mine/dig-up/extract etc. those essential chemical elements
 (iv) Transform the chemical elements into the required compounds
  (v) Fabricate the compounds into the required shapes and formats, etc. for components
 (vi) Assemble the resulting components
(vii) Provide the new machine with all the instructions, information, etc. that it, the first machine, possesses, together with anything else needed to complete its exact duplication
(viii) Send the duplicate off on its journey to the next exoplanet
 (ix) Start building the second duplicate.

A slight variant on this idea could enable terrestrial-type life-forms, though probably not precise copies of humankind, to spread throughout the galaxy as well. That variation (below) should be inserted between items (vii) and (viii) above and would comprise:

(vii #2) Carry on board the seeds and eggs (etc.) of some terrestrial life-forms under conditions for their long-term preservation
(vii #3) Manufacture the nutritional and other requirements for those seeds and eggs to be germinated/hatched
(vii #4) Nurture the resulting terrestrial life-forms until they are well established and thriving by themselves
(vii #5) Harvest fresh supplies of seeds and eggs
(vii #6) Add the new seeds, eggs, etc. to the payload of the duplicate machine

Until quite recently, we might have thought that, whilst von Neumann machines were possible in theory, their practical realisation would not be able to occur until very far into the future. The extraordinarily rapid developments in the capabilities of AI, though, which have happened in just a few recent years, could mean that the machines might be a reality *much* sooner than that ✿✿✿.

Either of the von Neumann variants is essentially a recap of the processes in the 'normal' origin and evolution of life-forms (Chapters 9 and 10), which has been jump-started to the point where *they* start with the levels of attainment *we* shall perhaps reach a few decades from now. Indeed, were we to have life expectances of a few tens of thousands years, Humankind would be an excellent example of a von Neuman machine; there is even the possibility that mistakes in the machines' duplication processes will mimic evolution.

In use, and assuming that the complete von Neumann machine could be accelerated to ~300 km/s, say,[520] then the machines could be present throughout the whole of the Milky Way galaxy within a few million years. The reason that this is possible is that a von Neuman machine already has all the facilities and the information that it needs for its duplication. When it lands on a new exoplanet, then it may take a century or even ten centuries to build all the industrial infrastructure which it needs, but not the 3,800 million years (Chapters 9 and 10) it has taken us to get to (or near to) that same point.

Thus, say a von Neumann machine is launched from within the Solar System at 300 km/s towards an exoplanet 100 ly away. It will take about 100,000 years to get there. But after that and probably within less than 1,000 years it will be sending out its own von Neumann machines to exoplanets 100 ly onwards from its position. We will now assume that not just one von Neumann machine will be sent out from the Solar System but several, or even many, over, say, a thousand-year period and in various directions. About 100,000 years after the first von Neumann machine was launched therefore, a sphere of the galaxy ~100 ly in radius and centred on the Solar System will have most, if not all, of its suitable exoplanets occupied by hard-working von Neumann machines, each sending out their duplicates to exoplanets another 100 ly further out.

About 200,000 years after the first von Neumann machine was launched then, a sphere some 200 ly in radius will be chock-full of them. After ~300,000 years, that sphere will 300 ly in radius, after 400,000 years it will be 400 ly in radius, after … and so on.

The Solar System is well out from the centre of the Milky Way galaxy so the spiral arm here is ~1,000 ly thick. Hence after ~500,000 years, the machines will have reached the 'top' and 'bottom' of the spiral arm and will then mainly be moving in towards the centre of and the far side of the galaxy. The disc of the galaxy is ~100,000 light years in diameter, so, the von Neuman machines will have occupied it *all* in a time of about 100 million years from when the very first one was launched.

If the speed at which the von Neumann machines were travelling could be increased to 3,000 km/s, then the 'conquest' of the galaxy would be accomplished in ten million years, whilst if only 30 km/s could be managed, it would take 1,000 million years. All these times are long compared with a human life span, but the Milky Way galaxy was formed soon after the Big Bang and so *its* age is ~ 12,000 million years.

We see, then, that interstellar travel *is* a possibility; if you are a machine capable of functioning for tens of thousands of years and more, or, with less certainty, you are an equally long-lived life-form.

We also now encounter a puzzle. The Milky Way galaxy is ~12,000 million years old. Even if it takes life-forms ~4,000 million years to achieve a technological standard close to ours at present, that still leaves up to 8,000 million years for those life-forms to have produced and sent out their von Neumann machines.

So, the puzzle is: Why are we not *inundated* with such machines from the thousands or millions of ETIs who had their First-Living-Molecule-moments thousands or millions of years before it happened on Earth? This puzzle has the name 'The Fermi Paradox' and it will be encountered again in Chapter 16.

---

[520] This is currently beyond our capabilities, but laser sail launch system and/or fusion reactors could alter this soon (see Chapters 15 and 18).

FALSE ALARMS

UFOs

> "… *Horatio.*
> Oh day and night; but this is wondrous ſtrange.
>
> *Hamlet*
> And therefore as a ſtranger giue it welcome,
> There are more things in Heauen and Earth, *Horatio,*
> Then are dream't of in our Philoſophy. …"[521]

There are indeed '… more things in Heauen and Earth … Then …' *any of us* can dream of, or know about, or experience or see. Which means that we are continually encountering things about which we have no prior awareness.

Mostly, the new 'things' are quickly and easily identifiable on the basis of other 'things' of which we already *do* have experience; go into a museum which you have never visited before and, although you have never seen the exhibits previously, they will mostly be recognisable because you have seen something similar in a different museum; a dinosaur skeleton can be recognised as such even if the dinosaur skeletons you have seen previously have been of different species.

Sometimes though, we see something and it is unrecognisable and then (because we *are* used to recognising things) we mis-interpret the new object as something totally different from its real nature. This phenomenon will probably be familiar to most readers and generally goes by various names such as trompe l'oeil, optical illusion, déjà vu, trick of the light, etc. (Figure 14.2).

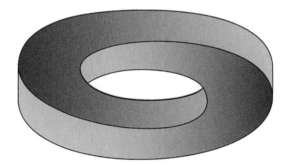

FIGURE 14.2 First impressions are not always right – an example of a Trompe l'oeil – look carefully and it is *not* a thick washer/squared-off doughnut. (Reproduced by kind permission of Pixabay and OpenClipart-Vectors.)

So, when we see something unusual, we may 'see it through rose-coloured spectacles' – i.e., see its appearance as we expect/wish to see it and not as it really is.

Which brings us to Unidentified Flying Objects or UFOs as most people know the term.[522] At this point most Ufologists will throw this book away, if they have not done so already, because Ufology:

---

[521] Shakespeare W. *The Tragedie of Hamlet, Prince of Denmarke.* Folio 1, 1623. Wording, spelling and lettering as in the original. "Philoſophy" here, probably may be taken to mean 'human knowledge / study'; however, Shakespearean scholars will argue over its meaning for days, given the chance.

[522] NASA is currently trying to change the name to 'UAPs' – Unidentified Anomalous Phenomena. For some reason, though, I find the words Against', 'Brick', 'Batting', Head' and 'Wall' keep recurring to me, though not always in that order.

"… is the investigation of unidentified flying objects (UFOs) by people who believe that they may be of extraordinary origins (most frequently of extraterrestrial alien visitors …"[523]

Whilst a good deal of the material on Wikipedia should be treated cautiously at times, the crucial word in this quotation is 'believe'. There is a large group of people who *believe* that UFOs exist and who *believe* that ETIs (their preferred word is usually 'Aliens') visit the Earth in flying saucers and the like.

Now the main characteristic of a *belief* in something is that it can exist in the complete absence of any evidence that it is valid. No amount of rational argument will persuade a believer to change her/his beliefs (although irrational persuasion may sometimes do so).

Thus, for a Ufologist, Area 51 (Edwards Air Force Base, in California) is a hive of UFOs and alien contacts. Similarly, the Wright-Patterson Air Force Base (Ohio) hides flying saucers (sometimes crashed), the cadavers of alien life-forms and perhaps even surviving aliens. Whilst abductions of humans from lonely roads or remote camp sites continue apace (why is it never from the centres of Harrods or Macy's on a busy day?), as do telepathic messages (*not* SETI), various sightings of unidentifiable flying things (Figure 14.2) by the pilots of (normal) flying aircraft and aliens already hiding within our midst.[524]

To a Ufologist, any denial of these beliefs is just the result of Bigotry, Blind prejudice, Conspiracies, Government cover-ups, Injustice, Lack of vision, Mental incapacity, Perfidious frame-ups, Put-up jobs, Scientific jealousy, Warped thinking, Wishful thinking, Xenophobia and probably Getting-out-of-bed-on-the-wrong-side that morning as well.[525] The Ufologist remains a true believer, come what may.

Half a century ago, J. Allen Hynek tried to bring some scientific organisation to at least some types of UFO experiences. He defined three types (out of many more possibilities) of UFO experience, the last of which will be known to many readers already; *Close Encounters of the Third Kind* from the 1977 blockbuster film of that same name. Only encounters inside a distance of 150 m are included in the scale, with the:

- first kind encounters being light(s) in the sky at night
- second kind encounters being daytime observations of UFOs

and

- third kind being UFO sightings with radar confirmation.

Now, there are also people who think that it *is probable* that non-terrestrial-types of life-forms *do exist* within the Universe and *could visit the Earth*[526] – and I include myself in that group (or I would never have written this book!). This group, however, has yet to see *any* evidence, which reaches even the *barest minimum standards for scientific acceptance*,[527] that such ETIs/Aliens do actually exist and have been/are visiting the Earth.

---

[523] https://en.wikipedia.org/wiki/Ufology, 2023.

[524] An internet search for, say, 'UFO Sightings List' will supply a plenitude of additional material on this topic.

[525] I might mention that I *have* seen two UFOs myself. Whilst I am sure that neither contained green-skinned, multi-headed aliens, I do not have definitive explanations for my observations. The first may have been an escaped barrage balloon or blimp (high, hardly moving, not an aeroplane and seen a long time ago when intentional sky balloons were very rare). The second was a glowing fuzzy light moving at erratically at a moderate speed at night and seen only briefly. This might have been a reflection of a distant light by an atmospheric reversing layer (i.e., a type of mirage). Despite these observations, however, I remain a UFO-non-believer.

[526] Subject, of course, to their overcoming all the problems reviewed in Chapters 14, 16 and 18.

[527] 95% confidence level, independent verification and publication in a reputable peer-reviewed journal. Repeatability would also be nice, but the nature of UAP sightings may make this difficult.

In the hope of finding some of the latter type of evidence, the Galileo project was founded by Abraham Loeb in 2021 and aims to be a 'Systematic Scientific Search for Evidence of Extraterrestrial Technological Artifacts' (it still only has minimal funding at the time of writing, nonetheless, its objectives are laudable). Whilst in 2022, NASA initiated a study into the feasibility of a future programme for UAP research, whose report concluded (*inter alia*):

> "To date, in the peer-reviewed scientific literature, there is no conclusive evidence suggesting an extraterrestrial origin for UAP." [528]

and

> "It is increasingly clear that the majority of UAP observations can be attributed to known phenomena or occurences *(sic)*. When it comes to studying such phenomena, our overarching challenge is that the data needed to explain these anomalous sightings often do not exist; this includes eyewitness reports, which on their own can be interesting and compelling, but are not reproducible and usually lack the information needed to make any definitive conclusions about a UAP's provenance. Thus, to understand UAP, a rigorous, evidence-based, data-driven scientific framework is essential."[529]

At the time of writing, any practical results from the NASA initiative are still awaited.

Thus, in respect of UFOs and/or UAPs, I remain highly sceptical with regard to their being due to aliens or to aliens' activities, but persuadable of those possibilities, if reputable evidence *ever* does emerge.

Meteorite ALH 84001

'Sceptical but persuadable' is also my position on the discovery of fossils of ETs. Unlike the UFOs, though, whilst ALH 84001 (Chapter 9) was a false alarm, its evidence was scientifically plausible; just not quite sufficiently plausible enough, in the end.

The evidence found in ALH 84001 is of possible fossilised nano-bacteria, which perhaps originated on Mars. Scanning electron microscope images of samples of the meteorite show segmented, chain-like and other structures up to 200 nm in size which are reminiscent, in some ways, of the structures and forms found associated with modern terrestrial bacteria. Magnetite crystals were also found and these can be of biological origin, but can also be produced abiotically. PAHs (Box 9.1) were also present, though these, too, can be biotic or abiotic in origin.

Today, whilst some workers still support a biological origin for the ALH 84001 structures, most feel that the evidence is insufficient to support such a radical and far-reaching conclusion, citing the Sagan Standard's[530] requirement that:

> "Extraordinary Claims Require Extraordinary Evidence".

WINDING-UP

The title of this Chapter posed the ethical question of whether or not we should try to find or even to try to contact ETIs. The conclusions are largely unresolved but may be summarised as

---

[528] https://web.archive.org/web/20230915112132/https://science.nasa.gov/science-pink/s3fs-public/atoms/files/UAP%20Independent%20Study%20Team%20-%20Final%20Report_0.pdf
[529] Ibid.
[530] An aphorism frequently employed by Carl Sagan within his speeches and writings.

- It is too late now to stop the process of ETIs discovering/contacting us, so we might as well continue with our efforts to discover/contact them.
- If ETIs want to visit us, or we them, it will be very expensive and may not be possible at all.
- We may hope that ETIs will be friendly and altruistic, but the alternatives should always be kept in mind.
- Whatever we may think, plan, anticipate or hope for, we may be sure that at least some ETIs will have totally different ideas, so we should always expect the unexpected.
- Maybe they are not there anyway (this is one of the possible solutions to the Fermi Paradox – Chapter 16).

15 SETI

BEWARE OF HUBRIS

You may be expecting that, when you have announced that you have detected the first *ever* ET/ETI (by any means at all), then the world and all its inhabitants will:

- lay itself at your feet,
- shout paeons of admiration and praise,
- beg you to accept this special achievement prize of $1,000,000,000,000,000 (with *lots* more to come),
- offer to mow your lawn, clean your shoes, polish your car, do your shopping,
- place Crowns on your head, Cadillacs around your feet, bathe you in Holy Oil and Gild your Toenails
- buy lorry-loads of dog biscuits for your dogs,
- pay off completely your Nation's National Debt,
- … … … well, I'm sure that you can add a few more possibilities of your own.

However – It ain't going to happen that way.

Many people, it may surprise you to know, have already made just such an announcement as that which started this section – and what did they get?

- Scepticism
- Being sent to Coventry,
- Being Totally Ignored by any Reputable News Sources,
- Death threats,
- Disbelief,
- Headlines in rubbish radio, TV, Internet channels and Newspapers,
- Malicious Stalkers,
- Ostracisation,
- Outright Derision,
- Scam attempts,
- … … again, there is space here, for your own additions, should you so desire.

These events may have escaped your notice because they have happened within two main groups of people:

- Ufologists

and

- Geologists, meteoriticists, micro-biologists, micro-botanists and a few other specialised scientists.

The announcement by a well-known Ufologist of her/his meeting with an alien visitor somewhere out in the boondocks, during which meeting, secrets were disclosed which are too dangerous for him/her to reveal to more uneducated peoples than herself/himself (until the right time comes), is still greeted with adulation and admiration by the coterie of UFO-believers. Most of the rest of the world now just ignores the statement, if, indeed they ever became aware of it in the first place.

DOI: 10.1201/9781003246459-18

The second groupings of people and of ET/ETI discoveries have only started to occur recently, although we might also put the Martian canals (Box 15.1) under this classification as well. The situation then is such that one group of well-qualified specialists thinks that they have evidence of the existence of ETs/ETIs, either now or in the past. Another group of equally well-qualified specialists thinks that the evidence has been mis-interpreted in some way and so they are not persuaded that the existence of ETs/ETIs has been proven. The case of the meteorite ALH 84001 is a well-documented example of this latter situation and is more extensively discussed in Chapter 14.

Here then, it just remains to point out to any existing or future persons, imagining that they may stand a chance of being the first human to find an ET/ETI, that they will probably need to persuade the rest of the world that they *have* actually done so and that the fighting off of millions of adoring admirers beating a path to their front door may *not* actually be the problem at all.

INTRODUCTION

The SETI Institute[6] was founded by Jill Tarter and Tom Pierson in 1984. It is currently based in Mountain View, California, 50 km SE of San Francisco, although many of its facilities lie elsewhere. Thus, the Allen Telescope Array (see below) is housed at Hat Creek, 350 km North of San Francisco. The SETI Institute employs around 100 active scientists and has many more affiliates who contribute to the enterprise whilst also having 'day-jobs' in other research institutes, observatories, universities and the like. The SETI Institute has three main divisions: Research at the Carl Sagan Center for Research, Education at the Center for Education and Public Outreach at the Center for Outreach. We will encounter the SETI institute again during our discussions in this and later chapters.

SETI[6] itself (i.e., the Search for Extraterrestrial Intelligence as an activity in general) though has a much longer history. Many early astronomers thought, long before there were any supporting observations, that not only would we find that the Solar System planets were inhabited but that the stars would also mostly have planetary systems and that those planets would additionally be inhabited.

The actual acronym 'SETI' probably originated from a NASA radio astronomy research project of that name which commenced in 1992.

THE BEGINNINGS

As mentioned earlier (Chapter 3), around 300 BCE Epicurus of Samos was teaching that there were infinitely many inhabited worlds within the Universe. This seems to be the first clear statement about the possibility of ETIs, but it may have had precursors about which we know nothing, and there could have been many other ancient cultures around the world wherein similar thoughts were being expressed and where no records have survived to the present day and/or where those cultures did not record their ideas in any lasting manner.

Following Epicurus, many others thought there might be inhabited planets within and beyond the Solar System. These suggestions were usually fictional (sometimes even intentionally so) and frequently included vividly imagined inhabitants based upon grotesque variations of terrestrial life-forms. A very incomplete and personal selection would include:

- Lucian of Samosata (*Verae Historiae* or 'True History' – c. 160 CE)
- *A Thousand Nights and a Night* – Middle-Eastern folk tales originating from c. 800 to c. 1200 CE
- Johannes Kepler (*Somnium, seu opus posthumum De astronomia lunari* or 'A dream, or a posthumous work on lunar astronomy' – 1608 – Published 1634).
- Cyrano de Bergerac (*L'Autre Monde: ou les États et Empires de la Lune* or 'The Other World: or the States and Empires of the Moon' – 1657 and *Les États et Empires du Soleil* or 'The States and Empires of the Sun' – 1662)

- Bernard de Fontanelle (*Entretiens sur la pluralité des mondes* or 'Conversations on the Plurality of Worlds' – 1686)
- Voltaire (*Le Micromégas* or 'The Small-Large' – 1752 and *Songe de Platon* or 'Plato's dream' – 1756)
- Camille Flammarion (*La Pluralité des Mondes Habités* or 'The Plurality of Inhabited Worlds' – 1862 and *Les Mondes Imaginaires et Les Mondes Réels* or 'Real and Imaginary Worlds' – 1864)
- Jules Verne (*Autour de la Lune* or 'Around the Moon' – 1869)
- Herbert G. Wells (*The War of the Worlds* – 1897)
- Edgar Rice Burroughs (*Barsoom* series – 1934–1948 and *Amtor* series – 1934–1946)
- Robert Heinlein (*Red Planet* – 1949)

and

- Ray Bradbury (*Martian Chronicles* – 1950)

Patrick Moore's 1955 book *Guide to the Planets*[531] could still talk about Venus possibly being:

> "… a moist humid world … a world of tropical vegetation … and perhaps primitive forms of life … insects, fishes such as the coelacanth and even great reptiles …"

Which, whilst it may sound as though it ought to be included in the fiction list, was not fictional, Moore's book was based upon the best scientific evidence available at the time. Moore's book and the earlier list of fictional imaginings, though, were not truly a part of SETI because these astronomers were not actually searching for extraterrestrial life, just using their vivid imaginations to suggest possibilities. The first active search[532] was thus based around the infamous Martian Canals.

MARS, SCHIAPARELLI AND LOWELL

The mistaken ideas about the canals on Mars and their implication that life is inhabiting Mars *now*, or at least until very recently, is an oft-told-tale. Nonetheless, it may still be new to some readers and so it is summarised in Box 15.1.

BOX 15.1 – MARTIAN CANALS

It all starts with the mis-translation of an Italian word into an English word.

During the close-range 1877 opposition of Mars, the Italian astronomer, Giovanni Schiaparelli, along with many other astronomers, spent a great deal of time observing the planet. These observations, of course, were all visual; photography was still in its infancy then. His telescope was a 218 mm (8¾ inch) refractor, a respectable size for that time. His observations included faint streaky linear or slightly curved markings on Mars which were sometimes seen to be double. It is worth noting here that Schiaparelli was colour blind, suffered from other eye problems and that few other observers were able to see this doubling of the streaky features. He used the Italian words 'canale' (channel) and 'fiume' (river) when describing these features, i.e., implying naturally produced phenomena.

---

[531] Eyre and Spottiswoode.
[532] 'Active' in the sense that *we* are *doing* something/anything, such as sending out signals, looking for incoming signals, mapping canals, sampling, undertaking chemical analyses, etc., towards finding ETs or ETIs. Not just getting on with our own business and then saying 'Oh, Hi There' when the ETIs unexpectedly land next door (see for example, Steven Spielberg's 1977 movie, *Close Encounters of the Third Kind*).

Unfortunately, some translations of Schiaparelli's writings into English rendered 'canale' as 'canal', i.e., a trench for flowing water (or liquid) implying an artificial product made by some type of life-form.

The implication that life-forms were living, now or until recently, on Mars naturally caused a storm of interest in the planet, and it was observed intensively over the next few decades by many scientists. Percival Lowell was a wealthy American businessman who became obsessed with observing Schiaparelli's 'canals'. He was wealthy enough to establish his own observatory (naturally called the 'Lowell Observatory') at Flagstaff in Arizona and to equip it with a 610 mm (24½ inch) refractor. It is sited, alongside several more modern instruments, on what is now called 'Mars Hill', at an altitude of 2,100 m.

Lowell used his telescope and other astronomers used other telescopes to observe Mars over the next several decades. Some observers saw 'canals' and some did not. Amongst those who did see canals, their recorded positions, shapes and sizes often did not agree with each other.

To cut a long story short, these observers of 'canals' were simply wrong – there are no 'canals' on Mars. Today's explanation of their observations (when it was not entirely due to their imaginations) is that 'vision' is a process which is a combination of what the eye sees and how the brain interprets that information. What the observers were actually (almost) seeing was several small, point-like features which were not actually seen or resolved properly, but the brain then interpreted these as single linear(ish) features.

That the 'canals' did not exist was shown conclusively by the first Martian spacecraft to return reasonably sharp images of Mars to the Earth (Figures 3.3 and 15.1), although by then many astronomers had come to that conclusion anyway. Incidentally, many (natural) dried-up river beds *have* been found to exist on the Martian surface, but no canals – and the observed river beds do not match the positions of Schiaparelli's and Lowell's features.

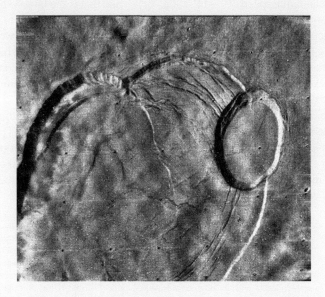

FIGURE 15.1 The central caldera (volcanic crater) of the largest known volcano within the Solar System, Olympus Mons. The main caldera is about 80 km wide. Volcanoes, dried-up river beds, dried-up flood plains, impact craters and sand dune fields are the actual main features of Mars' surface, not, 'canals'. (Reproduced by kind permission of NASA/JPL-Caltech.)

But the damage has been done and the Martian 'canals' concept still lives on today in the expectation amongst many people that if/when we do find extraterrestrial life, then it will be found on Mars first.

The 'SETI' associated with the Martian canals is rather different from most other forms of SETI. Schiaparelli, Lowell and the others were observing what they thought to be the products/remnants of ETIs. They had not discovered the life-forms themselves (which, of course, were not there) nor were they trying to observe those life-forms or to communicate with them. Doubtless, though, had the 'canals' really been there, those stages would have followed-on later.

A few other 'happenings' took place around the beginning of the 20th century which were linked to possible Martian life in one way or another. Thus, there were several suggestions in the early days of radio broadcasting that signals might be sent to the Martians and Marconi thought at one time that he had picked-up radio signals from Mars. During the 'good' Martian opposition of 1924,[533] quite widespread efforts were made to pick up any Martian radio signals which were coming to us. None of these 'happenings', as you may guess, gave any indication that Martians might exist.

MODERN SETI #1 – SETTING THE STAGE

WHY IS IT ALWAYS SETI RADIO SIGNALS FOR WHICH WE SEARCH?

Well, it is not *always* SETI radio signals for which we search, but that search does take up about 99% of the efforts being put into SETI.

Any active search by humankind for ETs and ETIs faces many problems. For the reasons partially discussed in Chapter 14, we cannot easily go and visit them and neither can they easily come and visit us. All forms of electromagnetic radiation, in any case, travel at 299,792,458 metres per second within a vacuum (the speed of 'light', a.k.a. 'c'[10]) – and that is the fastest speed through space attainable by anything.

At the same time, ourselves sending out light speed signals to a place where we think there might be ETIs and hoping for a reply would be a very long waiting game (even so, it has been tried – see below). We would have at least an 8 ½ year wait for a reply from the (probably non-existent) inhabitants of the nearest exoplanet (Prox Cen b – Chapter 8) and a 2,000-year minimum wait for any reply from an exoplanetary system 1,000 ly away from us.

Such figures would seem to imply that we should not waste *our* time sending out signals, but instead look for the signals sent out by others. On the other hand, if every ETI were to have this same thought, then *no one* would be sending out signals[534] (see the Fermi Paradox, Chapter 16).

Nonetheless, for some time to come, our serious efforts at active SETI are likely to be limited to trying to receive e-m radiation emissions[535] from those ETs and ETIs who started sending out their signals at least 4 ¼ years (or 1,000 years) ago.

---

[533] When Mars was only ~55,800,000 km away from the Earth. The planets at a 'poor' opposition could be over 100,000,000 km apart.

[534] Notwithstanding this, we and the ETIs may still be able to detect each other through our unintentional emissions.Consequently, I dread to think that ETIs might get their first impressions of *our* level of intelligence from 1950s TV shows. Those broadcasts will, by now, have travelled some 70 to 80 ly out into the Universe. Even worse would be the example chosen by Carl Sagan for his SF novel *Contact* (1985, Simon Schuster) wherein the ETI's first impressions of ourselves were obtained from the TV broadcasts of the 1936 Berlin Olympics. The latter, of course, being used by Adolf Hitler to promote his and the Nazi party's racial supremacy propaganda.

[535] Gravitational waves also (seem to) travel at the speed of light and some cosmic ray particles (Chapter 14) and neutrinos can move at speeds only a very little slower than speed of light. Although ETIs might send out signals via one or more of these entities or emit them unintentionally in other ways, our detectors for these types of outflows currently have to be enormous to be able to detect even the strongest of examples. This, to such a degree, that they may be many kilometres

Whilst any appropriate e-m radiation detections at any wavelength would suffice to indicate the presence of ETs or ETIs, in theory. In practice, certain wavelength regions of the e-m spectrum may be more suited to those purposes than others. This is because much of the e-m spectrum is absorbed by the gases and dust particles occupying the interstellar medium and the Earth's atmosphere. GMCs (Box 9.2), for example, appear as dark patches in the sky because they are absorbing the light emitted by those stars which are behind them from our point of view. Other absorptions mean that almost no radio waves reach us at wavelengths longer than about 10 m (30 MHz frequency)[536] and much of the UV light with wavelengths shorter than ~122 nm is absorbed completely by hydrogen atoms within the normal interstellar medium.

For reasons like those outlined in the previous paragraph and below and also from practical motives (e.g., Big radio telescopes cost *less* than Big optical telescopes) our active SETI activities began by observing at radio and microwave wavelengths and only now are they starting to operate within other parts of the spectrum (Box 15.2).

OK, So, the Sky Is Impenetrable at Some Wavelengths, but Why Not Use Light, which Does Get Through?

When it comes to sending or receiving messages across interstellar distances, there are optimum and less-than-optimum ways of doing it. Costings and the clear windows in parts of the e-m spectrum have already just been mentioned above, whilst the possible advantages of the 21-cm wavelength are discussed below.

A less obvious factor, perhaps, is the angular resolution for receiving signals and/or its equivalent for sending out signals; the beamwidth.

For sending out signals, it is relatively easy and cheap(ish) to send out a single radio signal with a wide angular beamwidth so that it will travel towards many hundreds, thousands or even millions of potential host stars of exoplanetary systems. Similarly, the fields of view of radio telescopes are generally much broader than those of optical telescopes, so that in searching for radio SETI signals, many potential sources can be covered in a single observation.

Now, at least within the Earth's atmosphere, there are two spectral windows – the radio window and the visual window and most SETI work has so far been in the former.

So why no Visual SETI work yet? After all, visual telescopes were invented long before radio telescopes?

The answer lies in the angular beamwidth of powerful light emissions. Unless we (or the ETIs) want to use energies enormous even by astronomical standards, the best light sources for sending out a visual signal would be lasers and, as is widely known, lasers have very narrow angular beamwidths[537]; indeed, for many terrestrial applications, that is the reason for their use. It is also the reason for their apparently high-power levels, since whatever radiation may be emitted, it is all concentrated into that narrow angle. Plainly, the sender of an interstellar signal would want to use the minimum power for the distance that the signal has to travel and that means using as narrow an angular beamwidth as possible.

---

long for gravitational waves and many cubic kilometres in volume for cosmic rays and neutrinos. Accordingly, we have little hope of finding ETIs via these possibilities. See the author's books *Astrophysical Techniques* 7th Edn., CRC Press, 2020, and *Understanding Gravitational Waves*, 2021, Springer, for further information. If the ETIs communicate by other means, such as hand (or tentacle) gestures (see Steven Spielberg's 1977 movie; *Close Encounters of the Third Kind*) or semaphore or telepathy (Chapter 12) or smoke signals or colour changes on their skins (octopuses) or … whatever, then we and they are probably doomed to remain unknown to each other forever.

[536] This cut-off, though, is due to the Earth's ionosphere and so radio telescopes in space *can* observe to longer wavelengths (lower frequencies). The interstellar medium, nonetheless, for wavelengths longer than about 3,000 m (100 kHz) is then completely opaque.

[537] A.k.a. the laser's Divergence.

Theoretically, the narrowest possible angular beamwidth (it will be larger in practice) is determined by diffraction[538] and that in turn depends upon the diameter of the laser's aperture and the wavelength of the emitted radiation. Currently the best practice is around 200 seconds of arc.[539]

We cannot know and would have difficulty in guessing sensibly, the levels of performance and sophistication of the optical equipment which ETIs, desirous of sending us a signal, might have available. Accordingly, we will look at the problem the other way around:

> with our present telescopes and lasers etc., could we send a signal to them?

For this discussion we will ignore the time lag problem (above) and also assume that we would use our own today's-best-available-means for the task (see also below and Chapter 18) – perhaps even, assume capabilities foreseeable a decade or two into the future.

The core of the signalling apparatus will be the laser. So, how powerful can it be and how narrow a beam can it produce? The answer to that is complex and rather uncertain, but a 'Back of the Envelope' calculation is shown in Box 15.2 (see also Box 18.3); you can skip to the conclusion, if you so wish.

BOX 15.2 – A BRIEF ESSAY ON LASERS AND OTHER MATTERS

1. Laser #1 – Power
   1.1. Pulsed lasers with incredible power levels, $\geq 10^{16}$ W, are available, resulting from the research into developing fusion power generators. But those emissions only last for around 10^{-14} seconds.
   1.2. For interstellar communications we need lasers which will operate continuously for long periods, preferably for decades. CW (Continuous Wave) lasers naturally have much lower instantaneous emission levels than the pulsed varieties. Furthermore, at their high-power end, many of the precise details of CW lasers are secret because of their military applications. Anyhow, the situation as of a decade or so ago was that megawatt (10^6 W) power levels could be reached, but only for a minute or two. CW lasers for industrial applications, though, can now be bought, off-the-shelf, with power levels starting to reach towards 1 MW.
   1.3. It may be that CW power levels in the region of 100 MW can now be achieved, but the duration of the output at that level is probably still quite short, although not as short as the pulses of a pulsed laser.
   1.4. For the purposes of this book, we will therefore assume that to have uninterrupted emissions lasting at least several months, the maximum power level is 1 MW. Any readers' privy to better information on the CW laser current-state-of-art will certainly be qualified to repeat this calculation with their better data, should they wish to do so.
2. Laser #2 – Angular Beamwidth/Divergence
   2.1. This depends, basically, upon the optical quality and the size of the laser's resonant cavity and upon the quality and size of any other optical components being used to produce the outgoing beam. Currently the best practice is around 200 seconds of arc (see text above).
   2.2. Angular beamwidth also depends upon the operating wavelength of the laser and this latter is usually fixed by the equipment and by the manner of the laser's operation.

---

[538] For the observational astronomers amongst this book's readers, this is the same diffraction limit as that of your telescope. For other readers, the Author's books *Astrophysical Techniques* 7th Edn., CRC Press, 2020, and *Telescopes and Techniques* 3rd Edition, Springer, 2013, give further details or try looking in any book on optics or on the internet.

[539] An interference phenomenon leads to 'Bessel Beams', which are sometimes said to have zero divergence, but in fact, this only means that they have a long depth of focus. Outside that focal region, those beams do diverge.

3. Telescopes
   3.1. Feeding a laser into telescope should reduce the angular beamwidth down to where it is comparable with the optical resolution of the telescope.
   3.2. Optical telescopes can currently achieve angular resolutions of 0.05 seconds of arc, for ground-based 10-m telescopes with atmospheric compensation and the HST (the JWST's resolution is only ~0.1 seconds of arc because it operates in the IR).
   3.3. Several multi-ten-metre diameter optical telescopes are currently under construction. The largest of these, with a 39-metre diameter, is the European ELT and that is due for completion in 2028. Observing at a 500 nm wavelength, this telescope will have a theoretical angular resolution of ~0.003 seconds of arc. Given that the ELT is to be sited high in the Chilean Atacama Desert and that it will have very sophisticated optics for compensating for atmospheric distortions, its actual resolution, in practice, may come quite close to this theoretical value.
   3.4. Optical interferometer instruments might be able to improve on such a resolution, but if ground-based, then ~0.001 arc-seconds is their current limit (for the VLT using the PIONIER[540] system) and phased arrays are discussed further below and in Chapter 18.
   3.5. Optical interferometers in space should be able to achieve their theoretical angular resolutions, provided that their optical quality and alignments are adequate. For example; 0.0001 seconds of arc for a 1 km interferometer operating in the green/cyan spectral region (500 nm). These, though, are significantly further into the future than the ELT.
   3.6. Unfortunately, it seems highly unlikely that the E-ELT astronomers would be happy to have a 1 MW laser shining into their precious instrument for months or years on end. But there are now, quite a few 8-metre and 10-metre optical telescopes around and some larger instruments constructed from several mirrors each of which is of a similar(ish) size. Consequently, in a few years from now you might be able to acquire a second-hand largish mirror/telescope for a song.[541]
   3.7. For practical purposes, at the time of writing, the best obtainable optical telescope resolution is about 0.05 seconds of arc.
4. Lasers AND Telescopes AND ETIs
   4.1. For the purposes of this section, then, we will assume that what is currently possible, or will be possible within a few more years, is a laser power of ~1 MW and an angular beamwidth of ~0.05 seconds of arc.
   4.2. Hence, with the ETIs at a distance of 1,000 ly, the laser will illuminate a circle ~2.3 x 10^{12} m in diameter and the intensity of that illumination will be:

$$\sim 2.4 \times 10^{-19} \text{ W m}^{-2} ****$$

To put this figure in perspective; the power received by the Earth from the Sun is ~ 1,340 W m^{-2}, a factor of ~5,500,000,000,000,000,000,000 higher than the brightness of our laser at a distance of 1,000 ly ®.
5. From the ETIs' Viewpoint
   5.1. We do not know what level of equipment an ETI civilisation might possess in order to receive our signal (any more than we knew what they might have in order to broadcast a signal).
   5.2. For that reason and for this calculation, we will therefore make the same assumption, i.e., that the ETIs are at a similar technological level as ourselves. In particular, we will

---

[540] Precision Integrated Optics Near-infrared Imaging Experiment.
[541] Actually, it might need around ten songs and even then, each would have needed to gross about the same as Lennon and McCartney's *Yesterday*, i.e., about $60 million.

assume that they have ~10-metre optical telescopes and optical detectors comparable to ours.

5.3. The faintest source detectable via a telescope is called its 'limiting magnitude'.[542] The numerical value of the limiting magnitude varies with the size of the telescope, the telescope's optical quality, the detector being used (eye, photographic plate, solid state detector and so on) and with the quality of the observatory's night sky. A fairly widely accepted value for the limiting magnitude of a (humankind's) 10-metre telescope, on a good observing site, using a modern solid-state detector and with a one-hour exposure time is 'magnitude 27'.

That number will not mean much to most readers,[543] but it translates into an incoming energy at the telescope of:

$$\sim 4 \times 10^{-19} \text{ W m}^{-2} \text{ ****}$$

Compare the above two numbers marked **** and you will see that our signal would be just *too* faint for detection by the ETIs.

5.4. However, for ETIs *750 ly* or less away from us, the illumination is greater than or equal to:

$$\sim 4.3 \times 10^{-19} \text{ W m}^{-2} \text{ ****}$$

and you will see that the third starred figure (the observed intensity of our laser) is now slightly larger than the second starred figure (the minimum detectable brightness). That being so, then:

EUREKA!

the ETIs *will* detect our signal!
Reverse the situation and

We would detect the ETIs!

HOORAY!! – We are there!!

BUT (you knew that this was coming, didn't you), the detection took a ten-metre telescope and, more importantly, an exposure time of an hour (Point 5.3, Box 15.2). We had not sent (or received) a signal at all – just sent a light which was very faint and contained no information except for the direction towards its origin and that it might be a laser source and hence of artificial[544] origin.

For our 'signal' to become a 'SIGNAL', it needs to carry a message. Carrying messages is exactly what all our internet, radio and TV transmissions do and they carry the message by rapidly varying

---

[542] The brightnesses of stars are measured via units called their 'magnitudes'. In this fashion, the star, Betelgeuse (α Ori), has a magnitude of 0.45 and Polaris (the Pole star, α UMi) has a magnitude of 1.97. Sadly, because of its long history (dating back over 2,000 years to Hipparchus), the stellar magnitude scale is illogical, confusing, non-user-friendly and a real pain in the neck, Also, sadly, a replacement for the stellar magnitude scale is unlikely any time soon. Happily, *we* will revert to SI units shortly within the main text.

Further details of the stellar magnitude scale may be found in the author's book *Telescopes and Techniques* 3rd Edition, Springer, 2013, most basic astronomy books and via the internet. You can, in any event, already see one of the scale's idiosyncrasies; Betelgeuse is a much brighter star than Polaris, yet the stellar magnitude of Betelgeuse has a lower numerical value than that of Polaris.

[543] Except for those readers who are also amateur astronomers.

[544] Masers, the equivalent of Lasers in the Microwave region, on the other hand, *can* occur naturally. Hence the ETIs might well interpret our 'signal' as a naturally occurring laser beam.

the strength of the signal. Most readers will know that such intensity variations are in a digital (binary) code and can be represent as a string of zeros and ones, with '1' being a strong emission and '0' a weaker emission.[545] The letter 'H' for example in this digital binary form would be 01001000[546], whilst 'h' would be 01101000.

In our example in Box 15.2 it took an hour's exposure for the laser signal to be detected. Assuming, therefore a rate of one bit per hour[547] for the message to get through successfully, then it would take 8 hours to transmit 'H' and 40 hours to transmit 'Hello'. Whilst to transmit

'Sorry, we've just made a mistake. We meant to say; Hi there! Neighbours!'

would take over three weeks.

Given that even a moderate speed internet system, at the time of writing (it will probably be regarded as a very slow system by the time that you are reading this), would operate at some 100 Mbps,[548] the message 'Hello' would be expected (by ourselves) to go through in 0.4 micro-seconds, or less.

Now, if you have an internet system with the moderate speed of 100 Mbps, this means that in 1 second of time there are 100 million ones or zeros coming into your computer/mobile phone, etc. It takes, though, around 50 to 100 wavelengths of the signal to be at the required intensity for a 'one' or a 'zero' to have been clearly represented. The frequency of the basic underlying signal (a.k.a., the carrier wave) has accordingly to be around 50 to 100 times higher than the speed of the message, i.e., $\geq$ ~5 GHz (wavelength $\leq$ 60 mm) for a 100 Mbps signal.

In this manner, if you wanted to send a message to the ETIs at 100 Mbps, then they would need to be able to detect a single pulse accurately in a time of about 0.001 microseconds, whilst, from the example above (Box 15.2), they needed an hour (at least). To achieve a speed of 100 Mbps (assuming now that they are $\leq$ 750 ly away), the intensity of our signal would need to be increased by a factor ~4 x 10^{12}.

Quite simply, such an advance is currently completely out of Humankind's reach.[549]

But, there are currently scientific and technological developments underway which *could* soon enable the above sort of levels of signal performance to become feasible. Those developments are centred upon the use of optical phased arrays and are discussed in more detail in Chapter 18 and, especially, in Box 18.3. Here we will just take the parameters reached in the worked example in Chapter 18 and apply them to the problem of sending/receiving optical signals from ETIs.

Thus, those developments which are relevant to the concerns of this section are that it may soon (within a decade or two or three):

- be possible to produce an optical laser signal generator with an output power of 100 GW or more

and

- which has a beamwidth of 0.000 1 seconds of arc or less.

---

[545] This is amplitude modulation. There are other methods of coding data, such as frequency modulation.

[546] The eight ones or zeros making up this code are called a Byte, and for many purposes an '8-bit byte' suffices since it enables 256 different symbols to be represented. If the ETI's though, communicate via individual symbols, like the Chinese, Japanese or the ancient Egyptians (hieroglyphics), then 16 bit bytes (65,536 symbols) or even 32 bit bytes (4,294,967,296 symbols) could be needed.

[547] 0.000 28 bits per second (bps).

[548] Mega bits per second.

[549] It might be accomplished by building a 20,000 km diameter optical telescope or by using 4,000,000,000,000 one-megawatt lasers.

With those levels of performance, the illumination would reach an intensity at a distance of 1,000 ly of:

$$\sim6 \times 10^{-11} \text{ W m}^{-2}$$

That is 150 million times the threshold detection level (Box 15.2) and whilst the improvement is not by the 4×10^{12} factor needed for 100 Mbps speeds, it *will* permit a speed of ~4 kbps (~7 kps at 750 ly).

Accordingly, since contact with distant ETIs will be a case of sending and/or receiving long monologues and will not be a conversational exchange, a speed of ~4 kbps will be adequate for many purposes.[550]

As always, though, hovering over all these calculations, but unmentioned for a few pages, there is the time lag; even if we do send a sufficiently adequate signal to the ETIs 1,000 ly away from us, it will *still* be 1,000 years before they can receive it and 2,000 years before we can hope for a reply.

Phased arrays may thus enable us to send signals to distant ETIs using the NIR and visible parts of the e-m spectrum. Whilst phased arrays can also operate as receivers by exchanging the lasers for optical detectors, the gains which follow are smaller than was the case for emission versions. The improvement in the angular resolution is identical to that of a similar emitting system. However, this will enhance the sensitivity of the receiving phased array mainly by reducing the background noise. The gain to the signal strength will only be by the number of telescopes being used compared with the signal from a single telescope. Only if the whole area of the array is covered cheek-by-jowl with telescopes which have the narrowest possible spatial gaps between them, will the *receiving* signal strength improvement approach towards matching that of the *transmitting* phased array.

Then, of course, there is always the possibility that the ETIs may be using equipment, engineering techniques, physical principles and so on, which are quite unsuspected by ourselves at present, so that *they* can send *us* signals, but not vice versa.

Nonetheless 'every little bit helps'.

So finally (and there is no 'but' to follow *this* sentence) we should, in a decade or two or three, really be able to claim:

EUREKA!

the ETIs *will* detect our signal!

Reverse the situation and:

We would detect the ETIs!

HOORAY!! – We are there!!

PULSED LASERS RECONSIDERED

At the beginning of Box 15.2, pulsed lasers were dismissed as being useful for SETI because of their very short pulse lengths and the assumption that interstellar signalling would require broadcasts to be years long. Now, if you are sending out a *Broad*cast, i.e., a signal sent to many potential recipients in the hope that some will notice that it is there, then that assumption is likely to be correct.

But, if you are aiming at one, or a few, recipients (sometimes called *Narrow*casting), then that is a different kettle of fish. You may still need to send out a 'Heads-Up' signal to get the recipient's

---

[550] The New Horizons Pluto mission, for example, used a data transmission rate of 38 kbps as it flew past Jupiter, but then, it was in a hurry to get to Pluto.

attention, but once you know, or at least suspect, that you have that attention, then pulsed laser sig-nalling might come into its own. There are two reasons for this change of heart:

- The *much* higher peak power levels available from pulsed lasers

and

- Making full use of the fact that we are using *light*, not *radio* wavelengths

The first point is relatively simple and has already been mentioned in Box 15.2, the power from a pulsed laser can reach 10^{16} W. There is naturally a trade-off between peak power and pulse dura-tion, but 10^{16} *is* 100,000 million times the power level which we earlier assumed for the CW lasers.

The second point is more subtle, but we have already covered the science background needed to explain its significance.

Earlier then, we saw that because sending a single bit *reliably* needs some 50 to 100 wavelengths of the carrier wave to have the intensity required to represent that bit (one or zero). Hence an internet system operating at 100 Mbps needs a carrier wave frequency of ~5 GHz (~60 mm wavelength). 5 GHz is, of course, a somewhat higher frequency (shorter wavelength) than most TV signals, but it is still well within the radio/microwave region of the e-m spectrum. We though, are now working in the *visual* region of the e-m spectrum where (using the earlier example) the wavelength is 500 nm and the frequency, 600,000 GHz. An internet system or transmitter using this cyan light as its carrier wave could thus potentially reach messaging speeds of ~10,000,000 Mbps. Such a speed translates into ~10^{11} words per second or to sending an entire copy of this book to the ETIs within a single laser pulse lasting one microsecond. Assume a pulse repetition rate of one pulse per second of time and then the whole Library of Congress[551] could be sent in about three years.

If the ETIs follow a similar line of argument to that above and so try sending us messages via optical laser pulses, what would we see? Well, we might *discover* the signal during routine spec-troscopic monitoring of the exoplanet's host star, assuming that the exoplanet is not seen separately from the star. The spectrum would show an intense and very narrow emission line at the laser's wavelength. We would not, though, detect the underlying message, because the exposure/integra-tion time used to obtain the spectrum would be measured in minutes or hours.

If though, it was realised that this was an ETI signal, the spectrograph would have to be replaced with a detector having a response time in the region of 0.000 000 1 microseconds (0.1 picoseconds). Now, the fastest detectors which we appear currently to possess (who knows what is hidden behind the military's closed doors) have response times of ~2 to ~5 picoseconds. So, we are actually quite close to being able to read the ETIs' message, providing that its signal is powerful enough.

We see, therefore, without trying to be too precise about the details, that pulsed lasers may, after all, be a better option than CW lasers for interstellar signalling within the visual region and that we might be moderately close to being able to receive and decipher them, if they are there[552] (under-standing what the ones and zeros mean is another matter – Chapter 15).

Optical SETI in Action

Like Radio SETI, Optical SETI has yet to make any discoveries of ETIs. The problems of Optical SETI, just discussed, may of course, indeed are very likely to be the reason for this null result. One stray thought, though, is that they *may* be avoiding us *because* we *might* send them the entire Library of Congress within three years of their doing so.

Joking apart, some optimistic *terrestrial* intelligences *are* trying quite hard to detect optical SETI signals, if they are there to be found.

---

[551] Equivalent, by my reckoning, to very roughly 100 million books similar to this one.

[552] A more in-depth, though now quite dated, analysis of optical SETI's prospects may be found at http://seti.harvard.edu/oseti/tech.pdf.

The Breakthrough Listen project (see below) mostly operates in the radio part of the spectrum, but it also covers the optical region by searching the spectra obtained by the Automated Planet Finder[553] (ATF) for the presence of intense narrow band CW laser-type emission lines (see above). It is thought that a 100 W laser at the distance of Prox Cen (4.25 ly) might be detectable although a different analysis suggests that an 84-kW laser at ~80 ly would be found (implying a ~250 W laser for Prox Cen b). Over its 10-year lifetime, the ATF is expected to examine about 1,000 stars within its primary mission.

A separate NIR SETI programme at the Lick Observatory, using a 1-metre telescope, is currently searching for pulsed laser emissions. Also currently in operation is the SETI Institute's partially crowd-funded LaserSETI which has two sets of four commercial cameras located in California and Hawaii and is searching the whole sky for laser pulses. Other similar projects have been based at Harvard (1.5-m telescope), whilst the Planetary Society[554] has had several Optical SETI projects, including the 1.8-m Planetary Society All-Sky Optical SETI telescope which started operations in 2006 in association with Harvard University.

Somewhat different from any of the above schemes is the optical SETI programme based upon the VERITAS[555] array. The VERITAS array comprises four 12-m diameter light 'buckets',[556] and it searches for the Čerenkov[557] light flashes coming from very high energy γ rays passing through the upper Earth atmosphere but which originated in outer space. Čerenkov radiation is seen as a very brief bright light flash. The VERITAS instruments are thus also almost purpose-designed for detecting light flashes from ETIs. In 2016 Breakthrough Listen started a search for SETI laser-flashes amongst VERITAS archive data. The star KIC 8462852 (a.k.a. 'Tabby's star'[558]) was selected, even though it is nearly 1,500 ly away from us, because of its anomalous brightness variations in its past. Nine hours-worth of observations were analysed. Unsurprisingly no artificial signals were found. VERITAS' data and that from other optical gamma ray observatories continues, though, to be monitored for possible SETI signals.

In the future, the planned PANOSETI[559] project envisages eventually having many individual stations sited in many places around the globe in order to monitor the whole night sky continuously for rapid light flashes. Each station would use two well-separated instruments so that local effects could be eliminated. Each instrument would comprise (perhaps) 24 individual telescopes with 0.46-m f1.32 Fresnel lenses.[560] Around 2,500 square degrees of the sky (~6% of the visible sky) would be imaged every nanosecond.

Optical SETI then may be possible, or the difficulties may be just too much for us at the moment. It is, though, very much the junior partner within the overall SETI effort when compared with SETI in the radio and microwave spectral regions.

OK. TELL ME ABOUT RADIO SETI, THEN

As mentioned above, the radio and microwave spectral region is one of the two main clear parts (windows) within the whole e-m spectrum, the other being the optical region (above). When we first

---

[553] A 2.4-m optical telescope sited at the Lick observatory which searches for exoplanets via the radial velocity method (Box 6.1).

[554] https://www.planetary.org/

[555] Very Energetic Radiation Imaging Telescope Array System.

[556] Low-optical-quality telescopes which can be constructed cheaply in large sizes. Used when high-quality imaging is not needed.

[557] For further information see the Author's book *Astrophysical Techniques* 7th Edn., CRC Press, 2020, or search the internet for 'Čerenkov radiation'.

[558] After Tabitha Boyajian who discovered the star's anomalous behaviour.

[559] Panoramic SETI, a.k.a. Pulsed All Sky Near Infrared Optical SETI. The first two prototype instruments have been installed at the Lick observatory at the time of writing.

[560] Low-optical-quality lenses which can be constructed in large sizes as flat plates, very cheaply. Used when high-quality imaging is not needed. For those readers who remember them, large(ish) Fresnel lenses were used in overhead projectors.

looked-out into the Universe we did so using our eyes and so we saw the Universe at optical wavelengths. When radio transmitters and receivers were developed, we started to see the Universe via the radio window. Seeing the Universe outside these two windows had to await the space age and the ability to put telescopes and detectors into orbits above the Earth's atmosphere. The potential for detecting radio signals from Extraterrestrial Intelligences thus built-up alongside the development of our own radio broadcasting.

Even for just the radio spectrum, though, some wavelengths can be identified as possibly being better suited to a SETI search than others. As already mentioned above, Earth-based radio telescopes are limited to wavelengths shorter than 10 m (30 MHz) although space-based telescopes can receive out to longer wavelengths. As we go to shorter wavelengths, the Earth's atmosphere and the interstellar medium become almost completely clear. The transparency starts to deteriorate again at about 30 mm (10 GHz) whereafter the absorption starts to rise sharply. The radio window is effectively shut from wavelengths shorter than 3 mm (100 GHz) where the absorption has increased by a factor of 20 and more over its level at 30 mm (10 GHz).

Some preferred wavelengths for a radio SETI could therefore be around 30 mm (10 GHz), 10 m (30 MHz) and 1,000 m (300 kHz – for radio telescopes in space), representing the shortest and longest wavelengths which we can monitor effectively.

Several specific regions within the radio window though are frequently selected for monitoring, for quite different reasons than just for their accessibility. The best known of these is an emission line arising from hydrogen atoms – of which there are huge numbers in the interstellar medium. This line is at a wavelength 211.061 mm (1.420 41 GHz) and is frequently called the '21-cm line' or the 'Hydrogen line'. Within 'normal' (non-SETI) radio astronomy, the 21-cm line is very important because it originates from the interstellar medium and then penetrates through interstellar dust clouds. It thus allows the structure of the Milky Way galaxy (and other galaxies) to be mapped and studied in great detail. Because of this application, the 21-cm part of the radio spectrum is intensively and frequently observed by many telescopes and across the whole of the sky. A signal broadcast at that 21-cm wavelength should therefore have a much better-than-average chance of having an observer looking in its direction at the same wavelength than if the signal wavelength had just been selected at random.

The natural 21-cm line, though, is an *emission* line, so it might seem that if the ETIs were to send out a signal at exactly that wavelength or if we were to observe at exactly that wavelength, then the ETI signal would be swamped by the galactic emissions.

Nonetheless, 21-cm line *is* used extensively for SETI work by terrestrial astronomers. Potential ETI signals are then not swamped by natural signals because the sources of the two emissions will usually have different line-of-sight speeds with respect to the Earth. Hence, the emission lines are Doppler shifted (Figure 6.2) by differing amounts and thus they are not superimposed upon each other after all.

Because the interstellar medium is cold and very nearly a vacuum, its 21-cm line is very sharp, i.e., its width is very small and a velocity difference with respect to the Earth as little as 15 or 20 km./s would more than suffice to change the wavelength of the signal enough for it to be seen separately from the galactic emissions. Now even 15 km/s is very fast by humankind's normal experiences. The Earth, however, is moving around its orbit at about 30 km/s,[561] the Sun is moving around the galaxy at ~230 km/s, the ETIs will have similar though different speeds and the source(s) of the natural emission(s) will also have yet more differing line-of-sight speed(s). In practice, therefore, it would be quite unusual if the various relative velocities were to differ by *less* than 15 km/s for most of the time.

---

[561] Therefore, since the Earth's speed within the Solar System reverses every 6 months as it moves around its orbit, its velocity *changes* by up to 60 km/s over a period of one year.

Since, in this situation, the ETIs are sending out a deliberate signal, *they* will be wanting it to be observed by other ETIs (i.e., *us*). They may well therefore choose the 21-cm line for their signal in the expectation that we will be observing the whole galaxy at that same wavelength and could therefore pick up their signal by accident. Conversely, *we* having followed the same line of reasoning, will be looking out for ETI signals at 21 cm *as well as* observing the galaxy, a win-win situation, if ever I heard of one. Be that as it may, however, the circumspect SETI researcher will want to scan his/her/its radio telescope across a small range of wavelengths centred on the 21-cm line and not just monitor the line itself.

A cluster of radio emission lines from the interstellar hydroxyl radical (OH) near 180 mm (1.665 GHz) offer much the same advantages to ETIs and SETI researchers as does the 21-cm line. Since a hydrogen atom and the hydroxyl radical can combine to produce water (H_2O) and since these spectrum lines occupy a particular clear part of the interstellar medium's radio spectrum, the spectral region between the 21-cm line and the hydroxy radical lines is often called the 'Water Hole'.[562]

Other suggested wavelengths, which might be favoured by ETIs for their signals, are based upon various 'famous' numbers in mathematics and physics, since these should be universally (literally) recognisable (Box 15.3). Thus, π written out begins; 3.141 592 653 ... and so, 211.061 mm (the '21 cm' line's accurate wavelength) multiplied by π gives a wavelength of 663.068 mm (452.129 MHz). Some ETIs might thus think that is a good choice of wavelength for their broadcasts. Given the abundance of such 'famous' numbers in mathematics and physics, though,[563] a SETI radio astronomer might just as well observe the whole of the radio window anyway, rather than attempt to guess what special wavelengths an ETI might choose on this sort of basis.

BOX 15.3 – PHYSICS – SOMETHING WHICH WE HAVE IN COMMON WITH THE ETIS

In case you are now thinking "Oh but the ETIs' units and numbers will be quite different from ours, so they will get different wavelengths from us". Well, you would be wrong, that is the beauty of mathematics and physics; they are the same everywhere, even if the labels (letters, numbers, etc.) being used differ.

Example

A very computer-literate ETI society is living on the (yet-to-be found) 10th exoplanet of the TRAPPIST-1 system: TRAPPIST-1 k. The Kayians use a base of '16' for their number system,[564] instead of *our* normal base of '10' (the decimal system). Now the Kay-ians have their own symbols for their 16 basic numbers with our '8', for example, having the symbol ËÞ. However, to use the Kayians' symbols for their numbers would unnecessarily make this example even more confusing than it may be anyway because the symbols themselves do not matter.

Thus, we will use humankind's symbols for our version of the hexadecimal system. The first ten of these symbols are just the same as our normal decimal numbers, i.e., 0, 1, 2 ... 8, 9. However the human hexadecimal system then uses letters so that 10 = A, 11 = B, 12 = C, 13 = D, 14 = E and 15 = F. The decimal number '1,010' then becomes the hexadecimal number '3F2'.[565]

Now, insofar as we can tell, physical laws are the same wherever you may be – and certainly they are valid over trivial distances of a few tens of thousands of light years. So, the

---

[562] Hence, an in-joke amongst SETI researchers is "that they hope to meet the ETIs at the Watering Hole".

[563] For example, Euler's number, e == 2.718 281 ... , the golden ratio, $\pi \lambda$ = 1.618 033 ... , the square root of 2 = 1.414 213 ... , the cube root of π = 1.464 592 .... and so on and so on.

[564] A.k.a., the Hexadecimal system, when used by human computer programmers.

[565] That is, 2 ones plus 15 sixteens plus 3 sixteen-squareds = (in decimals) 2 x 1 + 15 x 16 + 3 x 256 = 2 + 240 + 768 = 1010.

physics of the Kayians will be the same as ours. When we come to wavelengths and frequencies, we find that multiplying the two together gives us the speed of light, i.e.

$$\text{wavelength x frequency} = c.$$

Since the Kayians' physics is the same as ours, they will also find the same relationship, and we can test this using the values we have just been looking at for the 21-cm line.

Re-arranging the relation slightly, for convenience of calculation, we get:

$$\text{frequency} = c \text{ / wavelength}.$$

On Earth, therefore, for the 21-cm line we have:

Wavelength = 0.211 061 m
c = 299,792,458 m /s
(Frequency = **1,420,410,000 Hz**)

and thus the example relationship becomes:

c / Wavelength = 299792458 / 0.211061 = <u>1,420,406,000</u> = 21-cm line frequency (given the accuracy of the input figures).

Going back to the Kayians, not only do they have a different number system from us, but all their measuring units differ as well. Their unit for length is the Yin (= 0.3333 m) and their time unit is the Yang (= 23 seconds). For the same spectrum line which we have just been discussing in the terrestrial context (the 21-cm line), the Kayians would find:

Wavelength = 0.A21 847 Yin
c = 4D0 F66 BF0 Yin per Yang
(Frequency = <u>79B 3FF 0F0 per Yang</u>)

and thus the example relationship becomes:

c / Wavelength = 4D0F66BF0 / 0.A21847 = <u>79B 3F2 7D4</u> = frequency (given the accuracy of the input figures)

If we now convert the Kayians 79B 3F2 7D4 per Yang back to our units, we get 1,420,410,000 Hz for the frequency of the Kayians' 0.A21 847 Yin line. Which, since it is what we call the 21-cm line, this is exactly as it should be.

Thus, if you have managed to follow all this:

(a) You deserve to open a bottle of Champagne,
(b) You have shown that despite the differences in notation and units between the Kayians and ourselves, our physics and their physics give the same results,
(c) We are talking about the same things despite our differences

and

(d) If the Kayians broadcast signals on their 0.A21 847 Yin line, then we will receive them as 21-cm ETI signals.

However, although our physics is the same as that of the Kayians, it could all still come to naught if our atmospheres are different. Thus, if the Kayian atmosphere absorbs radio waves around the 21-cm region, then the Kayians would not, after all, be able to broadcast their 0.A21 847 Yin signals to the Universe (though, they could do so from space-based radio telescopes, if they so wished and had space travel).

PROJECT OZMA

In the late 1950s, the construction of a massive radio interferometer, which we now know as the Karl G. Jansky Very Large Array (VLA), was under consideration. As a test bed for that instrument, three 85-foot (26 m) and one 45-foot (14 m) radio telescopes were built at Green Bank, a small town in West Virginia about 250 km West of Washington. The first of these telescopes started operations in 1959 (Figure 15.2).

FIGURE 15.2   The 85-foot Howard E. Tatel radio telescope at Green Bank Observatory. Used for the first 21-cm line SETI investigations in Project Ozma. (Reproduced by kind permission of the NRAO/AUI/NSF.)

Frank Drake at that time was already working at the Green Bank Observatory studying Venus and Jupiter at radio wavelengths. He managed to obtain permission to use some of the Tatel telescope's observing time to look for artificial signals coming from the vicinities of the stars τ Cet and ε Eri. The program was named Project Ozma after Princess Ozma in L. Frank Baum's *Land of Oz* books.[566]

For four months in 1960, Drake therefore observed his two stars for about six hours each day. He monitored a region of their radio spectra centred on the 21-cm line for the reasons discussed in the previous section. At that time there were no known exoplanets (Chapter 4), but his chosen stars were fairly similar to our Sun. τ Cet is a little cooler than our Sun but at a distance of only 12 ly, it is the nearest star to us that is reasonably Sun-like. ε Eri is relatively cool (~5,100 K, 4,800 °C) and slightly closer to us than τ Cet at a distance of 10.5 ly.

---

[566] She does not, though, appear in the *Wizard of Oz*.

Drake did not detect any artificial signals from either τ Cet or ε Eri, but τ Cet *is* now known to have up to seven exoplanets and two of those exoplanets might be within the host star's habitable zone. ε Eri has a single known Jovian-type exoplanet, orbiting some 3.5 A.U. out from the star.

Thus, for Drake, whilst he did not hit his hoped-for ETI-signal jackpot during Project Ozma, his aim was remarkably good and Project Ozma was the first serious and scientifically well-based attempt to search for ETIs.

REPORT ON SETI OBSERVATIONS FROM 1960 TO THE PRESENT DAY

Results – 0/10
Efforts – 10/10

RESULTS

A mark of 0/10 for SETI's results to date is appropriate, but not quite fair. There is *some* information to be obtained from the fact that we have *not* detected any evidence of ETs or ETIs so far (see the Fermi paradox, Chapter 16). That is because, as already discussed, *we* should be detectable by some ETIs out to a distance of some 70 or 80 ly, from our radio and TV broadcasts. Any ETIs out to a distance of 35 to 40 ly would therefore have had time to receive our broadcasts and to send us back a reply. But they have not done so.

There are several possible explanations for this unresponsiveness:

- The ETIs are there but do not want to contact us (a.k.a., the Zoo Hypothesis)
- The ETIs are there but are keeping quiet until their invading army reaches us next month
- The ETIs are there but are still awaiting their politicians voting on a course of action
- The ETIs are there but have not quite invented radio receivers yet
- The ETIs are there but are now so advanced that they no longer bother with radios, etc.
- There are no ETIs closer to us than ~40 ly.

The last possibility would seem quite likely to be the correct one. But one fairly definite conclusion which *can* be drawn from our non-observations is that there are probably not *two or more* ETI groups living within ~40 ly of us, otherwise one at least would surely have contacted us by now, if only for one-up-ETI-ship reasons. A recent statistical estimate based upon this lack of contact estimates that there can be no more than five galaxy-wide broadcasts by ETIs made per century and that the odds are evenly for or against our detecting even one of those within the next 1,800 years.

THE LGM AND WOW! SIGNALS

Twice in the last several decades, researchers have detected radio emissions which they really thought for, a short while, meant that a signal from some ETIs *had* been detected. Although neither signal did change the SETI result's score from 0/10 to 10/10, those two detections are worth summarising in their own rights.

The LGM Signal

In late November 1967, a graduate student at the Mullard Radio Astronomy in Cambridge was patiently inspecting metre upon metre of chart paper produced at the rate of 30 metres a day by a radio telescope[567] operating at a wavelength of 3.67 m (81.5 MHz). This was a part of her work

---

[567] The radio telescope was of a design called a 'phased array' and it resembled none of the more familiar radio dishes (Figure 15.2). Instead, it looked like a field covered in dozens of criss-crossing washing lines and Bell had helped to construct it during the first two years of her doctoral studies. Further details of phased arrays may be found in Box 18.3.

towards obtaining a Ph.D. degree and Jocelyn Bell (for that was her name[568]) was hoping to solve some of the mysteries posed by Quasars. She noticed that one small (25 mm) section of this huge output seemed to be different from most of the rest of the signal.

Close examination of that unusual section of the output showed that it was a series of bright radio flashes occurring at very regular intervals[569] of 1.3 seconds, i.e., just exactly how some distant ETIs might send us a signal. She and her doctoral supervisor, Anthony Hewish, even labelled it 'LGM-1', standing for 'Little Green Men -1'.

Were Bell's Little Green Men really ETIs? Had she made the discovery of the century?

Well, it soon became clear that she had *not* detected the first signal from alien life-forms, but she *had* made the discovery of the century, or at least the discovery of quite a lot of decades. For instead of being an artificial signal, LGM-1 was the natural radio emission from a pulsar (Chapter 5) and Bell had thus discovered the first example of one of *those* quite incredible objects. Bell's pulsar now goes by the name PSR B1919+21.

The Wow! Signal

To begin with, the Wow! signal's story closely resembles that of the LGM signal: a short anomalous radio signal embedded in masses of output from a radio telescope. The details and outcomes, though, differ somewhat from those of the LGM signal.

Jerry Ehrman, when he found this anomalous signal in 1977, was already working voluntarily on a SETI project using the Ohio State University's Big Ear telescope. The Big Ear radio telescope, after initially producing the Ohio Sky Survey was then devoted to SETI work for the next 20 years.

The Big Ear, like Bell's instrument, was not of the standard radio dish design.[570] It resembled, (before being dismantled to make way for a golf course in 1998), two very large metal-mesh fences. The first 'fence' was flat and tiltable and about 30 x 100 metres in size. It reflected a radio signal, without focussing it, into a horizontal beam of radiation. The second 'fence' had a similar size to the first but was fixed in its position and was a segment of a sphere in its shape. The second 'fence' focussed the horizontal beam onto the radio detector which was situated between the two 'fences'. The telescope could point to objects at different altitudes by changing the tilt of the first 'fence' and observe them for about 70 seconds whilst they were in the detector's field of view.[571]

By 1977 somewhat more sophisticated output devices were available than Bell's chart recorder, but reams of recordings still had to be gone through by hand and eye. In fact, Ehrman would seem to have had a *less* user-friendly output than a chart recorder, because he had to look through line upon line of printed letters and numbers.

The Big Ear was operating then by observing at 50 different wavelengths simultaneously, all close to the 21-cm line. The strength of the signal at each wavelength was measured for 10 seconds and then printed-out, giving a line of 50 symbols. The equipment being used at the time, though, could only print *one* symbol to represent the strength of each wavelength. This was OK for the strengths[572] from 1 to 9, but what if the signal reached strength 10? The solution which the researchers used was to print a letter instead of a number; '10' thus comes out as 'A', '11' as 'B', … '17' as 'H'[573] … and so on.

---

[568] A.k.a., Jocelyn Bell-Burnell.

[569] The interval between the pulses is very stable indeed, but it is increasing slowly. Nonetheless, it is sufficiently regular that it can be measured to 13 significant figures. By 1991 it had reached the value of 1.337 302 088 331 s.

[570] Its design is named after its inventor, John Kraus. It is sometimes called a transit design because the objects are observed as they 'transit' the due-South meridian. This, though, should not be confused with the transit method of observing exoplanets (Chapter 7).

[571] A few years later the detector was mounted on a movable trolley, thus allowing for longer periods of observation – but that was too late for the WOW! signal.

[572] The strengths were measured as multiples of the average of the background changes (multiples of the background's standard deviation if you are acquainted with some statistics – see also Box 9.4).

[573] At first sight, this seems to be similar to the hexadecimal numerical system (Box 15.3). However, only a single number or letter is used and you just hope that the strength does not rise above 'Z' (35 x the standard deviation of the background).

After a few minutes of observing, the output would thus be a set of 50 columns of these symbols (Figure 15.3 Left) occupying a whole page – and there were hundreds of such pages for Ehrman to go through. It probably does great credit to his tenacity, therefore, that he found the Wow! signal only a 'few days' after it had been received.

The signal was received on the 15th August 1977. As can be seen from Figure 15.3 Left, it appeared at only one of the 50 wavelengths under observation and comprised six measurements over about a minute of time. The signal itself printed out as:

6, E, Q, U, J, 5

Translating these symbols in decimal notation, the successive strengths were:

6, 14, 26, 30, 19, 5

and these figures are shown in a more accustomed format in Figure 15.3 Right.

So, was the Wow! signal from some ETIs or not? Half a century later and the answer to that question is still debatable (and debated). No comparable signal was received by the Big Ear at any time or at any wavelength throughout whole of the remainder of its working life (21 years).

Thirty years after the Wow! signal had been received, Ehrman published a 17,000-word review of the very numerous explanations, ideas, analyses, simulations, suggestions and guesses which had been put forward relating to the Wow! signal, concluding:

"Thus, since all of the possibilities of a terrestrial origin have been either ruled out or seem improbable, and since the possibility of an extraterrestrial origin has not been able to be ruled out, I must conclude that an ETI (ExtraTerrestrial Intelligence) *might* have sent the signal that we received as the Wow! Source."[574]

By general consensus, though, the Wow! signal cannot be regarded as an ETI signal based upon the normal standards of proof used in science (see footnote[527]).

Still, you can assess the evidence for yourself and make your own mind up – and perhaps, one day, a second Wow! signal will come through and decide the matter for us for certain.

Efforts

We have essentially covered the efforts going into optical SETI earlier. Here, therefore we examine the current and possibly the future state of radio SETI.

A great deal of effort is now going into the type of observations which the Big Ear was undertaking, although, as already remarked, much of the funding comes from private individuals convinced of the value[575] of SETI. More conventional funding sources do, though, also support some of the efforts. Indeed, even that notorious critic of wasteful (US) government spending, William Proxmire, could be persuaded of its importance. As early as 1978 Proxmire had awarded NASA's efforts towards finding ETs and ETIs one of his deprecating and damaging Golden Fleece awards. A little later though, Proxmire, following discussions with Carl Sagan and others, did become convinced that ETI searches, at least, were worth supporting on the grounds that we, as an infant technological civilisation, might well be able to learn from more advanced civilisations.

SETI has continued to have a chequered history in terms of its official support, with, for example, NASA losing funding for its programmes in this area for a time in the early 1990s, following criticisms by another US senator, Richard Bryan. One way and another, though, the work has continued and many other studies in many different nations have flourished. There are numerous excellent

---

[574] *The Big Ear Wow! Signal (30th Anniversary Report).* North American AstroPhysical Observatory May 2010.
[575] See Chapter 1, curiosity and blue-sky research.

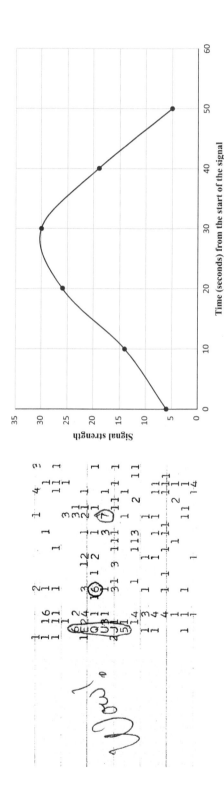

FIGURE 15.3 Left – The output from the Big Ear as inspected by Ehrman. The Wow! signal is circled in the second-from-left column. The word 'Wow!' was written onto the printout as Ehrman's instantaneous reaction upon seeing the signal – and that is how the event received its name. (The image is in the Public Domain, but I would like to thank the Big Ear staff anyway, for making it so generally available.) Right – A more conventional graph showing the rise and fall of the Wow! signal. The intensity variation is due to the varying responses of the detector, as the source crosses its field of view. The source itself was almost certainly much more constant in its intensity and probably lasted for a much longer time. (© C. R. Kitchin 2023. Reproduced by permission.)

sources[576] for SETI's history over the last three or four decades, but all SETI programmes have so far failed to find any ETs or ETIs. However, to give a flavour of this activity, some details of a few of these programmes will be discussed further here.

China has recently completed the 500-m diameter FAST[577] radio dish which is currently the largest filled-aperture telescope in the world. FAST has SETI as a part of its core scientific mission and had a false alert in June 2022 – which later turned out to be a terrestrial source. If used for CETI with a 1,000 GW transmitter, then it is claimed that that signal could be received at any exoplanetary systems around the nearest 1 million stars; that figure though is dependent upon the sensitivity of those ETIs' radio receivers.

Iosif Shklovskii inspired many SETI researchers and promoted much early interest in the subject with his 1962 book, *Universe, Life, Intelligence*.[578] So, the Soviet Union/Russia made a good start towards having an interest in SETI. However, funding problems have subsequently limited their practical observational work severely. Nonetheless, an Israeli/Russian entrepreneur, Yuri Milner provided $100 million in 2016 to fund the Breakthrough-Listen project. The latter is based at the University of California, Berkeley, using the (now, independently owned) Green Bank radio observatory, the Australian Parkes radio telescope and, for optical observations (Box 15.2), the Automated Planet Finder[579] at the Lick Observatory.

The SETI Institute (see above) together with the Radio Astronomy Laboratory (University of California, Berkeley) operates the Allen radio telescope array. This telescope was initially funded with donations totalling, now, over $30 million from Paul Allen.[580] It comprises an array of 42 six-metre radio dishes and operates automatically at several wavelengths between 27 mm and 600 mm (11.2 GHz and 500 MHz – which includes the 21-cm line). Half of the Allen array's observing time is devoted to SETI observations.

The US-based SERENDIP[581] programmes make use of other radio telescope research work. They are sometimes called 'Piggy Back' or 'Commensal' programmes since they use radio observations obtained for other purposes and then search them for possible SETI signals. Various 'editions' of SERENDIP have evolved since it first started in 1979 and 'SERENDIP VI' began in 2014 using the Green Bank radio telescope and the, now destroyed, Arecibo telescope. The programme has detected some hundreds of questionable signals, but none have yet proved to have ETI origins. As of 2019, SERENDIP was still functioning on the Green Bank telescope alongside another piggy-back programme called GREENBURST. COSMIC[582] is currently using the data acquired by the VLA in a piggy-back fashion in a joint programme between the SETI Institute, Breakthrough Listen and the NRAO.[583] The FAST telescope had a citizen-science project called 'SETI@home' to analyse its piggy-backed data for SETI signals for a while. That project is now 'in hibernation'.

The European Low Frequency (radio) array (LOFAR) operates at much longer wavelengths than the SETI instruments discussed so far: 1.3 to 30 metres (230 MHz to 10 MHz). It comprises around 20,000 simple dipole antennas (rather similar to some old-style TV aerials) which are distributed over a wide area of Europe ranging from Sweden to France and from Eire to Poland. It works alongside Breakthrough Listen and has so far monitored 1.6 million stars for any SETI-type signals at long wavelengths including stars in quite a number of galaxies outside the Milky Way. I probably

---

[576] Paul Shuch's *Searching for Extraterrestrial Intelligence: SETI Past, Present, and Future*, 2011, Springer, or; https://en.wikipedia.org/wiki/Search_for_extraterrestrial_intelligence, would be good starting points for any reader wanting further details.

[577] Five hundred metre Aperture Spherical Telescope.

[578] Printed in expanded form, in English with contributions from Carl Sagan, in 1966 as *Intelligent Life in the Universe*, Holden-Day.

[579] A 2.4-metre robotic telescope whose main mirror blank was cast in Russia.

[580] One of the founders of *Microsoft*. Other smaller, but still substantial, donations have helped to operate the instrument in recent years (see also footnote [6]).

[581] Search for Extraterrestrial Radio Emissions from Nearby Developed Intelligent Populations

[582] Commensal Open-Source Multimode Interferometer Cluster.

[583] National Radio Astronomy Observatory

hardly need to add (but will anyway) that no SETI-type emissions have been detected by LOFAR to date.

CETI

As previously mentioned (Chapter 14 and above), the sending out of signals deliberately intended as messages to ETIs (CETI) has already been attempted. Of course, there have been no replies to date. The practice has been criticised (Chapter 14), but it *has* occurred. Here, therefore we briefly review those 'experiments'.

In many ways, the most famous act of CETI was/is not a signal at all. The Pioneer 10 (Jupiter flyby mission, launched March 1972) and Pioneer 11 (Jupiter and Saturn flybys, launched April 1973) spacecraft were the first human artefacts to be launched with high enough speeds to leave the Solar System. Since they just might (in many, many, many years, Pioneer 10 is expected to pass within ~0.75 ly of the cool star HIP 117795 in Cassiopeia about 90,000 years from now) be found by ETIs, each spacecraft carried a plaque with information about ourselves, the Earth and Solar System (Figure 15.4).

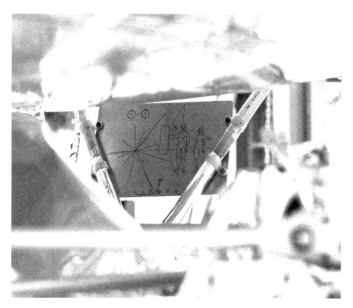

FIGURE 15.4   The 'Greeting to Aliens' plaque on Pioneer 10, seen in position on the spacecraft.

This image is probably familiar to most readers, but in summary, it shows (top left) the transition of molecular hydrogen which produces the 21-cm line (see above), the figures of a man and a woman with the Pioneer spacecraft to the same scale behind them, the position of the Solar System with respect to 14 pulsars and a schematic of the Solar System with Pioneer 10's path through it on its way out of the Solar System. (Reproduced by kind permission of NASA.)

The two Voyager spacecraft, which will also leave the Solar System in due course, similarly carried greetings for ETIs. The New Horizons mission to Pluto and beyond (launched January 2006), however, does not carry any such message.

Next, in order of fame (or infamy) is the three-minute radio signal sometimes called the 'Arecibo message', sent in November 1974 from the Arecibo radio dish towards the globular cluster, M13.[584] It was a 450 kW, frequency-modulated signal, at a wavelength of 126 mm (2,380 MHz) comprising 1,679 binary bits. If the number '1,679' seems to be a slightly eccentric choice, then that is not

---

[584] About 25,000 ly from the Earth, so we might get a reply 50,000 years from now.

actually the case. It is the product of two prime numbers, 23 and 73, and the originators of the message hoped that any ETIs receiving the message would realise that this meant that it was a rectangular image with sides 23 units and 73 units long.

Primary numbers are universal (see also Box 13.3) whatsoever the form of mathematics that the ETIs may have developed. Most of humankind's discussions of interstellar signalling now automatically assume that any messages will have a similar basis. An ordinary 2D image thus being the product of two prime numbers, a 3D model or a 2D movie would be the product of three prime numbers and a sequence of 3D models would be the product of four prime numbers and so on.

The Arecibo message itself purports to tell the ETIs of the numbers one to ten, the principal chemical elements making up DNA and that molecule's structure, what humankind looks like, the then number of humans on the Earth and details of the Solar System and of the Arecibo telescope. Whilst it is a miraculous achievement to get that amount of information into such a brief message, it would be a greater miracle if even the most super intelligent of ETIs could decipher it successfully (Figure 15.5). Similarly, in 2023, the SETI Institute's 'Sign In Space' project simulated the receipt of

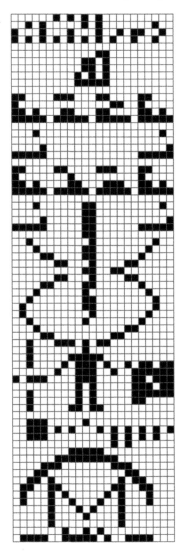

FIGURE 15.5 A pictorial representation of the Arecibo message. (Reproduced with kind permission of Arne Nordmann under the Creative Commons Attribution-Share Alike 3.0 Unported License; https://creative-commons.org/licenses/by-sa/3.0/deed.en.)

an ETI radio message on the Earth, by transmitting a coded message to the Earth from the ExoMars Trace Gas Orbiter in orbit around Mars. Although the message was received and interpreted as a 2D pattern quite quickly, in so far as I can ascertain at the time of writing, the actual meaning of the message is still undeciphered.

The Yevpatoria 70-m planetary radar dish, sited in Ukraine, has been used several times in sending signals, some on a commercial basis, to potential ETIs. Starting in 1999, two sets of messages, called Cosmic Calls 1 and 2, were directed towards some nearby stars, such as 15 Sge and 47 Uma,[585] which had some non-zero possibility of hosting life-forms. The messages were each sent three times at ~150 kW, at a wavelength of ~60 mm (5 GHz), were some 400,000 bits in length and had speeds between 100 and 2,000 bps. According to one report (on the BBC), members of the public could add their own messages for $15 a time to the signals. The same radio dish was used in 2001 to send messages containing, amongst other things, some musical pieces, to six more nearby stars. This programme was called 'Teen Age Message' since the contents of the messages were suggested by Russian teenagers. The same messages had been offered to the Arecibo observatory for transmission a year earlier but had been declined because of the dangers (Chapter 14) of revealing our existence to ETIs of unknown character.

The potential danger inherent in sending out messages, which led to the initial reluctance to transmit the 'Teen Age Message', seems to have been quickly forgotten. In 2008, NASA played the Beatles' song *Across the Universe* using a 63-m dish of its Deep Space Network to any ETIs residing in the general direction of Polaris. Whilst in 2018, a private organisation with the name 'METI', used an EISCAT radar dish in Tromsø to send a 2 MW message towards Luyten's star, a red dwarf ~ 12 ly away, with two (perhaps four) exoplanets which are of Earth-masses or somewhat more.

In 2013 a very brief message (144 characters) was sent towards the star GJ 526 (which is a cool flare star about 18 ly away from us and which may be seen through a 150 mm or larger telescope, a few degrees from Arcturus). The, now non-operational, 30-m radio dish at the Jamesburg Earth Station in California was used to send the message, in what was called the 'Lone Signal' project.

An old concept, known as a Bracewell probe, might on long time scales, be able to satisfy both those worried about revealing the Earth to ETIs and those wanting to send signals to ETIs. In the original concept, the Bracewell probe would be what we would now class as a high-end AI machine with extensive records of all aspects of the Earth, Terrestrial life and anything else we might like to tell the ETIs. It would be sent through space (so we are back to very long travel times again) to somewhere close(ish) to the ETI's home system. It would then communicate with the ETIs and tell them all they wished to know about us *except* where the Earth is in space. A related concept would be to send a probe out a *long* way from the Earth and use it as a relay. The signals reaching the ETIs would then come from the probe and not from the Earth's direction.

Whatever the merits or demerits of CETI, though, as remarked several times already 'It is Too Late Now'; the cat has potentially been out of the bag since 1886 and Hertz' discovery of the radio spectrum. It could still, though, be advisable not to stir the wasps in their nests by sending out signals deliberately aimed at where we think ETIs might be found (see Louis XIV quote, Chapter 14).

Incidentally, if you want to send your own (definitely harmless) message to a few ETIs, it is very easy to do so. Go out on a clear night and point a torch anywhere that you fancy in the sky. You can also say 'Hello' using the Morse code if you so wish. A one-watt LED torch emits ~10^{18} visual photons every second. If you point the torch in one direction for 100 seconds, that will produce 10^{20} photons heading outwards from the Earth all in generally that same direction. Each one of those photons, unless it is absorbed by an atom or molecule somewhere and somewhen, will carry on moving out into the Universe forever, or at least until the Universe ends in some fashion. Perhaps one of those photons will end up in an ET's eye, so delivering your message. But if not, you will still have the satisfaction of knowing that there are ~10^{20} photons travelling through parts of the Universe where none would have been, had *you* not pointed *your* torch *just so*, ten billion years

---

[585] These two stars are solar analogues. 15 Sge is ~58 ly away and is a binary system with a brown dwarf whilst 47 Uma is ~45 ly away and is (now) known to host three Jovian type exoplanets.

earlier; a memorial (if rather an anonymous one) vastly longer-lasting than anything the Pharaohs might have dreamt about.

1010000000111101001110111001110111010100010100 … … … …

GREAT – at long last, after X decades of work by XX thousands of people and by spending $XXX billions we have finally got a signal from those nice ETIs, called the sneilAehT,[586] who live on that gorgeous exoplanet just 77.2 ly beyond Sirius.

Isn't it all WONDERFUL!!!

Errr …, Um …, Uh … Yes … But … … … Well, what do We *DO* next??

Fortunately, 'What do we do next' is already being thought about. Generally, it is assumed that the signal will be in binary format, although this could well be wrong. However, as this is a hypothetical scenario, we will assume, to keep things moderately simple, that it *is* a binary signal. In which case, the title to this sub-section shows what it could look like, except that it will almost certainly continue in this same fashion for page, after page, after page, after ….

So, where do we start?

If we have inadvertently intercepted an ETI interbank message in 120 different languages, notifying their corporate associates on 540 other exoplanets, of a change in the sub-prime base rate from 110010100 to 110010101 and which message is encoded at three different quantum levels in 120-bit bytes, with the originating language is based upon a sexagesimal numerical system and an hieroglyphic alphabet comprising 216,000 basic characters and 3,600 modifiers … then …

No Chance.

We have, then, to assume/hope that the message is *intended* by its senders to be received *and* understood.

Since we have no idea of the capabilities of the message's senders, we ask instead and as usual 'how would *we* go about send a message which was intended deciphered and understood by intelligent beings with absolutely no knowledge of ourselves or of our circumstances?'

Fortunately for this author, one approach to this problem has been extremely well described by Carl Sagan in his SF novel *Contact*.[31] So, the next few paragraphs are based upon his ideas.

We have already seen[587] that any type of conversation is quite impracticable over light-year distances. It is almost certain, therefore, that the sneilAehT will simply send out a continuous message lasting months or years and then repeat it multiple times. When we first detect them, inevitably, of course, it will be in the middle of one of those repetitions. Somehow, the sneilAehT will have to indicate when one repetition ends and the next starts. One example of how this could be done is just to send (say) a string of 65536[588] zeros.

The first part of the message would then have to be a tutorial training ourselves to read the message itself. This would probably have to start by indicating the binary coding. In this example, humankind's fairly usual coding of 8 bits to a byte is being assumed. This could be indicated by, say, 8 zeros followed by 8 ones and then repeating *that* sequence 4095 times (another 65536 bits).

Thereafter, the tutorial proper might start with the numbers and basic arithmetic. Using our own 8-bit byte representations, then,

$$1 + 1 = 2$$

would look like:

00110001 00100000 00101011 00100000 00110001 00100000 00111101 00100000 00110010

| 1 | space | plus | space | 1 | space | = | space | 2 |

---

[586] I am sure that you already know that the sneilAehT read from right to left.

[587] Chapter 14, Section 'What If the ETIs Are Friendly?'

[588] 2^{16}.

except that the gaps in the sequence would not be there in reality; here they are left in for clarity. That sequence would be repeated several times. Any competent human code breaker would quickly notice the 72-bit repetition and also pick up the repetitions of the '1' (00110001) and the 'space' (00100000) sequences. A computer-based code cracker would do the same in microseconds.

The lesson might then continue with repeated sequences of:

$1 + 1 + 1 = 3$

00110001 00100000 00101011 00100000 00110001 00100000 00101011 00100000 00110001

1   space   plus   space   1   space   plus   space   1

00100000 00111101 00100000 00110011

space   =   space   3

and;

$1 + 2 = 3$

00110001 00100000 00101011 00100000 00110010 00100000 00111101 00100000 00110011

1   space   plus   space   2   space   =   space   3

This looks cumbersome, but even at only 100 bps and with ten repetitions, the transmission would take less than half a minute (~3 seconds at 1,000 bps and so on).

By this stage the symbols, 1, 2, 3, =, + and space may well have been identified and their meanings realised for what they are, because under the circumstances, human recipients of the message would be expecting some sort of progression like this. If not, then continuing the sequence up to, say, $1 + 2 + 3 = 6$ and so on should soon get the idea over.

With the basic numerals and '+' and '=' understood, the decimal notation can be introduced using (say), $1 + 2 + 3 + 4 = 10$. The multiplication and division operations then via (say): $2 \times 3 = 6$ and $9 \div 3 = 3$.

The alphabet would be a little trickier, but could, say, start off from the now known maths symbols by replacing those symbols by their word equivalents and going through another training progression. Thus '3' say would be replaced by 'three' and then become:

01110100 01101000 01110010 01100101 01100101

t   h   r   e   e

Probably within a few hours of such training, a computer program would have been written to convert the digital signal coming from the sneilAehT straight into our own familiar notations.

After that, might come an image (indicated as the product of two prime numbers – see above) of, say, a beautiful adult sneilAehT, labelled with the names of all 16 tentacles, the 3 green, warty heads and the 14 types of drool coming from its 7 mouths.

The sneilAehT, of course, would not know that we had received their message for another 85.8 years (Sirius is 8.6 ly away from us) – and then only if we sent them an immediate reply. Furthermore, they would not know if we had succeeded in deciphering their message within the signal.

Once the initial training is over, then, the message will probably continue with whatever the sneilAehT think that *we* might like to know. This could be a video from 85.8 years ago of the semi-final of the inter-planetary llabtoofecapS game (which lasts for at least 2 years, without injury time) or detailed instructions on how to cook their famous national dish, snoinodnaepirT.

Alternatively, as in Sagan's 'Contact', we could receive the building instructions for a vehicle which would travel (via wormholes,[589] not through space) the distance of ~26,000 ly from the Earth to the Centre of the Galaxy in just a few hours. It could even be the instructions on how to obtain unlimited energy from the vacuum of space, or … well, space is left here for your own 'Birthday gift' list:

……………………………………………………………………………………………………

The message imagined being received on Earth in the 'Contact' novel had another twist to it; it was actually two messages overlaying each other. The obvious message was an amplitude modulated signal listing the prime numbers. This however was just a 'heads-up' for the real message which was being transmitted within the same signal by changing the radio wave's polarisation (Chapter 11).

In fact, THREE independent messages could have been sent by changing the radio wave's wavelength/frequency as well, usually called frequency modulation, or FM. Many readers will probably have their own FM radios and so the concept should not be unfamiliar.

The idea of amplitude *and* frequency modulation, usually called 'Mixed Modulation' within a single broadcast, may though seem likely to be impossibly confusing. However, many (probably all) readers will also be familiar with this as well. If you have ever listened to music with more than one instrument playing simultaneously, then you have probably noticed that the violins (say[590]) could be heard quite distinctly and separately from the flute (say). However, the only signal actually going into your ears would be a simple air pressure wave, which at any given instant of time would only be exerting a single value of the pressure. Assuming, that the air pressure when everything is silent is 101,350.000 Pa, then at some instant during the music it might have risen to say 101,350.002 Pa. A millisecond or so later, that pressure may have changed to 101,350.003 Pa,[591] say, but it is still, at any of those instants, just a single pressure. Nonetheless not only will you hear the violins and flute as separate instruments, but they will become louder and softer (i.e., amplitude modulation) and the notes being played will change their pitch[592] (i.e., frequency modulation).

Thus, a single signal from (or to) an ETI could contain up to three independent messages and these could be about quite different topics and/or with different levels of sophistication and/or three different camera views of the semi-final of the inter-planetary llabtoofecapS game and/or …..

Sagan's approach to sending a message which (after working on it) could be deciphered without prior knowledge, except for basic mathematics, is one of several which various researchers have tried to develop. The Busch binary language,[593] for example, using a blind test message of ~75,000 bits repeated three times, was able to convey our scale of linear units by using the Planck length[594] as a starting point. This was a blind test, but since a convenient ETI was unavailable, the recipient was a human and would thus have had the same background knowledge, training and expectations as the sender. Its results may not, therefore, truly represent whether or not a real ETI could have understood the message.

---

[589] The wormhole is a concept from General Relativity which provides a short cut between two well-separated points in space. It can be imagined as an interstellar 'Panama canal' (which makes the journey from Rainbow City to Panama City about 80 km instead of the ~23,000 km it would be if you were to go around Cape Horn). However, wormholes, if they exist and if you can find one, are probably not a practical means of making journeys, since you may have to go through a black hole first in order to get to them.

[590] Please substitute you own favourite instruments.

[591] The sound which you hear would have increased in loudness from ~40 db to ~44 db with this change.

[592] Sometimes in the way that the composer had intended.

[593] https://www.lpi.usra.edu/meetings/abscicon2010/pdf/5070.pdf

[594] A basic unit of length derived from the universal constants of nature: the speed of light (c), the gravitational constant (G) and the Planck constant (h). Its formula is

$$\text{Planck Length} = L_P = (h\,G\,/\,2\,\pi\,c^5)^{0.5} = 1.616 \times 10^{-35} \text{ m}.$$

Since L_p is based upon physical constants which are the same everywhere and not, for example, on the distance covered by 1,000 paces of the marching Roman legionnaires (which is the original definition of a 'mile' – NB: a 'pace' is two 'steps'), then it will be the same actual distance in space between two points for the ETIs as it is for us.

16 That MHoQs Again; A.K.A. 'The Fermi Paradox'

INTRODUCTION

In the preface to this book, the Most Human of Questions (MHoQs) was introduced and it is

Is there life elsewhere in the Universe?

Since then, we have journeyed through a great deal of cutting-edge science to try and find the answer to that question.

We have thus seen that:

- There are many thousands of exoplanets already discovered, some reasonably well-suited to hosting life, others less so.
- It is highly likely that there are at least millions of exoplanets within the Milky Way galaxy and quite possibly far more exoplanets than the ~100,000 million known stars in that galaxy.
- There are perhaps 1,000,000 million galaxies in the visible Universe and most, if not all, seem likely to have abundant populations of exoplanets.
- There may be even more free-floating exoplanets than there are exoplanets which are gravitationally bound to stars.
- It may be possible for life to exist elsewhere than on the surfaces of 'conventional' planets. For example, in oceans buried deep below the surface, within hospitable niches in otherwise untenable circumstances (e.g., perhaps high in Venus' atmosphere).
- Life-forms with biochemistries very different from those of terrestrial organisms may be possible.
- That many of the molecular building blocks of life can form abiotically under a wide variety of circumstances.
- That once any type of First-Living-Molecule has come into existence, then Darwinian evolution will change and develop it to fit its environment better.

All of which suggests that, whilst the answer to the MHoQs is still '*Yes*', it may very soon change to '*No*'.

On the other hand, as we have also seen, some quite determined efforts to detect life away from the Earth, by any available means, have so far failed. Furthermore, the von Neumann machine concept (Chapter 14) suggests that, if there is intelligent life out there, then there has been plenty of time for it to have spread widely throughout the galaxy and perhaps throughout the whole Universe. Both of which lines of evidence suggest that the answer to the MHoQs is '*Yes*' and that it is likely to stay '*Yes*'.

The discordance between these two answers and their lines of evidence comprises the Fermi Paradox (see also Chapter 14).

DOI: 10.1201/9781003246459-19

THE DRAKE EQUATION

Before examining the Fermi Paradox further, there is another relevant input to be considered, although in the end it is not of much practical help. This is the Drake equation.

In 1961 Frank Drake tried to estimate the number of ETIs which we might expect the various SETI and related programmes to be able to find. He did this by identifying all the factors which might influence our ability to detect an ETI civilisation and then multiplying them together, hoping to get a numerical answer (e.g., '2,000,002', '31,415', '0', or whatever). Unfortunately, estimating the actual values of five (of the seven) factors has so far proved to be nigh on impossible. Nevertheless, the factors are listed below and they, at least, indicate areas of the subject requiring further investigation (the symbols are not those usually encountered in this context, but are, I think, easier to follow):

- The rate of formation of stars within our galaxy ($Rate_{Star\ formation}$)
- The proportion of stars which host exoplanets ($Prop'n_{exo\text{-}planet\ system}$)
- The proportion of exoplanets which could support life in some form ($Prop'n_{potential\ life}$)
- The proportion of potentially inhabitable exoplanets which do support life-forms ($Prop'n_{life}$)
- The proportion of inhabited exoplanets which develop intelligent life-forms ($Prop'n_{intel\ life}$)
- The proportion of intelligent life-forms developing technologies detectable from a large distance ($Prop'n_{tech}$)
- The survival time of life-forms with technologies detectable from a large distance ($Time_{survive}$).

The Drake equation is thus:

$$Number\ of\ detectable\ ETI\ civilisations = Rate_{Star\ formation} \times Prop'n_{exo-planet\ system}$$

$$\times prop'n_{potential\ life} \times Prop'n_{life}$$

$$\times Prop'n_{intel\ life} \times Prop'n_{tech}$$

$$\times Time_{survive}$$

The Drake equation, clearly, does not include any contribution from free-floating planets, and we have seen (Chapter 7) that free-floating exoplanets might outnumber those bound to host stars by a large factor. Neither does it take account of any ETI civilisations developing in some other way, independently of the presence of a star. Furthermore, first factor, ($Rate_{Star\ formation}$), does not seem to be appropriate. Life-forms could develop on old exoplanets as well as new when conditions change on the former as their host stars evolve. Moreover and especially for the lower mass stars, their post-formation life-times and their later slow evolutionary changes last thousands to millions of times longer than the times taken for their formations. The simple number of stars in the galaxy, N_{Galaxy} probably therefore should replace $Rate_{Star\ formation}$.

However, this last comment is immaterial when we have no sensible estimates of the values of $Prop'n_{potential\ life}$, $Prop_{life}$, $Prop_{intel\ life}$, $Prop_{tech}$ or $Time_{survive}$. Even the values of $Rate_{Star\ formation}$ and $Prop'n_{exo\text{-}planet\ system}$ are uncertain to within quite large factors, although the latter may be better pinned down soon as more and more exoplanetary systems are discovered (Chapters 4 to 7).[595]

Notwithstanding, then, the fact that the Drake equation would seem to be highly relevant to a discussion of the Fermi Paradox, we shall just have to make do without its solution.

---

[595] The value of N_{Galaxy} is known; it (probably) lies between 50,000 million and 200,000 million.

THE (NOT FERMI'S) PARADOX

The title of this section may seem confusing, but the 'Fermi Paradox' is somewhat unfairly attributed to Enrico Fermi, brilliant polymath though he was. It is always nice to give credit where it is due, but the Fermi Paradox actually seems to have occurred to many people, expressed in various ways, over quite a long period starting from the early 1930s. Fermi's only contribution to the topic was in 1950 when, whilst chatting over lunch at Los Alamos with, amongst others, Edward Teller, he asked the question "where is everybody?",[596] meaning why had the ETIs not been found? Various more equitable labels for the paradox have been proposed such as the 'Hart Paradox' and the 'Hart-Tipler[597] Conjecture, but the 'Fermi Paradox' seems to be here to stay and so that term will be used from now on in this book.

THE FERMI PARADOX (AFTER ALL)

THE QUESTION

The essence of the Fermi Paradox has been outlined already, but in a concise form we may express it as:

Why have we found no evidence for ETs or ETIs, given that the Galaxy / Universe is old enough for life-forms to have originated and evolved to interstellar messaging / space travelling levels many, many times over and so to have permeated the entire Galaxy / Universe many, many times over?

THE ANSWERS

Ask five people interested in SETI and associated matters 'What is the answer to the 'Fermi Paradox' and you will get ten (if not 100) differing responses. A brief internet search will give you dozens of 'answers'. Some of the answers are possible, some are possible for an individual ETI civilisation, but unlikely to affect billions of civilisations, some are unlikely and some are just barmy.

Table 16.1 lists some of the possible suggestions – you pays your money and you makes your choice. I have deliberately not indicated which answers come into which of the above categories for fear of being assassinated by the aliens on board the next UFO to land in my garden.

TABLE 16.1
Some suggested answers to the Fermi Paradox (in randomised order)

- Extraterrestrial life is rare or non-existent
- Humans have not listened for long enough
- Intelligent life may exist hidden from view
- We have seen the evidence but not recognised it
- Who knows? – Not me
- Alien species may have only settled part of the galaxy
- Earth is deliberately being isolated
- Information is cheaper to transmit than matter is to transfer
- Humans have not listened properly
- Civilisations only broadcast detectable signals for a brief period of time
- Everyone is listening but no one is transmitting

---

[596] Different remembrances of the lunch give slightly differing versions of what Fermi said.
[597] Michael Hart and Frank Tipler.

TABLE 16.1 CONTINUED

Some suggested answers to the Fermi Paradox (in randomised order)

- We have not looked hard enough yet
- No other civilisations have arisen
- There is nothing in it for them (i.e., for the ETIs)
- We ARE the first
- It is too expensive to spread physically throughout the galaxy
- Advanced civilisations may limit their search for life to technological signatures
- It is the nature of intelligent life to destroy itself
- Extraterrestrial life is rare or non-existent
- Alien species may isolate themselves from the outside world
- There is periodic extinction by natural events
- We exist – who cares about any others anyway
- They are too technological
- Extraterrestrial intelligence is rare or non-existent
- Alien life may be too incomprehensible
- Intelligent alien species have not developed advanced technologies
- It is the nature of intelligent life to destroy others
- No other civilisations currently exist
- UFOs exist, but the government or the military or who-knows-who keep them hidden
- We will detect them tomorrow
- They are too alien
- Alien species may not live on planets
- Earth is deliberately being avoided
- Colonisation is not the cosmic norm
- ETIs are too far apart in space and/or time
- Lack of resources needed to physically spread throughout the galaxy
- The Earth is quarantined
- Communication is dangerous
- The thought of having the Library of Congress sent to them keeps them hidden
- Space for your own ideas ....

Essentially, then, the Fermi Paradox remains a Paradox at the time of writing. Revert to the MHoQs and the entries in Table 16.1 would change little except that they could all be summarised as 'Your Guess is as Good as Mine'.

17 Signatures – Catching ETs and ETIs Unaware

INTRODUCTION

This topic looks at the possible inadvertent detection of ETs and ETIs via evidence from them which is *not* being sent out deliberately towards other known or suspected ETIs nor as just a wide-angle broadcast sent 'on spec' but still with the intent/hope of it reaching other ETIs at some time. This topic could have been placed as a sub-section within an earlier Chapter, but it seems sufficiently significant and distinct to deserve its own chapter. Additionally, if one or two of the 'answers' to the Fermi Paradox are correct (Chapter 16), 'signatures' may be the only way by which we will ever find any ETs or ETIs.

A signature,[598] then, in this context is not an inky scrawl on the bottom of a cheque, it is some phenomenon, event, change, circumstance or state-of-affairs in, on, around or near an exoplanet which could only have been produced, or is highly likely to have been produced, by some activity of some life-form(s). It is therefore a sign that those life-forms are currently present on the exoplanet or have been present there in the past.

An example from Chapter 10 may help to clarify what this usage of 'signature' is all about. Around 850 million years ago, ETI spectroscopists observing the Earth would have noticed a slow increase in the abundance of oxygen in the atmosphere. This rise was due to the Great Oxygen Event (GOE) and was due to photosynthesis amongst the abundant marine life-forms then inhabiting the Earth.

This increase in the Earth's oxygen levels is an example of a 'Biosignature', i.e., a change in the Earth which has arisen just because of the every-day living processes of the life-forms on the Earth at that time. The simple *presence* of oxygen, water and other necessities of life is called bioindicator. A bioindicator is an essential for the presence of life, but it does, of itself, necessarily show that life *is* present. Thus, it is the *changes* in the oxygen level on the Earth which would be the biosignature, whilst its simple presence alone would be a bioindicator.

If the ETIs had anaerobic metabolisms, then the rise in oxygen may have been a huge puzzle and they would certainly have decided immediately to avoid the Earth like (literally) the plague. But if the ETI astrophysicists were advanced enough to obtain high dispersion spectra of a very small and dim planet, X-hundreds of ly away, then the ETI exobiologists would surely have sussed-out the possibility of oxygen-based life and so they would know that the Earth had life on it, even if they regarded that life much as we regard gangrene.

A Biosignature is one of the two principal signatures. The other is the 'Technosignature'. There are quite a lot more varieties, such as Bio-technosignature, Chemicalsignature, Geo-chemicalsignature, Heatsignature, Isotopesignature, Prebiosignature, Spectral-biosignature and Spectralsignature. These, though, are mostly sub-topics of the two main signatures or are rarely to be encountered or both. We will therefore concentrate on just the two main signatures – 'Bio' and 'Techno'.

---

[598] A.k.a. 'Marker' as in 'Biomarker'.

DOI: 10.1201/9781003246459-20

BIOSIGNATURES

Biosignatures, then, result from life-forms' 'natural' activities, whilst technosignatures result from life-forms' 'technological' activities. ETs thus are likely to be found via biosignatures, whilst ETIs could be found via both biosignatures and technosignatures.

There is likely to be a grey area between the two signatures, when the life-forms are starting to evolve towards developing technical behaviours. Thus, the Earth's GOE (Chapter 10) was clearly a biosignature, whilst industrial pollutants appearing in the atmosphere would equally clearly be a technosignature. But where does (say) clearing jungles by burning and then running herds of animals in their place come? It probably does not matter much in reality, just as long as *we* can detect *some* signature of life.[599]

Biosignatures, though, are probably the only way in which we will be able to find inhabited exoplanets, *before* those inhabitants have developed intelligence and/or industrial-type activities.[600]

In this section we are dealing with biosignatures which can be detected from multi-light year distances. Obtaining samples, laboratory analyses whether in situ or on the Earth, going on safari and bringing back dragons' heads, picking up meteorites, finding crashed spacecraft on Europa, being kidnapped by UFOs and other direct intervention processes have been considered in Chapters 11, 13 and 14 (except for the UFO kidnappings), and some of those would be better described as technosignatures, anyway.

Remote sensing of exoplanets can potentially detect life-forms' biosignatures via direct imaging, photometry, spectroscopy and (perhaps) via polarimetry, and of these possibilities, spectroscopy is the most promising.

Clearly, for a biosignature to be detectable remotely, it must affect the whole or a large part of an exoplanet; the 'flap of a butterfly's wing is not going to be detected, but the defoliation of large forested areas (or their equivalent) by the larvae of butterflies (or *their* equivalent) following a population explosion might well be noticed.

The main current line of searching for biosignatures is by looking for constituents of an exoplanet's atmosphere which might have/must have arisen from some life-form's natural living processes. That, essentially, means analysing the atmosphere's composition via spectroscopy. We have already seen spectroscopy being used to detect exoplanets (Box 6.1) via using it to measure the exoplanet's radial velocity. Although Box 6.2 is primarily concerned with measuring radial velocities, Figure 6.2 (ii) does indicate some chemical elements which have produced absorption lines in the solar spectrum. Thus, lines C, F, G' and h arise from atomic hydrogen, D_1 and D_2 are due to sodium, E is due to iron and H and K are due to calcium.

Thus, we know, without going to the Sun and taking samples for laboratory analysis, that the composition of the Sun includes hydrogen, sodium, iron and calcium. Indeed, the second most abundant element in the Universe, helium, was discovered within the Sun[601] before it was even known that there could be such a chemical element from terrestrial analyses ®. The chemical composition (atoms, ions and, if the temperature is low enough, molecules) of a distant object may, accordingly, be determined from the pattern[602] of absorption and emission lines present in its spectrum.

---

[599] It may be of interest that the JWST should be able to detect bioindicators when observing the Earth from a distance of ~50 ly. Even biosignatures might be detectable.

[600] See the discussion on Dolphins (Chapter 13).

[601] Pierre Janssen and Norman Lockyer in 1868 noticed a previously unidentified yellow emission line in the spectra of solar prominences. Jansen's observation preceded Lockyer's by two months, but their papers announcing their discovery were read together at the same 1872 meeting of the French Academy of Sciences. Both Janssen and Lockyer are therefore credited with discovering helium (especially by the English).

[602] All chemical elements, the neutral atoms, ions, isotopes and molecules have *different* patterns of spectrum lines. Thus the 'C' line due to hydrogen (Figure 6.2) has a (at rest and in air) wavelength of *656.3 nm*, whilst the same line produced by deuterium (a.k.a. 'Heavy Hydrogen'; an isotope of hydrogen with a proton and a neutron for its nucleus instead of just a proton by itself) has a wavelength of *656.1 nm*. Hence all the chemical components of a star or an exoplanet can be separately identified, *provided* that the spectrum is of adequate quality *and* that the temperature of the object is suitable for the lines' production.

Furthermore, the strengths of those lines can be used to estimate the *quantity* of that substance which is present within that distant object.

For a few exoplanets which are sufficiently distant from their host stars to be resolved (i.e., seen separately) from those host stars, it may be possible to obtain their spectra by direct observation (see Direct Imaging and Free-Floating Exoplanets, Chapter 7). This will reveal the gross composition of the atmosphere, if it has one, and/or the spectrum of the solid or liquid surface. Now, unfortunately (at least for spectroscopists, most of us are actually quite grateful to be resting on the 'solid' earth) the spectra of solids and liquids are much less specific than those of atoms, ions, etc., within a gas. Hence, most of the time, the best spectroscopy you can do for solids is not much different from looking at them by eye, i.e., you can see broad colours, but not sharp spectrum lines (i.e.: limestone cliffs are white, sandstone cliffs are a light orange-brown, basalt cliffs are black).

The presence of photosynthesising vegetation, similar to that on the Earth, however, does produce one recognisable sharp spectral feature. This occurs just where the visible and NIR regions merge, at about 700 nm wavelength. Chlorophyll's spectrum changes at that point from strongly absorbing to strongly reflecting and so a 'Red Edge' is seen there in the spectrum. If ever observed, the red edge will not only be strong biosignature, but a strong biosignature for *Terrestrial-type* life.

The majority of exoplanets cannot be resolved from their host stars though, so any spectrum is an amalgam of the spectrum of the star and that of the exoplanet. Additionally, most of the time, the light from the star completely swamps any exoplanet emissions, so no separate exoplanet spectrum can be obtained. Howbeit, when an exoplanet transits its host star (Chapter 7), an opportunity arises for obtaining the spectrum of the exoplanet's atmosphere separately from that of the star. This is because during a transit the exoplanet is silhouetted against the brighter surface of the star (Figure 7.1). If we were able to see closely what was happening, then whilst the star's surface is effectively white and that of the back of the exoplanet is effectively black, there is a *thin grey circle* just around the edge of the exoplanet. This is where the star's light is percolating through the semi-transparent out edges of the exoplanet's atmosphere.[603]

The spectrum obtained during a transit is still an amalgam, but now, a small proportion of the *star*'s light has passed through the upper reaches of the exoplanet's atmosphere. In doing so, the components of the exoplanet's atmosphere will produce absorption lines of their own superimposed upon the stellar spectrum lines. As a result, the *differences* between the stellar spectrum during a transit and the stellar spectrum outside a transit are the spectrum of the exoplanet's atmosphere – and, Hey Presto – we can see the composition of the exoplanet (or at least that of the outer reaches of its atmosphere).

With massive new space telescopes, such as the JWST and Euclid instruments now in operation and soon to be joined by the Roman space telescope and then perhaps by LUVOIR (Large Ultraviolet Optical, Infrared Surveyor, 2039?), the amount of stellar spectroscopic data becoming available is increasing exponentially over very short time scales. Even though most of the data acquired by these new and future telescopes will not be of known exoplanet host stars, by piggy-backing equipment (Chapter 15) onto the main instrument, each spectrum can be checked anyway on the off-chance that a transiting-exoplanet-atmospheric-spectrum has been caught.

I have, though, missed out an important caveat in that last statement so, it should read:

> "… a transiting-exoplanet-atmospheric-spectrum has been caught … *by the AI / Machine-Learning programs powering the piggy-backed equipment.*"

At the time of writing, therefore, much effort is going into developing the equipment and the programs capable of searching 100,000s (Millions? Billions?) of opportunistically obtained spectra of stars, for the very, very occasional anomaly which indicates that an exoplanet with an atmosphere was transiting that star at that moment. The rate of such discoveries is likely to be tiny, but they will

---

[603] Naturally, no such effect occurs for exoplanets without an atmosphere.

almost always be discoveries of exoplanets which no other approach would ever have found – and we should get an idea of its atmospheric composition, quickly and in every case.

So, after all that effort, have we got any positive identification of ETs or ETIs from biosignatures?

Well, … No, … But … the principle of the technique has been established, with several recent such observations from the JWST as well as the earlier results. Also, the following elements and compounds, which are potential biosignatures and/or biomarkers, *have* been detected in some exoplanets' atmospheres:

- Argon
- Dimethyl sulphide
- Isoprene
- Methane
- Nitrogen
- Oxygen
- Ozone
- Phosphine

(NB: not *all* of these have been found in *all* the exoplanets – in fact only one, two, occasionally three atoms/molecules are detected for any individual exoplanet).

Other, not yet detected, but likely, possibilities for biosignature atmospheric gases include ammonia, methyl bromide and nitrous oxide. Bioindicators could also include argon, carbon dioxide, methane and ozone. There are also 'prebiosignatures', such as acetylene, carbon monoxide, cyanoacetylene, hydrogen sulphide, nitric oxide, sulphur dioxide and others, which might perhaps be interpreted as indicating that some form of life, with luck, could emerge in one million … 10 million … 100 million … 1,000 million years … from now.

Whilst biosignatures can intrinsically indicate the presence of life on an exoplanet (finding the spectrum signature of chlorophyll, for example, would be as close to being a sure thing as we are ever likely to get), changes in the biosignatures could also show the presence of life, as for the Earth's changing oxygen levels (above). Similarly, a change in the biosignature following (perhaps years later), a major flare or other change in the host star could be an indicator. The exoplanet change, perhaps resulting from an environmental disaster on the exoplanet or some other major comparable effect.

Although this really comes under the heading of technosignatures, another example, would be for an ETI monitoring the Earth's radio and microwave activities, to notice a drastic drop in their levels following a Carrington-event level solar flare[397] as our communications networks are wiped out and (one hopes) are then followed by a recovery back to previous levels over the next decade or so.

Another 'change' which might support the actual presence of life on an exoplanet with a carbon biosignature, is if the carbon abundance on *that* exoplanet is significantly *lower* than that on other exoplanets within the *same* exoplanetary system. The concept behind this suggestion being that provided the life-forms are fairly closely similar to those on the Earth, the atmospheric carbon will have become depleted through the formation of carbonate rocks. Whilst the ratios of various chemicals' abundances (stoichiometry) could also help establish the actuality of life-forms producing a biosignature, since those ratios will (probably) have different values when resulting from life's action, from when they result from 'natural' processes.

Spectroscopy then has yet to discover any ETs or ETIs for us. But it is not the only possible route to biosignatures. Monitoring the brightness of an exoplanet, where this is possible separately from that of the star, could detect some bioindicators at least. In particular, large areas of water (liquid or ice) or any other specularly reflecting material could produce glints. That is to say, brief flashes of intense light when the host star is reflected from those areas. The other biosignature possibilities, for all that they were mentioned at the start of this section, direct imaging and polarimetry do not seem to be very hopeful, at least not until our telescopes become capable of watching the exo-trilobites emerging from the exoplanet's seas.

TECHNOSIGNATURES

FIGURE 17.1   The Nile Delta at night – a clear Technosignature for the presence of Intelligences on Earth. (Reproduced by kind permission of NASA.)

Figure 17.1 shows what technosignatures are all about; an action (switching on the lights at night) of the intelligences (Humankind) living on the planet (Earth) as a part of their everyday lives which is *not* an intended signal to ETIs living elsewhere, but which, nonetheless, acts exactly as such.

Humankind's lights at night (called light pollution by those wishing to see the night sky) are just the visual equivalent of the inadvertent radio signals discussed as one of our means of eventually detecting ETIs (Chapter 15). As such, for ETIs searching for us, they are a beacon saying 'There is intelligent life right here'. If ETIs similarly illuminate parts of their exoplanet in the visual region, then that is just as good a technosignature for us that ETIs exist there. A recent current estimate, though, suggests that for the JWST to detect the exoplanet, Prox Cen b which is just 4.24 ly away from us via the light pollution caused by its (hypothetical) ETIs, that light pollution would have to be ~500 times worse than it is on the Earth at present. Hence, although our light pollution is a tech-nosignature for ETIs, they would have to be a lot more advanced than ourselves before being able to detect us in our present state of technology.

The beauty of technosignatures, on the whole though and as opposed to biosignatures, is that they *always* indicate life; sometimes the life is there now, sometimes it may have died-out, but the evidence for its past existence is still there.

So, find a definite technosignature and you:

- have found an ETI civilisation
- SETI is completed

and

- the MHoQ's answer is known (at last) to be 'NO'.

So, how do technosignatures differ from biosignatures? Well, some of them do not. The spectroscopic discovery of chemicals naturally produced by life-forms; we have seen above to be a biosignature. Take the same process, but discover chemicals which can only be produced by an ETI's technological activities and you have a technosignature. The chemicals which might constitute such a discovery include:

- CFCs (Chlorofluorocarbons)
- Nitrogen dioxide (though this can be produced naturally)
- Nitrogen trifluoride

and

- Sulphur hexafluoride

Other technosignatures which have already been mentioned include light pollution (above) and inadvertent radio and microwave signals (Chapter 15). Another type, which we will deal with in Chapter 18, is technology on stellar to galactic scales, such as Dyson swarms, which are large enough to be seen directly. Another, possibly even larger-scale signature might be the Ultra-High Energy cosmic rays for which natural production methods have yet to be found (see below).

THE 'OH-MY-GOD' PARTICLE

The 'Oh-My-God' particle is the highest energy cosmic ray received to date (Box 14.3). Its energy (~50 J) was so high that it must have originated within ~100 million ly of the Earth. That is because cosmic ray particles are continually colliding with the photons of the CMB (Chapter 8) and cosmic rays above an energy of ~8 J[604] lose energy rapidly in that process. This is because in those collisions, pairs of sub-atomic particles called pions[605] are created. This creation requires around 4×10^{-11} J per pair of pions. So, some 200,000 million such collisions would reduce the 'Oh-My-God' particle's energy to ~8 J. On average collisions occur about five times a day, which gives us the above ~100 million ly for the maximum(ish) distance to whatsoever produced the particle.

Now, not many natural events *can* produce 50 J cosmic rays. Something, as yet unknown, occurring within active galactic nuclei is generally expected to be their source, but no one knows for certain.

So, if there is a possibility that the 'Oh-My-God' particle cannot be produced naturally, there must also be a possibility that it has been produced 'un-naturally', i.e., by ETIs somewhere. To date, the highest energy sub-atomic particles produced in our accelerators is around 10^{-6} J. Surely, though, inside a sphere 200 million ly in diameter, there must be *one* ETI civilisation which is ~10^7 times better at making particle accelerators than we are?

It does not seem likely that the 'Oh-My-God' particle was sent to us as part of a deliberate signal, since the next cosmic ray particle of comparable energy was not detected until 30 years later. So, at a bit rate of ~10^{-9} bps (Chapter 15) it would take ~1,200 years to say 'Hello' to us. However, it might count as an 'ultra-technosignature'.

TAXONOMY

Producing lists and organising them is a feature of many human activities. It seems a little premature to do this, however, within a subject which yet has to contain a single item to be classified and

---

[604] This is known as the Greisen–Zatsepin–Kuzmin limit (GZK). Further details may be found in the Author's *Astrophysical Techniques* 7th Edn., CRC Press, 2020, and via the internet.

[605] There is insufficient room here to expand on this; see books on sub-atomic physics and/or the internet for further details.

wherein it could be years, decades, centuries, millennia (or never) before there *is* anything to be classified.

Nonetheless, there are now, at least two proposed classification systems for ET and ETI signatures: the Balby scale and the Ichnoscale.[606]

The Balby scale classifies technosignatures on the basis of their length of existence as identifiable technosignatures:

- Type A – Lifetime ~1,000 years
- Type B – Lifetime ~1,000,000 years
- Type C – Lifetime ~1,000,000,000 years.

The Ichnoscale compares a possible external technosignature with the level of the same phenomenon as it occurs for the Earth today (i.e., the equivalent terrestrial technosignature).

- Ichnoscale level 1 – technosignature is the *same* as that of the Earth
- Ichnoscale level 2 – technosignature is *twice* that of the Earth
- Ichnoscale level 3 – technosignature is *three* times that of the Earth
- …
- Ichnoscale level 10 – technosignature is *ten* times more intense than that of the Earth
- …

and so on. Thus, as we have seen above, for Prox Cen b to be detectable from the Earth via its light pollution, that pollution would need to be ~500 times the Earth's current level. In other words, we could detect the light pollution technosignature of ETIs on Prox Cen b *if* it were at an Ichnoscale level of 500.

---

[606] The name is derived from 'Ichnos'; the Greek word for 'Footprint'.

18 The Wilder Speculations: From the Almost-Sane, through the Three-Quarters-Crazy to the 'You Really Must Be Totally Out of Your Mind' Level

Carl Woese, the discoverer of the archaea life-forms (see Archaeon Eon, Chapter 9), is recorded as having a variant on Newton's famous epigram,[607] which reads as follows:

> "If I have seen further than others, it is because I was looking in the right direction."[608]

Well, this chapter is where we are going to be looking very far indeed, but, quite possibly, not always in the right direction.

INTRODUCTION

Aliens, Androids, BEMs, C-3POs, Cybermen, E.T., Insectoids, Klingons, Kzinti, LGMs, Martians, Monoliths, Na'vi, Outsiders, Ramans, Space invaders, UFOs, Vulcans, Wookiees, Xenomorphs, … (this list could go on for some time) are all popular names for what, in this book, have generally been called Extraterrestrial life-forms, ETs, Extraterrestrial Intelligences and/or ETIs.

The fact that they are *popular* names is due to the fascination which many of us humans have with the idea of having surrogate pseudo-humans who work out our fantasies using scenarios wherein *we* remain safe and need not feel guilty about mowing down 10 million of the enemy with our Zap guns. At the same time and in many cases, the scenarios have some pseudo-scientific background, so that we may also feel that it is not *just* an idle fantasia within which we are wasting our time but that it is educational as well.

Be that as it may and some of these fictional confections are, in truth, quite competently scientifically based, they all remain as daydreams (or nightmares). Sometimes, though, ideas may be mooted amongst scientists and others, which might *seem* to be little different from daydreams, but which *have* been conceived in the expectation that they might actually be put into practice in real life. That expectation might be over-optimistic, or if not, it might be a century or a millennium before they can be realised, but the ambition behind them is that they should be more than *solely* horsefeathers.

Here, then, we look in a little more detail at some of these 'Daydreams-with-Corporeal-Aspirations'.

---

[607] 'If I have seen further it is by standing on the shoulders of Giants'.
[608] *Carl Woese Tribute.* J. Ngeow, C. Eng. Personalised Medicine, 231, **10**, 2013.

DOI: 10.1201/9781003246459-21

PHYSICALLY TRAVELLING TO SOME OF THE NEARER EXOPLANETS ✺✺

Much of this book has, up to this point, been plugging the line that we cannot travel interstellar distances, or if we do, it will be very difficult/expensive/time consuming/fatal or worse.

The discussion of von Neumann machines (Chapter 14) has, though, shown that such travel may not be quite impossible, although it would remain very expensive and time consuming.

Until not many years ago, this point is about where this chapter would have stopped. As far as sending humans or getting visits from ETIs (Chapter 14) or even sending out Voyager-type space probes and getting results within a human lifetime is concerned this is *still* where we stop.

However, when it comes to sending itsy-bitsy-teenie-weenie space probes to the stars, the situation has changed radically within the last decade. We will look into this shortly (indeed, the basic fundamental development, phased arrays, has already been encountered in Chapter 15 since Bell-Burnell's 'LGM' discovery was made using such an instrument). First, though, we need to fill in some background and even before that there is a legal point to consider (Box 18.1), who owns things in space?

BOX 18.1 – WHO OWNS THINGS IN SPACE?

This may seem a ridiculous question when we still struggle to get spacecraft as far as the Moon.[609] It may even seem a ridiculous question under any circumstances whatsoever.

Indeed, the concept of 'ownership' may be quite unknown to every ETI in the Universe or perhaps some/most have a quite different concept(s) of ownership, such as:

> "I have now walked thrice widdershins around your city, chanting the mantra '*yenom*' continuously and I have jumped over yonder broomstick, so, now *your* city belongs to *me*".

Nonetheless there are already earnest international committees debating whether, for objects within the Solar System, when a space tourist brings back a souvenir rock from the Moon; does she/he own it? does the tourist firm own it? does the launch agency own it? does the tourist's nation own it? does some world organisation own it? Then there is the separate question of who owns the scientific data, or the mineral rights, held within that rock? Then again should taking souvenir rocks from the Moon be made a criminal offence? If so, what can be a suitable punishment? This list seems likely to go on forever and probably very soon will.

Whether is 50 years, 500 years or 5,000 years from now, similar questions seem likely to arise eventually, even with just sending gram-scale spacecraft (see below) to nearby exoplanet systems. After all, the tradition of claiming a new land for your country just by planting a flag on it is not *that* old.

Could then, someone sending a gram-scale spacecraft laden with a million micro-dots to be released into the destination exoplanet's atmosphere, with each micro-dot carrying the legend:

> "I, Prof. A. Bcdef, hereby claim complete ownership in perpetuity of this exoplanet, its host star and any and all other Celestial Bodies, whatsoever maybe their nature, which are gravitationally bound to this exoplanet's host star, on behalf of my Illustrious Nation; the Great Land of ZYX!"

really make such a claim stick? I do not know, it would seem though, that it ought to cleared-up *before* the problem actually arises. I leave this as a thought for readers of this book who may also have some legal training. Of, course, everything *could* already be owned by those grasping ETIs from just beyond Polaris.

---

[609] The average success rate of lunar missions (including the ones which only just managed it) is ~50%.

Magic Carpets

We have seen (Box 14.2) that present-day rocket fuel is totally inadequate for reaching relativistic speeds. Not only does it not provide sufficient energy, but the fuel has to be carried on the rocket. Several alternatives to conventional rocket fuel have therefore been suggested over the last few decades.

We shall almost certainly want to have gravity on our magic carpets; for, although it is a pain to lift that bag of cement into the mixer, there are other situations where is useful (if only to stop your reading glasses and sandwiches disappearing off to the other end of the spacecraft). There are only two realistic solutions,[610] though, if we want gravity: rotation and acceleration.

The former must be familiar to most readers from widely available and enchanting, but imaginary paintings by various space artists, each seeking to show how one of humankind's future homes, might appear from a few hundred metres' distance. Usually it is a silvery cartwheel, with additional appendages added here and there. The cartwheel hangs there in space, shining magnificently against the starry black backdrop of the outer Universe. The artist cannot show this next part (unless it is a movie), but we all understand that the cartwheel is slowly and oh-so-irresistibly revolving around its centre point/axis/hub and thus providing a sufficient amount of simulated gravity (centrifugal 'force') for those inhabitants occupying the outer rim regions. The 'gravity' then conveniently decreases to zero-g at the hub (where the sporty-types fly around on their human-powered feather wings inside the vast air-storage tanks).

Constant acceleration is much simpler – you simply turn your rocket motor on and forget to turn it off, you will then feel this apparent gravity constantly thereafter throughout the spacecraft. Of course, as we have already seen, this tends to use up a lot of fuel and you *do* have to remember to turn the spacecraft through 180° when half-way towards your destination so that you slow down again.

Nuclear Pulse Propulsion 🌀🌀

This is a rather terrifying idea. Basically, you get a large plate of steel, mount your spacecraft on one side of the plate and explode hundreds, even thousands, of nuclear bombs behind the other side of the plate. The blast from the bombs provides the propulsion for the spacecraft while the plate protects the spacecraft from the bombs.

This may sound like an appallingly bad joke, but it was considered seriously in the 1950s and 1960s, indeed even the use of fusion bombs was contemplated. Experiments with conventional explosives showed that the idea *did* work and Project Orion was a concept study into its viability which showed that payloads of hundreds or thousands of tonnes could theoretically be sent out to the (Solar System) planets. Interstellar journeys were investigated by Freeman Dyson and one of his designs was for a 400,000-tonne spacecraft, powered by 300,000 one-megaton bombs which would reach 10,000 km/s. It would take ~130 years to reach α Cen (which is about 0.1 ly further from us the Prox Cen) – and there would be no return. Other similar concepts were Project Daedalus (to go to Barnard's star) and Project Longshot (intended to orbit α Cen B).

Fortunately, the 1963 treaty[611] banning nuclear explosions in space, estimated costs (up to a year's US GNP[612]) and, perhaps, some common sense have meant that serious attempts at nuclear pulse propulsion have yet to be made.

---

[610] Additional options might be to invent artificial gravity generators 🌀🌀🌀🌀🌀, or to suspend a small, well-secured, black hole below the floor of your spacecraft 🌀🌀🌀🌀🌀, but, I class these as 'unrealistic'.

[611] "... Parties to the Treaty undertake not to place in orbit around the earth any objects carrying nuclear weapons or any other kinds of weapons of mass destruction, install such weapons on celestial bodies, or station such weapons in outer space in any other manner.

The moon and other celestial bodies shall be used by all ... Parties to the Treaty exclusively for peaceful purposes."
– Article IV, 1967 *Outer Space Treaty*. A similar statement was adopted unanimously by the UN General Assembly in 1963.

[612] Gross National Product.

Anti-Matter[613] 🌼🌼🌼🌼

This one is a favourite of the SF authors, but not realistic at the moment, since we can only produce anti-matter a few sub-atomic particles at a time.

Larry Niven's characters, Beowulf Schaefer and Gregory Pelton, though, do discover an anti-matter planet,[614] so all you have to do is work out a way of mining on that planet without becoming yourself a super-nova-sized explosion; and – 'Bob's your Uncle'. Thereafter, you add an electric charge to your kilo or two of anti-matter, so that you can suspend it safely in a vacuum using electrostatic charges. Then you drip ordinary matter atoms, a few at a time, onto your anti-matter and use the resulting energy to power the backward-facing lasers, which are your propulsion system. The starship 'Enterprise' in the *Star Trek* series uses an anti-matter drive to travel faster than light 🌼🌼🌼🌼🌼 – but that *is* over-gilding-the-lily.

Bussard[615] Ramjets 🌼🌼🌼

This is another popular SF concept which sounds as though it is an outside actual possibility (several centuries from now). Ramjets (on Earth) burn fuel to obtain their thrust as usual, but the air supply is forced into the engine ('rammed') by the forward motion of the aircraft/rocket, etc. Such a ramjet cannot work if it is stationary.

The interstellar (Bussard) ramjet obtains its fuel *and* its thrust via its forward motion through the interstellar medium. The ISM contains, *inter alia*, many H^+ ions (a.k.a. protons, ionised hydrogen atoms). The moving spacecraft projects ahead of itself a large (100s to 1,000s km across), funnel-shaped magnetic field. That magnetic field gathers up the H^+ ions and channels them in towards the centre of the 'funnel'. Once received by the spacecraft at the centre of the funnel, the H^+ ions are fed into a fusion reactor 🌼🌼. Energy is generated by the reactor and used to eject the reactor products (deuterium, tritium, helium-3 and helium-4) at high velocities out from the rear of the spacecraft. Thrust on the spacecraft is thus produced and the spacecraft is accelerated, or at least kept moving[616] at its already high speed.

Ion Drives (Solar-Electric Propulsion) 🌼🌼

The Bussard ramjet uses ions for its propulsion and so is a type of ion drive. Real ion drives have been used within the Solar System, but their thrust is low. They use electrical acceleration of ionised xenon to a speed of ~30 km/s for their propulsion. The xenon, though is carried by the spacecraft and so there is only a limited supply. The power for accelerating the xenon ions, though comes from the Sun, via solar panels.

This is rather similar to the Bussard ramjets concept but is probably only suitable for in-Solar-System use. However, it could be beefed-up to be an interstellar drive (perhaps) by using super-conducting magnets to gather atoms/ions from the ISM (*a la* Bussard) and supplying the required energy by laser illumination (see below) of the solar panels when the Sun is too far away and too faint to be usable.

Twisting Einstein around Your Little Finger – The Alcubierre Concept 🌼🌼🌼🌼🌼

An apparently FTL drive in which the spacecraft never actually travels faster than light but warps space-time around itself so that *space* itself travels past the spacecraft and so, to an outsider, the spacecraft would appear to get to its destination in less time than it would have needed had travelled

---

[613] For readers unfamiliar with anti-matter, it is a type of 'inverse' matter. Positrons (and you may have heard of, or even undergone, a Positron Emission Tomography (PET) scan) are the anti-matter version of electrons. Combine matter and anti-matter and they mutually annihilate each other, producing pure energy in the form of e-m radiation. A collision between an electron and a positron, for example, will produce two high-energy γ rays.

[614] *Flatlander* short story in *Neutron Star*, 1968, Futura Publications.

[615] The idea was proposed in 1960 by Robert Bussard.

[616] Because gathering up the H^+ ions causes drag and so slows the spacecraft down if it is not propelled continuously.

at the speed of light. The concept would require the availability of Negative Mass and/or Negative Energy and a few other necessities not to be found in your local corner shop.

Sailing ✿✿

OK – I know – now you are really laughing – and probably picturing a million-mast square-rigger (doubtless named Mayflower II and complete with cabin girl /boy and ship's cat) setting off from Plymouth on its mission to colonise some new New World, 1,000 ly away.

Well, laugh away as you will, Sailing *is* our *best* answer to date for going on our interstellar travels and it is the only one which may be realised within a few decades. As such, we need to devote a major section to its description (below).

ON SAILING THE GALAXY'S INTERMINABLE EXPANSES ✿✿

Our 'Magic Carpets', apart from the last mentioned, are mostly many decades, perhaps many centuries, away from realisation if, indeed, they are possible at all. Most are also **very, very, very, very** expensive.

The finally mentioned magic carpet though, which is merely **very** expensive, is a distinct practical possibility for ourselves in, maybe, a decade or five – and it is called the Photon-Sailing Micro-Spacecraft.[617]

The major reasons for the high costs of interstellar travel are the high masses and high speeds involved in getting humans or even just their larger artefacts across multi-light-year distances. But sending small spacecraft at slower speeds may be more feasible, or it could be so within a few decades.

Now 'small' in this context really does mean 'small'; maybe just a few grams, maybe a kilogram or two, but not much more than that. Already, there are many small spacecraft in orbit around the Earth, but these 'Cube-Sats' are based upon one or more 100-mm cubes each with masses of ~2 kg. For interstellar missions we need to be thinking of spacecraft 10 to 100 times smaller than Cube-Sats. The micro-spacecraft's speeds though still have to be quite high; but they need to be 'only' 0.1 c to 0.2 c and not the 0.999 999 ⋯ c speeds discussed earlier.

Even if we wanted to send off our 1 kg micro-spacecraft at 'only' 0.1 c, its kinetic energy at that speed will amount to ~450,000,000 million joules, i.e., the energy in ~10,000 tonnes of present-day rocket fuel (Box 14.2). If we were to try to launch our 1 kg micro-satellite via a present-day-type of multistage chemical rocket from the Earth's surface, we are instantly back again (see Chapter 14) looking at starting off with a system massing 10^n tonnes (where I leave you to calculate '10^n', if you so wish – but it will be *humongous*).

Even 10,000 tonnes of rocket fuel, though, is a substantial amount, but it is not an impossible amount; the Saturn V lunar rockets, for example, had an initial fuel load of ~2,000 tonnes. Now we cannot reduce the 10,000 tonnes requirement for rocket fuel (unless a far more efficient source for the energy were to be found ✿✿). So, if we are *really* determined to send something out to the stars, we have to find a way of *using the fuel, whilst leaving it behind* ®.

Leaving the fuel behind may sound more of a fantasy than the most absurd SF story yet written, but the way in which it (perhaps) could be done has already been mentioned twice earlier (Chapters 2 and 14) and demonstrated to work on small scales. The marvel of sailing ships, compared with those with engines, is, of course, that they *do* leave their fuel behind and so, *that* is why we are going to SAIL out to the stars.

Our sailing micro-spacecraft may not have any masts at all, it will not utilise the World's winds, it will more resemble a parachute than it does a Spanish Galleon and its sails may look like cooking foil, but it will be *sailing* and it will be *leaving its fuel behind*.

---

[617] A.k.a. Gram-scale spacecraft or Wafer-sats, especially the *very* small varieties.

It will be able to do this because, although there are no winds in space, there *is* light (and other e-m radiations; the NIR is a frequent choice in practice) and the micro-spacecraft will be riding the *Winds of Light* (Box 18.2) and not the winds of our atmosphere, for its interstellar voyage.

BOX 18.2 – THE WINDS OF LIGHT, OR, WHY DO COMETS' TAILS POINT AWAY FROM THE SUN?

The fact that the tails of comets point away, more or less, from the Sun in space (Figure 18.1), has been known to the Chinese since around 635 C.E.[618] In Europe, though, it was not noticed until Petrus Apianus, then working at the University of Ingolstadt, observed the 1531 appearance of Halley's comet. The phenomenon, rather unfairly, is now called Apian's law.

FIGURE 18.1   Comet Hale Bopp. The comet's tails are pointing away from the Sun, which is below the horizon, straight downwards on this image from the comet's head. (© C. R. Kitchin 1997. Reproduced by permission.)

Although the fact of Apian's law was known early, the reason for it was not confirmed until 1899. The fundamental cause of comets' tails pointing away from the Sun, then, was shown to be because sunlight (and all e-m radiations of any other wavelength/frequency) exerts a

---

[618] Joseph Needham. Quarterly Journal of the Royal Astronomical Society, 87, **3**, 1962.

pressure when it falls onto a surface. Kepler had suspected this as far back as 1619, and it was predicted to be a consequence of Maxwell's equations describing e-m radiation (1861/62), but only in 1899 did Pyotr Lebedev actually detect the pressure directly.

The pressure is tiny; about 10 μPa for direct sunlight, or about 1 kgf over a square kilometre – and that is when the radiation is 100% reflected (the values for the force are halved if the radiation is absorbed). Even so, it suffices to push the dust grains which are coming out from a comet's nucleus away from the Sun. This dust tail can be seen in Figure 18.1 as the brighter tail angling slightly to the right. The much fainter tail, which is close to being vertical on the image, is formed from ions which have been caught up in the Sun's particle emissions (the solar wind).

So, solar and any other form of e-m radiation exerts a force on any object which it encounters. From Box 18.2 we see that if we had a 1 kg micro-spacecraft out in space and attached to something like a parachute with a surface area of 1 square kilometre, the solar radiation pressure would exert a force of 1 kgf outwards from the Sun on the parachute. Assuming, for the moment, that the parachute has a negligible mass, then it would be inflated and pushed away from the Sun until the lines joining it to the micro-spacecraft drew taught. The parachute would then pull the (1 kg) micro-spacecraft away from the Sun with its 1 kgf and thus give the micro-spacecraft a 1 g acceleration away from the Sun.[619] A 1 g acceleration will increase the speed of the micro-spacecraft by 10 m/s for every second that it acts. That is, its speed will increase 1 km/s every 100 seconds and so it will reach 0.1 c (30,000 km/s) in ~3,000,000 seconds (~35 days).

So, apart from the minor problem of making a one-square kilometre parachute out in space, what is all the Fuss About? Why have we not sent small spacecraft out to the stars years ago?

The fly in the ointment is that the calculation suggesting that 35 days of sunlight will lift our micro-spacecraft up to a speed of a tenth of that of light was based upon the assumption that the 1 g acceleration was constant. Of course, the micro-spacecraft's acceleration is *not* constant, because, as it moves away from the Sun, out towards the stars, the solar illumination decreases and so also, therefore, does the acceleration. Thus, its actual progress is as shown in Table 18.1.

As can be seen from Table 18.1, the initial progress is great; out to Mars' orbit (~1.5 A.U.) in just over three days and at Pluto's orbit (~49 A.U.) in just over 50 days (compared with current journey times of ~7 months and ~9½ years, respectively).

Plainly, a micro-spacecraft cannot carry any fuel in order to slow down when it arrives at Mars or Pluto. Even so, it could still obtain images and send them back to us. Thus, if some intrepid ETIs do land on Mars, we could have close-up photos of them only three and a bit days later, provided only that everything was set up to launch the micro-spacecraft at a moment's notice.

On the other hand, it is about at Pluto's orbit (~30 A.U. to ~50 A.U.) that the fly in the ointment makes its presence really felt. The acceleration drops rapidly as the micro-spacecraft's distance from the Sun increases. By around a hundred days from the start, it is becoming clear that it will never achieve the desired 0.1 c velocity; indeed even ~ 0.006 c (~1,800 km/s) is probably too much to hope for. Still and all, the fastest speed achieved by a spacecraft to date is 163 km/s and that was reached by the Parker Solar Probe when its orbit took it close to the Sun, so 1,800 km/s is not to be sniffed at. Furthermore, from Earth orbit, the escape velocity from the Solar System is 'only'

---

[619] The fact that the micro-spacecraft is in an orbit will complicate its actual motions, but this does not change the final result much. The force of solar gravity on a 1 kg object, in orbit 1 A.U. distant from the Sun (i.e., its solar weight), is ~0.000 6 kgf. The solar sail is pulling at 1 kgf. So, for simplicity, we may ignore the tiny effect of the solar weight.

TABLE 18.1

Progress of a 1 kg Micro-spacecraft with a 1 Square-Kilometre Light Sail under Solar Illumination and Starting from Already Being in Space at the Distance of the Earth from the Sun (1 A.U.)

Time	Distance	Acceleration	Speed
(days)	Moved (A.U.)	(g)	(c)
0.0	0.0	1.0	0.0
1.0	0.227	0.664	0.002 5
2.0	0.775	0.316	0.003 8
3.0	1.484	0.162	0.004 5
4.0	2.274	0.093	0.004 8
5.0	3.112	0.059	0.005 0
10.0	7.60	0.013	0.005 4
20.0	17.01	0.003 1	0.005 6
50.0	46.0	0.000 45	0.005 71
100.0	94.6	0.000 10	0.005 743
250.0	241	0.000 017	0.005 761
500	486	0.000 004 2	0.005 770
1,000.0 (2.7 years)	975 (0.015 ly)	0.000 001 0	0.005 770
10,000.0 (27 years)	9,790 (0.15 ly)	0.000 000 010	0.005 772 6
25,000.0 (68 years)	24,500 (0.39 ly)	0.000 000 001 7	0.005 772 8

42 km/s[620] so the micro-spacecraft will head out to the stars, but it would take about 700 years to cover the 4.25 ly to Prox Cen.

Nevertheless, micro-spacecraft powered by solar sails could be real handy little runabouts for touring the Solar System (Figure 18.2).

Still, we really need 0.1 c speeds or higher, in practice, in order to send micro-spacecraft even to the closest exoplanet, Prox Cen b.

OK then, let us replace the one square-kilometre-sail with one ten times the size and see what happens (remember we are still assuming zero mass for the sail, but we will return to that in a moment). The results for a 10 square-kilometre-sail are shown in Table 18.2.

It is disappointingly clear that even a 10-fold increase in the sail area has failed to give us a speed of 0.1 c, although the speeds *have* increased. Looking closely at the figures in the two tables, the terminal speed values differ by a factor close to the square-root of 10.[621] Indeed, given that the increase in the sail area was by a factor of 10 and that the acceleration decreases as the square of the increasing distance, this is no coincidence.

Finally, then, we can estimate, by extrapolation, what size of sail would be needed to attain a speed of 0.1 c – and it is ~300 square kilometres, i.e., a parachute with a circular canopy ~20 km in diameter.

Now constructing a 20-km diameter circle out of an exceedingly thin film, whilst out in space, is probably a 🌀🌀 to 🌀🌀🌀 task, but it is not the factor which will stop us solar sailing to the stars; that will be the mass of the parachute. In 2010, the Nobel Prize for Physics was awarded to Andre Geim and Konstantin Novoselov, for their work on graphene, a material which takes the form of a

---

[620] The Pioneer, Voyager and New Horizons spacecraft were launched from the Earth which is already moving at 29.8 km/s around its orbit, so their launch speeds could be less than 42 km/s and still allow them to escape from the Solar System.

[621] 0.018 255 / 0.005 772 8 = 3.162 244. Square root of 10 = 3.162 278.

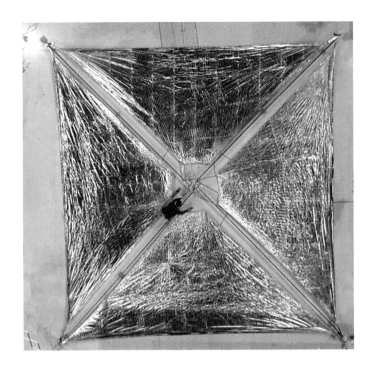

FIGURE 18.2   The solar sail for NASA's ACS3 (Advanced Composite Solar Sail System) technology demonstration spacecraft laid out in order to test its deployment. The sides of the sail are about 9-m long and it masses about 15 kg in total. Launch is currently expected for 2024. It should be able to change its altitude by up to two km per day using solar radiation pressure. Larger sails may have to be more like a parachute in shape. (Reproduced by kind permission of NASA.)

TABLE 18.2
Progress of a 1 kg Micro-spacecraft with a 10 Square-Kilometre Light Sail under Solar Illumination and Starting from Already Being in Space at the Distance of the Earth from the Sun (1 A.U.)

Time (days)	Distance Moved (A.U.)	Acceleration (g)	Speed (c)
0.0	0.0	10.0	0.0
1.0	1.6	1.51	0.014
2.0	4.3	0.359	0.016
3.0	7.15	0.151	0.017 1
4.0	10.1	0.081 5	0.017 4
5.0	13.1	0.050 8	0.017 6
10.0	28.2	0.011 7	0.017 9
20.0	58.8	0.002 80	0.018 1
50.0	151	0.000 431	0.018 20
100.0	305	0.000 106	0.018 23
250.0	770	0.000 016 8	0.018 24
500	1,540	0.000 004 18	0.018 250
1,000.0 (2.7 years)	3,090 (0.049 ly)	0.000 001 05	0.018 253
10,000.0 (27.4 years)	31,000 (0.40 ly)	0.000 000 004 24	0.018 254
25,000.0 (68 years)	77,000 (1.2 ly)	0.000 000 001 67	0.018 255

sheet of carbon which is just one atom thick. In the Nobel Committees' announcement, they wrote of graphene that:

"It is so strong that a 1 m² hammock, no heavier than a cat's whisker, could bear the weight of an average sized cat without breaking."[622]

Given that an 'average sized cat' weighs about 4 kg and a cat's whisker weighs about 1 mg, you will probably not be surprised that graphene is currently 'the strongest material ever measured'.[623] Apart, then, from graphene being almost transparent (when 100% reflectivity what is needed) and unavailable in square kilometre sheets, it would be difficult to find a better material for making solar sails.[624]

All the same, incredibly light in weight though graphene may be, it is not weightless (or in this context, massless). At 1 mg per square metre, a square kilometre of the material masses ~1 kg (and 300 square kilometres would be ~300 kg). Since we have been considering a 1 kg micro-spacecraft in Tables 18.1 and 18.2, adding the mass of the sail would reduce the speed figures in Table 18.1 by the square root of two (down to ~70%) and reduce those in Table 18.2 by a factor of the square root of 11 (down to ~30%).

In fact, after taking account of the masses of the sails, the terminal velocities are much the same whatsoever the sails' areas may be:

- Sail area – 0.10 square kilometres – Terminal velocity – 0.001 74 c
- Sail area – 0.25 square kilometres – Terminal velocity – 0.002 58 c
- Sail area – 1 square kilometre – Terminal velocity – 0.004 09 c
- Sail area – 10 square kilometres – Terminal velocity – 0.005 50 c
- Sail area – 300 square kilometres – Terminal velocity – 0.005 76 c.

So, the idea of sailing to the stars seems to be getting nowhere fast; although 'fast' is hardly the appropriate word here.

The first option at 0.10 square kilometres (a sail just 300-m x 300-m), which still has a top speed of 0.001 74 c (520 km/s), though, sounds ideal to enable your great grandchildren to send boxes of chocolates to their lovers on the other side of the Solar System, some-time around the year 2100.

Unfortunately, this is all still pie-not-in-the-sky, because for *actual* current light-sail materials, the density is around 1 g per square metre. So, all those sail masses would be 1,000 times larger than those just calculated and the actually achievable velocities would be reduced very much further.

LASERS TO THE RESCUE? 🌀🌀🌀

The reason that solar sails failed to get our micro-spacecraft up to relativistic speeds was mainly because the acceleration decreased due to the decreasing light intensity from the Sun as the spacecraft moved out through the Solar System. No laser beam has zero divergence (Box 15.2) and so the same phenomenon (decreasing light intensity) will occur along any laser beam sent out into space. A laser beam's divergence, though, can be made very small. In Box 15.2, we saw that ~0.05 seconds of arc might be possible sometime within a decade or two. We also saw in Box 15.2 that an individual

---

[622] The Nobel Prize in Physics 2010. Illustrated information

[623] https://en.wikipedia.org/wiki/Graphene#:~:text=Graphene%20is%20the%20strongest%20material%20ever%20tested%2C%20with%20an%20intrinsic,1%20TPa%20(150%2C000%2C000%20psi).

[624] A related material, aerographite, comprises a convoluted web of micron-sized carbon tubes. Its mass is around 180 mg per square metre for a 1-mm-thick sheet, but it can also be fabricated into very lightweight 3-D support structures as well.

laser might reach a power of 1 MW, if not now, then perhaps fairly soon. Combining these two estimates we may see just how well laser-driven sails[625] can do in the race for relativistic speeds.

When it comes to lasers, the acceleration which they can provide to a micro-spacecraft remains constant all the while that the laser-illuminated-spot on the sail remains smaller than or equal to the size of the sail. Once the laser beam diameter becomes larger than the sail, some of its light will spill over the edges of the sail and so, just like a solar sail, the micro-spacecraft's acceleration will decrease with increasing distance from the laser. The speed achieved by the micro-spacecraft when the laser beam size equals the sail size is thus the crucial factor. The speed of the micro-spacecraft *will* continue to increase just as long as the laser continues to shine. Nevertheless, its acceleration will decrease rapidly (c.f. the solar sail examples in Tables 18.1 and 18.2) and the beam-sail equality point is the point of diminishing returns; soon after that, the spacecraft team will likely cut their losses and turn their laser off (or use it for another purpose such as launching the next micro-spacecraft).

Earlier in this section, we saw that a 1 kg micro-spacecraft at 1 A.U. from the Sun would experience 1 kgf (~10 N) and so be given a 1 g acceleration by a massless, reflective sail with an area of 1 square kilometre. Furthermore, we saw that a constant 1 g acceleration would get the micro-spacecraft up to a 0.1 c speed in about 35 days.

Now, the radiation (at all wavelengths) at a distance of 1 A.U. out from it from the Sun (a quantity known as the Solar constant) amounts to ~1,380 W per square metre. The total radiant energy falling onto a square kilometre thus amounts to 1,380 MW and so gives the (1 kg) micro-spacecraft its aforementioned 1 g acceleration.

The light pressure from a 1 MW laser would thus give the micro-spacecraft an acceleration g/1,380, i.e., ~0.000 7 g (0.007 N). The laser beam width will be about 1 km across when the micro-spacecraft has travelled about 4 million km. Starting from stationary, travelling that distance will take about 30 hours and by then the velocity will have reached ~75 km/s (0.000 25 c). The solar sail version of the micro-spacecraft reached ten times that speed (0.002 5 c) in just 24 hours (Table 18.1).

Clearly, then lasers are *not* coming to the rescue.

LASERS (WITH A LITTLE BIT OF HELP) TO THE RESCUE? 🌀🌀🌀

At the beginning of this chapter, it was written;

> "Until not many years ago, this point is about where this chapter would have stopped."

and

> "… the situation has changed radically within the last decade."

The situation has changed radically because of new scientific and technological developments in an area called 'Photonics'[626]. This essentially is Electronics, but utilising light beams[627] instead of electric currents. In fact, the optical fibres and lasers used to provide high-speed broadband communications and related technology, are one aspect of photonics whose applications are already quite widely familiar; but there will be a great deal more to come from photonics applications going into the future.

For the areas of interest discussed in this book, photonics has changed the situation because it allows techniques which have been in use at radio wavelengths for many decades to be applied at much shorter wavelengths. In particular, the optical version of the radio Phased Array (above, Box

---

[625] A.k.a. 'Directed Energy Propulsion'.
[626] The subject of photonics is both highly technical and developing rapidly. For readers interested in looking into the subject further, an internet search for 'Photonics' is probably the best way forward.
[627] Actually, these are usually in the NIR, not the visible part of the spectrum.

18.3 and Chapter 15) enables laser power to be enormously increased by combining the outputs from many individual lower-power lasers and for the divergence of the resulting beam to be reduced by many orders of magnitude. It sounds like the 'answer to a maiden's/young man's prayer', but it does have one drawback; the micro-spacecraft have to become micro-micro-spacecraft. Proposed projects in this area (see below) are *genuinely* envisaging sending spacecraft with masses of one,

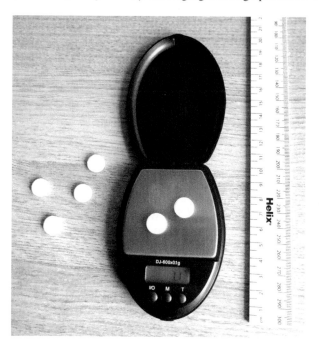

FIGURE 18.3 A gram-scale spacecraft would have a total mass similar to that of the two pills (which together mass 1.1 grams) on the scale. (© C. R. Kitchin 2023. Reproduced by permission.)

two or a few grams out to the nearest stars (Figure 18.3). Henceforth, therefore, our discussions will refer to 'Gram-Scale Spacecraft' and not to those enormous ~1 kg micro-spacecraft.

An outline of phased arrays is given in Box 18.3, which concentrates on their applications to the subject areas of this book. Further and other details may be found via an internet search and/or in the author's book, *Astrophysical Techniques*.[628]

BOX 18.3 – PHASED ARRAYS

The phenomenon behind phased arrays is that of interference between two or more beams of e-m radiation. An analogy of the way in which an array operates, which may help to understand it somewhat, although the analogy is not based upon interference effects, is as follows.

A SHAGGY DOG STORY

One of my dogs has normal hearing, whilst the second is completely deaf in one ear. If both are some distance away when I call them, the dog with full hearing knows instantly where I am. The partially deaf dog, however, has to look around in all directions and catch sight of me before knowing my direction. A signal in one ear therefore registers that an important sound has been received, but that is all. Two ears register the occurrence of the sound in the same

---

fashion, but the differences between the sounds going into the two ears can be interpreted by the brain to give the direction from which the sound originated. To make the analogy closer to the way the phased array operates, imagine the dogs are hearing a continuous noise. The normal dog will rotate her head until the sounds in both her ears match and her head will then point[629] towards the sound's origin. The deaf dog will gain nothing as her head moves; the sound does not change, so she still needs to use her eyes to establish its direction.

Thus, two ears (= antennae for a phased array) are required to determine a signal's direction.

BACK TO PHASED ARRAYS

Phased arrays were first developed to operate in the radio region and, as already mentioned, Bell Burnell (above and Chapter 15) used a radio phased array when she detected the first pulsar. Initially phased arrays were used because they were cheap and cheerful to construct. They are built up from two or more (usually many more) very simple radio antennae. Often, the antennae are little more than a couple pieces of wire cut to the right lengths. The antennae are mounted in fixed positions over an area which can be just a hectare or two or can be many square kilometres in size. Each antenna then picks up radio signals coming to it from all directions, so its output is a mish-mash of multitudes of garbled individual signals.

If we now connect two such antennae together, their two mish-mashed signals will mix together, but the signals will not be quite the same. They differ very slightly because there is a certain physical separation between them in the field (or wherever) within which they happen to be positioned. Now as we have seen many times already, the speed of light, or radio waves in this case, is fast, but not infinitely so. Two aerials form a line, and if they are separated by 100 m (say) and a signal comes in from the left, then it will arrive at the left-most antenna about 0.3 microseconds before it reaches the right-most antenna – and vice versa.

The two versions of the same signal combine then, when the two aerials are connected together and with a very small delay between the two versions. Normally the two versions will not be in-step with each other and they will, on average tend to cancel each other out. *But*, suppose the original signal had a wavelength of 100 m (3 MHz). When the two versions of *that* signal from the two antennae are combined, because they are separated by 100 m and the radio signal's wavelength is 100 m, one of the two versions will have slipped a single wavelength with respect to the other, but they are still *in step*[630] with each other and will remain so. The mixed versions of that particular signal will therefore reinforce each and it will *not* die away.

OK – that is it – now *you* know how *a phased array works*.

There are some bits and pieces left to deal with, naturally, but whether the two versions of the same signal cancel each other out (destructive interference) or reinforce each other (constructive interference) is at the heart of the phased array's operation. So now let us deal with those 'bits and pieces'.

The example just discussed (wavelength 100 m, aerial separation, 100 m) may not have seemed to advance our way towards travelling to the stars very much, so let us expand on the situation a little.

The two antennae being considered will still be absorbing many signals from all directions simultaneously. However, we will still concentrate on those coming from the left and along the line of the antennae. Suppose there are ten quite different signals coming from the left and being picked up by the antennae and which have the following wavelengths:

---

629 Hence the name 'Pointers' given to a breed of gun dog.
630 *In step* is more normally called *In Phase* and it is from this that the device's name is derived.

256 m, 255.55 m, 110.773 m, 100 m, 99.95 m, 83.02 m, 64.666 67 m,
33.333 333 m, 31.415 m and 31.417 m.

The output from the antennae might be expected to be a multiple mish-mash of all those signals. However, on the contrary, only those signals which are interfering constructively, i.e., those which emerge in step because their wavelengths are exact multiples/divisors, (or nearly so), of the 100 m separation of the aerials will be strong signals. The other signals will interfere destructively and become very much weaker, if not fade completely. Accordingly, only those signals with the wavelengths of 100 m and 33.333 333 m will emerge strongly, since these wavelengths fit into the 100 m aerial separation once and three times respectively.

So, our very simple two-antenna phased array is already acting as a filter and selecting, in this example just two out of the incoming ten signals. Were we to add two more antennae between the two originals at 33.333 333 m and 66.666 666 m from the initial left-hand antenna, then only the 33.333 333 m signal would emerge.

From now on it is simple. If we add more antennae and in a 2D pattern, then we can select out not just the example's 33.333 333 m signals but also eliminate signals of that *same* wavelength but which are coming from a *different* direction.

The phased array has thus become a radio telescope which can point to a particular spot in the sky and then only detect the signals coming from that spot at one particular wavelength. Furthermore

- the spot in the sky observed by the array is fixed with respect to the Earth and so scans along a band across the heavens due to the Earth's rotation
- the array's sensitivity is given by the area which it covers and is not that of a single antenna

And

- the array's angular resolution is given by the distance between the two most widely separated antennae and is also not that of a single antenna.

All that is great, but suppose we want to look at a part of the sky that is not within the band being scanned by the array, do we have to build a second array?

The answer to that question is 'almost, but not quite'.

The spot in the Earth's sky which is observed by the array is determined by the delays between the signals from each of the antennae and the antennae's physical distribution over the area of the array. Now, the latter is fixed (unless we want to dig up each antenna and move it around, or the antennae are mounted on movable platforms, like the individual dishes making up the VLA[631] and ALMA[632] radio (dish) telescope arrays).

So, to choose a new observing point, the signal delays will have to be changed. Luckily, although not yet mentioned, the delay for the signal from any particular antenna has two components:

- the first and already considered is due to the physical position of the antenna on the ground with respect to the other antennae forming the array
- the second is the time it takes for the electric current, which results from the e-m signal being received by the antenna, to travel along the cables to the instrument's processing centre.

---

[631] Chapter 15.
[632] Atacama Large Millimetre Array; ESO's instrument sited in Chile.

Now the speed of an electric current[633] along a cable is not quite as fast as that of light and varies depending upon the nature of the conducting cable, but a figure of 0.9 c (270,000 km/s) is reasonably representative. So, if we want to change the delay between the two variants of the signal considered in the two-antenna example above, we could

- physically change the actual separation of the antennae from (say) 100 m to 200 m

or (and *so* much easier)

- we could just clip in an extra 90 m[634] of cabling between the antenna and the processing centre.

In the very early versions of this type of instrument,[635] physically going around with a wheelbarrow full of cables and changing them all over (a tedious and lengthy task when 100 or more antennae were involved) *was* done. Today, changing the signals' delays can be accomplished electronically and so quickly that the array can track a specific source as it moves across the sky.

The principal features, then, of phased arrays operating in the radio part of the spectrum are summarised in Box 18.3. But that is not much help, so far, in getting a spacecraft to Prox Cen b. The remaining steps, though, are straightforward:

- There are no fundamental differences between radio waves and light waves (including the NIR), so an optical phased array will work on the same principles as a radio phased array
- Radio and optical receivers will operate as transmitters instead, once their receiving elements are replaced with appropriate emitters.

Optical phased arrays, then, have always been theoretically possible. That they are now becoming realisable in practice, or almost so, is due to the developments within photonics mentioned above.

Thus, although neither radio nor optical phased arrays have (yet) been applied to sending signals to or to receiving signals from ETIs (Chapter 15) they could well be the next major development in that area. For sending spacecraft to Prox Cen b they are the only current option that is anywhere near to being viable.

Phased Arrays and Gram-Scale Spacecraft: A Marriage Made in Heaven

The principal properties of phased arrays, which are now of importance, were mentioned in Box 18.3, but their significance was not stressed there. However, their moment has now arrived and those important properties are that:

- the array's sensitivity is given by the area which it covers and is not that of a single antenna

and

- the array's angular resolution is given by the distance between the two most widely separated antennae and is also not that of a single antenna.

---

[633] This is not the speed of individual electrons, which is around 0.001 m/s, it is the speed of the electric field.

[634] Not 100 m because of the slightly slower speed of the electric current.

[635] Readers interested in this should start their research by looking for 'Mills Cross' radio telescopes.

Thus, operating at a wavelength of 500 nm (say), which is in the green-cyan part of the visual spectrum, instead of needing a problematical multi-MW laser and failing to get much in the way of speed for a sail-driven spacecraft, we could use 100,000 one MW lasers and combine their output into a single 100 GW laser beam using an optical phased array. Furthermore, if those lasers were distributed over a 1 km diameter circle, their beamwidth after combination would be better than even a 1 km diameter normal optical telescope's angular resolution, i.e., it would be about 0.000 1 seconds of arc, an improvement by a factor of 500 on the 0.05 second of arc beamwidths discussed earlier.

The force from a 100 GW light beam would be ~70 kgf. Acting on a 10-gram spacecraft-and-sail-unit, this would produce an acceleration of ~7,000 g. A speed of 0.1 c would be reached in a time of about 7 minutes. The gram-scale spacecraft by then would have travelled a distance of ~6.4 million km and the beam width (= sail diameter) would be about 3 m. The sail area that would be needed would thus be about 7 m². Now reflective sail material currently masses about 0.001 kg m⁻². So that would place the sail mass at about 7 grams, leaving ~3 grams for the payload. However, a graphene (see above) membrane masses about 0.000 001 kg m⁻², so anticipating a sail membrane mass of ~ 0.000 1 kg m⁻² being developed within about the next decade is probably reasonable. That would leave us with ~0.7-gram sail and a ~9.3-gram payload.

In summary, then, a potentially viable system for sending out gram-scale spacecraft to nearby stars is *almost* within our grasp and should be physically possible[636] within a decade or two. The parameters (rounded somewhat since there are still many uncertainties in the calculations) of one such a system would be roughly:

- 0.01 kg – spacecraft mass
- 0.009 kg – payload mass
- 100 GW – laser power obtained from 100,000 one MW lasers operating at a wavelength of 500 nm and with their beams combined via an optical phased array
- 0.000 1 seconds of arc – beamwidth for the (1 km) optical phased array
- 3 m – sail diameter
- 0.001 kg – sail mass
- 7,000 g – acceleration when under illumination from the lasers
- 7 minutes – acceleration time
- 6.3 million km – distance travelled under acceleration
- 0.1 c – final speed
- 40 years – travel time to Prox Cen b
- 45 years – the time before we, on Earth, get the spacecraft's results

THE OPTIMISTS

The above considerations on how to get a gram-scale spacecraft to Prox Cen b or some other nearby star/exoplanet in less than a human lifetime are all back-of-the-envelope rough calculations. They can be tweaked,[637] there may be clever design changes, there may be new developments or some of the anticipated advances used in the calculations may take longer/be less effective/not happen at all; BUT, we can probably begin to see the light at the end of the tunnel and so:

the person who will push the button to launch Humankind's first inter-stellar space mission to a nearby exoplanet has probably already have been born.

---

[636] Whether or not it is economically possible is another matter.

[637] A recent Breakthrough Listen concept, for example, envisages using molybdenum disulphide for a three-metre diameter sail and a micro-spacecraft comprising just an electronic circuit board about 50 mm square (my estimate) and massing 50 to 100 g (my guess). However, the researchers confess that they are "… still figuring out how to retrieve information from the microchip probe …" (CNET Press Release 20th February 2022).

or, to quote Philip Lubin:[638]

> "Directed energy propulsion allows a path forward to true interstellar probes. … Shuttle technology cannot get us to the stars but directed energy can … This technology is NOT science fiction. Things have changed. The deployment is complex and much remains to be done but
> *it is time to begin*[639]."

and some people *have* begun.

(i) Workers at Breakthrough Starshot[640] suggest the possibility of sending gram-scale spacecraft to Prox Cen b, utilising a 100,000 MW laser array (this is quite similar to the calculated example above). They anticipate accelerations of 10,000 g being applied for 10 minutes, giving a final speed of 0.2 c, at a distance of ~18 million km. These figures would suggest a total mass of ~7 grams. The sail size is suggested at around 4 to 5 m diameter, giving a mass, if using graphene, of ~0.015 g, although it would, in practice with spars and reflective coating, be significantly more.

If operating as an interferometer using green-cyan light (500 nm wavelength) and with the maximum separation of the individual lasers of 1 km, the central spot of the interferometer pattern could have an angular diameter of ~0.000 1 seconds of arc. At the micro-spacecraft's then distance of ~18 million km, the beam width would be about 10 m. This is close enough to the suggested sail size of ~5 m that a bit more tweaking should get the two sizes to match.

Other problems would be addressed by sending many gram-scale spacecraft, so that most may be expected to avoid or survive collisions with interstellar dust, etc. Figures of a total of 1,000 micro-spacecraft being sent with individual launches at 10-minute intervals are suggested. Currently several 2-megapixel cameras are suggested as a payload and with a laser using the light sail as a reflector to send the data back to Earth.

Despite the apparent viability of this proposal (and the earlier one above), many of the technologies needed for its success are still years away from being adequate for the job and that applies to all current proposals in this area of interest. For example, any photo-cells used to power the spacecraft when it is near to Prox Cen b will need to be specially designed, perhaps they will even need new materials to be invented, since the star's surface temperature is only half that of the Sun and so its radiation peaks in the NIR (around 1 μm; the Sun's radiation peak is at ~500 nm). Even then, at a given distance of the spacecraft from the star, it will receive only ~6% of power which it would receive were it to be at the same distance from the Sun.

(ii) Very, very, very small indeed is beautiful according to George Church[641] who envisages sending out probes the size and mass of a microbe and which are essentially close analogues of von Neumann machines (Chapter 14). At a mass of 10^{-15} kg (one picogram) for the spacecraft, most of the earlier problems disappear. Just scaling from the earlier energy calculations, without considering any other factors, the laser power required to reach 0.1 c, reduces to 4 mW, about the strength of a lecturer's laser pointer.

Church notes that picogram-scale microbes are capable of reproducing themselves from basic chemicals, a process which is more complex than most/all of our current automatic machine activities. He suggests, therefore, that the pico-craft (= human-fabricated microbe) after landing on an exoplanet would set about producing some contraption to send messages back to the Earth.

(iii) The Starlight concept of the University of California is partially funded by NASA. It is very similar to the Breakthrough Starshot project (above). Its gram-scale spacecraft would be powered

---

[638] Physics Professor at the University of California, Santa Barbara and specialist in, *inter alia*, directed energy (i.e., lasers).

[639] My emphasis. Reference; *A Roadmap to Interstellar Flight.* JBIS, 40, **69**, 2016 (also at https://www.nasa.gov/wp-content/uploads/2015/05/roadmap_to_interstellar_flight_tagged.pdf). If you wish to research this topic for yourself, then this paper would be a good starting point.

[640] https://breakthroughinitiatives.org/initiative/3

[641] Leader of Synthetic Biology studies at the Wyss Institute.

by plutonium, be the size of a DVD disc and travel at 0.2 c. The intent is for the payload to be seeds and even living examples of the nematode, *Caenorhabditis elegans* or tardigrades (Chapter 12). This introduces the potential for contaminating exoplanets and other environments and so that part of the project is not funded by NASA.

Other similar proposals are being considered/thought-about/dreamt of – and doubtless more will soon come, but those you may research (or originate) for yourself.

Now, if you have the tenacity, read Box 18.4.

BOX 18.4 – IGNORE THIS BOX IF YOU HAVE ALREADY HEARD QUITE ENOUGH ABOUT INTERSTELLAR TRAVEL

OK – you have reached this sentence, so, you must be a glutton for punishment. The problems of interstellar journeying do not end once you can launch your spacecraft at a relativistic velocity towards its destination. You will also have to consider/solve/get around at least the following additional difficulties:

(i) The 100,000 one MW lasers will be pouring 10_{11} W of power into spacecraft and its sail, which, if its mass is 10 grams, will raise its temperature to about over 100,000,000 °C within a tiny fraction of the first second of its journey. That, of course, will only happen if all that energy is absorbed – and the sails are made as reflective as possible. Since a laser's radiation is emitted over only a very few wavelengths (i.e., it is close to being monochromatic), multi-layer reflective coatings and the more recently developed silicon nitride photonic nano-crystals may be usable, which can have reflectivities of ~99.999% over those few wavelengths. This, however, still leaves kilowatts of energy being absorbed and that will have to be radiated away *very* quickly if spacecraft is not to be damaged or destroyed.

(ii) The probe will need terminal guidance in some fashion when it gets to its destination – it will be of little use if it goes through the system 100 A.U. out from the star and when the exoplanet is on the far side of the star.

(iii) The acceleration of ~7,000 g produced by the 10,000 lasers would shatter any complex payload to smithereens.

(iv) Invent a means for a tiny micro-spacecraft to broadcast a signal which is strong enough to be received back on Earth at a distance ≥4.25 ly (see discussions in Chapter 15 on the problems in sending signals across multi-light year distances).

(v) The possibility of the gram-scale spacecraft hitting a grain-of-salt-sized bit of interstellar dust (Box 14.3) at a speed of 0.1 c, thus releasing ~0.5 MJ of energy. This is equivalent to hitting the spacecraft with about 300 bullets from a Colt-45 pistol simultaneously. After such an experience, we probably need not expect very much in the way of results from the mission even if the collision has not diverted the spacecraft's course so much that it misses its target by light years.

Launching numerous gram-scale spacecraft in the hope that some would get through is one already-suggested solution to this problem.

If (optimistically) we take the gram-scale spacecraft to be a thin wafer which has dimensions 5 mm x 50 mm x 50 mm, then edge-on, its cross-sectional area would be 0.000 25 m². In travelling, edge-on, a distance of 4.25 ly, it would sweep through a volume of ~10,000 km³ ®. From Box 14.3, the possible number of salt-sized dust particles in the ISM is estimated at one per 300 km³. So, the number of collisions of the wafer[642] with salt-grain-sized particles during

---

[642] Never mind the sail – we will just assume that the sail gets a tiny hole punched through it with each of its collisions and that is it.

the journey to Prox Cen B could be ~30 – and just one such collision would almost certainly destroy the wafer. The estimates of the density of salt-grain-sized ISM particles given in Box 14.3, of course, could be wrong by orders of magnitude either way. However, taking those figures as they stand, for now, it would seem that millions, not a thousand or so, of gram-scale spacecraft would need to be launched before even one survived to reach its destination.

(vi) How will ETIs throughout the Universe view our firing off of what amounts to a bullet with a kiloton of energy,[643] out into the Universe in a more or less at random direction? That 'bullet' will continue to travel through the Universe, at that speed, until the end of time – or until it annihilates some ETI city. In the latter case, we could then, *really*, face hostile invasions (Chapter 14), especially if it is to be thousands of such 'bullets' each time and many stars to be thus targeted, not one (Breakthrough Starshot, above).[644] Of, course if the problem (v) calculation is correct, then *this* problem will disappear.

(vii) Find the ~$10,000 million to ~$1,000,000 million to purchase the 100,000 one MW lasers, launch them into Earth orbits and then fund all the other parts of the project.

Spin-offs

Right, then, you have spent the 10^n billion and sent off your one (or thousands of) gram-scale spacecraft towards Prox Cen b. Now, you wait for about 45 years to get your results. In the meantime you have an 10^n billion miracle of modern engineering sitting around doing nothing.

Unless you are careful, your 45 years (and a lot more) are going to be spent inside a gaol gazing forlornly at its walls as you serve out your sentence for flagrantly wasting tax payers' money (you should, of course, have remembered that this practice is legal only for politicians).

Luckily your 10^n billion miracle of modern engineering has quite a number of uses;

- For a start, you can send another one (or thousands of) gram-scale spacecraft towards the next nearest star – and then another lot towards the next star after that – and then the next star after that – and then …. With nearly 60,000 stars within 100 ly of the Earth, this should keep you out of gaol for quite a while.
- The machine will be a nifty meteorite defence. Suppose a large meteoroid is found to be on a collision course with the Earth and it is already only 2 million km away from us. The 100 GW laser beam can be used to illuminate a one-metre circle on one side of the meteoroid. Almost instantly, some 50 tonnes every second of the meteoroid will flash into gas and boil off violently into space. However, you are not trying to melt the meteoroid; instead, the boiling rock acts as a jet (or a rocket) and the reaction to it pushes the whole of the rest of the meteoroid in the opposite direction. Within a few minutes the meteoroid's orbit will have been changed sufficiently for it to miss the Earth.
- If the hostile ETIs (Chapter 14) do make an appearance, then the launcher's ability to send off kinetic energy missiles every few minutes at speeds of 0.1 c with each packing a kiloton should be a significant deterrent.
- With slight adaptations, the launcher could send packages of a few kilograms anywhere throughout the Solar System at speeds of ~0.01 c (3,000 km/s; ~24 hours for the Earth-Mars journey time).

---

[643] The energy of a 10 gram mass travelling at 0.1 c is ~1.08 kilotons of TNT.

[644] On the other hand, we would still have the 100,000 one MW lasers and a society able to fire off 0.1 c, 1 kiloton bullets might be viewed as something to be avoided. The chances of hitting an ETI city are extraordinarily remote, but our ethical position is extraordinarily invidious whatsoever the eventual result might be.

- A few more adaptations and loads of a few tonnes could be sent anywhere throughout the Solar System at speeds of ~0.001 c (300 km/s; ~1 week for the Earth-Mars journey time).

N.B. The last two applications for the launcher would get the payloads to the destinations in the times suggested, but they would still be travelling at 3,000 km/s (or 300 km/s). Any destination wishing to use the technique would thus need some means of slowing-down and catching the payload. Perhaps some sort of electro-magnetic net orbiting the destination planet/satellite/asteroid might do the job, or if the aim can be made precise enough, a launcher at the destination working in reverse might slow the package down ✿✿✿? However, by the time that the launcher is in operation, this minor snag will surely have been solved.

In any case, the launcher is clearly going to be so useful that another dozen are likely to be needed within a century or two and you may cease worrying about gaol and start planning for your Nobel Prize.

TRACTOR BEAMS ✿✿✿

Another tempting notion from the SF lobby – and this one might *not* be insane. An SF tractor beam is generally a narrow beam of some (usually unspecified) force which can link two spacecraft and be used by one of them to tow the other around space without having any material linkage such as a rope, a cable or a chain.

Well, electrostatic (Figure 1.2) and magnetic fields might almost be persuaded to do just that. A recent study into how defunct Earth satellites and other such debris might be cleaned-up suggests that "… an electrostatic tug could pull a satellite weighing (sic) several tons about 200 miles in two to three months …". [645]

Once that system is operating, then all that needs to be done is to up-grade the beam until it can apply a pull sufficient to accelerate an object continuously at 1 g for about 3½ days and then our gram-scale spacecraft is off on its journey at 0.1 c towards that interesting-looking exoplanet just beyond Tabby's star.

As an alternative way for launching gram-scale spacecraft, this method seems to me to have a lot going for it. There is one smallish drawback to it, although that should be solvable. The system would not be one spacecraft towing another, for then, we would still have the problem of getting the towing spacecraft up to a speed of 0.1 c. The pulling end of the tractor beam would be mounted on a smallish asteroid situated some ~30 AU out from the Sun (Neptune's orbital radius) because that is the distance the spacecraft will have to travel under acceleration by the tractor beam to reach 0.1 c. The tractor beam would then simply pull the gram-scale spacecraft towards it for the required 3½ days.

The aforementioned drawback would then be that, if we did exactly as has just been described, the spacecraft would hit the tractor beam generator at 0.1 c, destroying it and probably the asteroid as well. The two obvious, simple solutions though would then be either to drill a hole through the centre of the asteroid (which should be easy-peasy by the time that all this can happen) so that the spacecraft goes straight though the asteroid or to have a second tractor beam which, at the last moment, pulls the spacecraft slightly to one side (so missing the asteroid).

Before your hopes rise too much though; it is just *such* a pity, though, that the strength of an electrostatic field decreases as the square of the increasing separation/distance.

Still, a relay system might even get over that last problem; a chain of tractor beams (of the hole-in-the-centre variety) with small separations and perhaps using greater accelerations (say 1,000 g) might get a gram-scale spacecraft up to 0.1 c in ~50 minutes, so the required length of the system would be 'only' about 50 million km.

---

[645] Strain D., CU Boulder Today, 2023

PLASMA WINGS 🌀🌀🌀

Figure 13.6 shows a terrestrial albatross gaining its flying energy via dynamic soaring close above the ocean waves. Albatrosses are remarkably good at this process and can remain air-born without flapping their wings for months on end. Dynamic soaring requires the air to have a layered structure, with the wind speed higher in one layer than in the other. The apparently miraculous soaring effect is basically due to the low, but not zero, friction between the albatross and the air layers and to the conservation of momentum.

As a numerical example, take the lower air layer to be at ground level and stationary. The higher layer is moving across this ground layer at (say) 10 km/h. The albatross is currently gliding through the ground layer at 10 km/h. The sequence might then be:

(i) Albatross' ground speed, 10 km/h (Let us call this the + direction). The Albatross' air speed is therefore + 10 km/h.

(ii) Albatross does a quick upward turn into the upper layer (which is moving at 10 km/h in the opposite direction to the Albatross) and then completes the 180° loop so that it is now flying down-wind within the upper layer.

(iii) The Albatross retains its initial ground speed (because of momentum) although this is now in the opposite direction. The drag of the moving air layer soon brings the Albatross' air speed up to match the air-layer's speed (i.e., air-speed = 0 km /h).

(iv) Albatross' ground speed, − 20 km/h; Albatross' air speed, 0 km/h.

(v) Albatross does a quick downward turn into the lower layer again and then completes that 180° loop again.

(vi) Albatross' ground speed, + 20 km/h; Albatross' air speed, + 20 km/h.

(vii) Quickly (so that it does not lose much air speed), the Albatross loops up into the upper layer again and after again being accelerated by the moving upper layer, we have;

(viii) Albatross' ground speed, − 30 km/h; Albatross' air speed, 0 km/h.

(ix) Down into the ground layer again and

(x) Albatross' ground speed, + 30 km/h; Albatross' air speed, + 30 km/h.

(xi) Upper layer again and

(xii) Albatross' ground speed, − 40 km/h; Albatross' air speed, 0 km/h.

and so on.

After each complete looping cycle, the Albatross' ground speed increases by 10 km/h. A seeming miracle; but the energy for the increasing speed is actually coming from the kinetic energy of the moving upper layer of air (in this example). If you are worried about the '0-km/h-air-speed-periods' and the Albatross then falling into the ocean, the sensible birds just sacrifice a bit of energy-extraction-efficiency and keep their speeds *above* stalling level throughout all the cycles.

But, you are doubtless thinking, why are we discussing Albatrosses in the final chapter of a book about exoplanets and ETs?

Well, Albatrosses are very beautiful and well worth looking at. However, if you insist on being relevant, then it is not so much the Albatrosses which we now need to consider, but their dynamic soaring abilities.

The Albatross obtains its energy by a careful alternation of its presence between two regimes (air layers), gaining energy from one and trying to lose as little as possible whilst in the other.[646] But the air above ice-cold oceans is not the *only* place where such regime types may be found.

What is relevant to interstellar travel is that conditions for dynamic soaring may exist at the boundary where the Solar Wind finally slows down to sub-sonic speeds and intermingles with the

---

[646] Incidentally, this is probably instinct at work and not due to teaching by its parents or even to intelligent thought. Other birds, including gulls, can perform the same feat.

interstellar medium (the Termination Shock). The materials are still plasmas and their interactions lead to rapid changes in magnetic fields, compressions and to turbulence.

In this maelstrom, a suitably equipped small spacecraft might quickly find a couple of appropriate plasma streams to use for dynamic soaring, maybe even getting up to the magic 0.1 c and so to be on its way to another nice-looking exoplanet and with all the 'fuel' (except what was needed to get out to the termination shock) having been provided free of charge. The plasma 'wing' to be used might be a plasma wave antenna.

GENERATION SHIPS 🌀🌀🌀🌀

The discussion, so far in this and earlier chapters, has shown the huge barriers which there are even to just sending or receiving signals across interstellar space and the even larger problems of sending physical items on multi-light year journeys. In fact, we *might* now be able to send and receive e-m radiation signals with our present facilities, but sending even minute quantities of material across interstellar space is still for the future, albeit perhaps, the not-too-far-distant future.

So, it would seem ridiculous to contemplate sending vessels massing billions of tonnes and populated by humans and other terrestrial life-forms off to a distant exoplanet with the intent of colonising it.

Well, it 'seems ridiculous' because it *is* ridiculous; nonetheless the concept *is* being contemplated.

It is fair to say, though, that the concept is one that so far occurs only in fiction; nobody has any serious proposals or requests relating to this idea submitted to funding agencies or which might have any other real intent behind them. Video games based upon the concept do abound though, the first written story dates back over a century and formulating the requirements for gigantic interstellar spacecraft is a popular topic for science students' undergraduate theses.

The basic concept is usually called a 'Generation Ship'[647] for reasons discussed below or a 'Space Ark' and it grows out from the very long journey times required for any form of interstellar travel. The Voyager 1 spacecraft is travelling fast enough to leave the Solar System and head out to the stars, although it is not aimed at any particular star. It is currently moving at a speed of almost 17 km/s or, as is more useful in this context; ~1 ly per 18,000 years. If it was heading that way, it would require over 75,000 years to get to Prox Cen b. Even the most optimistic SF writers are inclined to think that this is on the slow side and opt for journey times of a thousand years or so. If travelling, say 10 ly, this would imply a speed in the region of ~1 ly per 100 years (~0.01 c, 3,000 km/s), which the gram-scale spacecraft should achieve with ease, but getting a billion-tonne spacecraft up to that speed would be an entirely different matter.

Ignoring speed problems, the Generation ship acquires its name because it is envisaged as housing a completely self-contained and infinitely (or nearly so) self-sustainable environment for a human population which is large enough to be genetically stable over many generations. The individual humans forming the population of a generation ship are born (Box 18.5), live and die on board it and after many such generations have passed, it arrives at its exoplanet destination. The SF writers usually throw in some animals, plants, seeds, frozen embryos and all the equipment to build a viable colony in a hostile environment as well, not forgetting the fuel to slow down and land on the exoplanet in thousands of years' time.[648]

---

[647] A.k.a. 'Bernal sphere', although this latter concept was started (in 1929) as a possible means of providing additional habitats within the solar system.

[648] I find it difficult to understand why the population of a generation ship would not just carry on living on it after arrival at its 'destination'. Why should they leave behind their accustomed environment, which must obviously have been good enough for them for many centuries, to go and land on and colonise some totally unfamiliar and possibly hostile exoplanet - but then, perhaps I do not have the right adventurous spirit.

BOX 18.5 – SEX IN SPACE

The viability of generation ships, amongst other factors, will depend upon the possibility of humans having sex in space, bringing any resulting pregnancies to successful fruitions and then raising the babies to the point where they can start the next generation. Of course, by the time that generation ships are viable, test-tube babies with robotic carers and AI-powered upbringings and educations[649] may be the norm.

Somehow, though, one feels that the more natural processes will have solved any problems in this area long before generation ships are being built. After all the 'Mile-High-Club'[650] reportedly dates back to 1916 (in aeroplanes, perhaps to 1785 in balloons – peer-reviewed research papers, though, do tend to be rather few and rather vague on this topic).

Given the development of large space stations (the International Space Station and the Tiangong Space Station at present, but many more are likely to be constructed in the next few decades), plus the start of space tourism, asteroid mining and the probable construction of habitats on the Moon/Mars, also within a few decades, it would seem inevitable that some enterprising couple will soon find a quiet little hideaway in which to become the first members of the '1,000-mile-high club'.

Whilst such private-enterprise contributions will doubtless be condemned by all and sundry, there may be some who will be glad to study the results; the ethical problems of carrying-out such studies on an official basis are probably quite insurmountable.

Unsurprisingly, then, a Generation ship has to be BIG.[651]

- Robert Heinlein in *Orphans of the Sky*[652] envisages a constructed ship with "… miles and miles of metal corridors …"
- Arthur C. Clarke's Ramans (who are ETIs visiting the Solar System, except that they ignore us and carry on their journey to some other destination) are on board a craft "… at least forty kilometres across …".[653]
- Anne McCaffrey requires three Generation ships for her 6,000 colonists (and 25 dolphins), although most of the life-forms are in some form of cryogenic sleep, so the 'generation' part is a bit of a misnomer. Furthermore, it only took the ships '15 years' using their matter/anti-matter drives to travel to their destination, so it was hardly worth the bother of going to sleep anyway.[654]
- Larry Niven takes the biscuit though, with his *Ringworld*[655] – this is a solid ring encircling a star. The ring is some 940 million km in circumference and ~1.5 million km wide with a population of trillions of various life-forms, some humanoid, many not. The whole lot (star included) is fleeing from a chain reaction of supernovae in the galactic centre. Although this is a generation ship of sorts, it is not *going somewhere* in order to land its occupants onto some exoplanet; the Ringworld already *is* the home to its life-forms. We shall therefore return to this concept in the 'Mega-structures section' later in this chapter.

---

[649] Isaac Asimov's book *The Naked Sun*, 1958, Michael Joseph, explores an interesting, imaginative society based upon such a type of environment.

[650] A reclusive society for people who have had sexual intercourse whilst in flight.

[651] One estimate, which may help you to visualise sizes, is that 50 Ha (~120 acres) of farmland might feed 500 people, although this would clearly vary widely depending upon the details of the agricultural methods to be used. 50 Ha would be the surface area of a cube-shaped spacecraft with ~300 m long sides.

[652] 1963, Victor Gollancz.

[653] 1973, Victor Gollancz.

[654] Numerous books within the *Dragon Riders of Pern* series.

[655] 1972, Victor Gollancz.

The authors of Generation ship novels are generally fairly circumspect in providing practical details of those ships and usually choose to explore the interactions between the life-forms involved in such an exotic environment as the main part of their novel. So, I can give little information on how these ships are persuaded to start their travels through the galaxy.

One suggestion as to how the ships themselves could be *made*, though, may be of some interest, (although I can no longer remember the reference) and is this:

- Take a smallish asteroid
- Cut a shaft down to its centre and create a cavity there
- Fill the cavity with water
- Seal off the shaft and any other holes in the asteroid
- Place lots of large thin-film mirrors around the asteroid aligned so that sunlight is reflected onto the asteroid from many directions
- Continue heating the asteroid until it starts to melt
- Continue heating the asteroid further until the heat percolates right to the centre
- Stand well back
- Watch as the water at the centre boils and the steam pressure blows up the largely molten asteroid into a gigantic balloon
- Allow to cool and then fit out the interior with all the mod cons required for a population of thousands to live out next few centuries until they arrive at their new home.

I can think of many problems likely to occur during this procedure, most of them fatal, but at least the author had given some thought as to how a Generation ship *might* be built.

TERRAFORMING WITHIN AND WITHOUT THE SOLAR SYSTEM ✿✿✿

We have seen (Chapter 13) that nowhere within the Solar System, except for the Earth, is intrinsically inhabitable by humankind and/or most other terrestrial life-forms in its current state. We have also seen (Chapter 13) that there have been suggestions for the improvement of the Martian environment by enhancing Mars' magnetic field until the planet develops a magnetosphere to shield the surface somewhat from Solar charged particles. Those suggestions are highly speculative, but not quite Science Fictional.

Such a change to Martian magnetic field is an example of terraforming and that process may be defined as:

> Making large scale and permanent (or nearly so) changes to an environment which is currently hostile towards the survival of terrestrial life-forms so that that environment becomes less hostile or even benign and hospitable for those same life-forms.

It is likely that terraforming is a level ✿✿✿ or even a level ✿✿✿✿ task for humankind at present. The details of what is needed for a terraforming task, though, will depend upon the nature of the problem. Thus, if we were to find an exoplanet quite similar to the Earth, but too cold, then seeding that exoplanet with methane-generating bacteria in order to spark off a runaway-greenhouse effect could be within our reach. The suggestion in Chapter 13 to induce a magnetosphere around Mars by applying static electric charges to large globs of material moving around its outer mantle layers will probably be beyond our reach for a long time to come.

Terraforming, whatsoever the problem may be, is likely to involve planetary-scale manipulations and changes. Thus, in Chapter 13, we saw that, should we wish to do so, moving Ceres to be conveniently near to the Earth would require the fusion conversion of some 440,000 million tonnes of hydrogen into helium. Humankind will hence probably need to have developed to C-Factor [10]✖ (Table 13.4) or higher before it becomes feasible for ourselves.

Even within the Solar System, it is likely that colonies/industrial activity/mining/farming/ etc., away from the Earth, will only be able to be undertaken within protected environments (i.e., space-suits, spacecraft, larger-scale self-contained bases on the Moon, Mars, Ceres, Europa and the like). The self-contained bases, though could become quite large if, say, lava tunnels on the Moon or Mars can be sealed off to retain an atmosphere and then be heated. On small scales one could imagine that some device akin to a diving bell might work; a geologist (say) dons a large plastic bag, seals its edges to the surface and could then fill it with air and so work directly on the now openly exposed surface digging up sample minerals.

MEGA-STRUCTURES 🍬🍬🍬🍬🍬

The Ringworld has already been mentioned above. It is one example of a mega-structure, but it is not the earliest to have been conceived, nor is it even the largest. A Dyson sphere could be a good deal bigger than the Ringworld (unfortunately it is un-buildable until anti-gravity 🍬🍬🍬🍬🍬 is discovered – but then you cannot have everything). Ringworld itself is also un-buildable until we discover/invent materials hundreds or thousands of times stronger than those we have at present. A Dyson swarm, though could be a definite possibility. Other, increasingly improbable suggestions abound (usually, though, they are just even more fanciful variations of one of the Dyson variants), but I leave to you the pleasure of searching for those.

DYSON SPHERES

Let us start this section with the Dyson sphere which will then set-the-scene for later ideas, even though it is un-buildable (certainly by ourselves, possibly by any ETIs anywhere howsoever advanced they might be). In fact, we already have covered the essential background to Dyson Spheres in Chapter 17, since the idea occurred to Dyson whilst (in today's terminology) thinking about ETI signatures.

In 1960, Dyson published a short article entitled "Search for Artificial Sources of Infrared Radiation".[656] In this, he assumes that any technologically based civilisation will need more and more material resources and more and more energy supplies as that civilisation progresses towards more and more advanced levels (whatsoever forms those advanced levels might actually take). He then follows that assumption to its logical conclusion; that any such civilisation which does not have cheap and rapid interstellar travel (and we have seen how difficult that could be) will eventually need to exploit *all* the resources available within its home host-star-and-exoplanetary-system. For our-selves that would mean that eventually we will be using all the energy from the Sun and will have dismantled and re-used all the material forming all the Solar System's planets, moons, asteroids, interplanetary medium, etc. For any sceptics, he then explains:

"The following argument is intended to show that an exploitation of this magnitude is not absurd. ... the time required for an expansion of population and industry by a factor of 10^{12} is quite short, say 3000 years if an average growth rate of 1% per year is maintained. ... the energy required to disassemble and rearrange a planet the size of Jupiter is about 10^{44} ergs[657], equal to the energy radiated by the Sun in 800 years ... the mass of Jupiter if distributed in a spherical shell revolving around the Sun at twice the Earth's distance from it, would have a thickness of ... 2 to 3 meters, depending on the density ...".

---

[656] Science, 1667, **131**, 1960.
[657] 10^{37} J.

This spherical shell would absorb all the Sun's energy (except, perhaps, for small amounts from x-rays, γ rays and long wavelength radio waves which might be able to penetrate the shell) and its inner surface would have an area some ~10^9 times the surface area of the Earth.[658]

Probably you will have noticed some obvious problems with this concept already. However, we will come back to those in a moment. Firstly, we will complete Dyson's argument, since that is relevant to Chapter 15. The shell will absorb all the energy coming from the Sun, but with only a 2- to 3-m thickness, this energy will quickly percolate through to the outer surface of the shell from where it will radiate out into space. That radiated energy will have an intensity of ~340 W m^{-2} and this corresponds to a temperature of ~278 K (~ 5 °C). While most of us would regard 5 °C as quite chilly, the outer surface will be radiating at MIR wavelengths (peak emission at ~ 10 nm) and so will stand-out like the proverbial sore thumb as being an intense IR source when viewed via IR telescopes. Thus, Dyson's article concludes that; to find the most advanced ETI civilisations we should be searching at IR wavelengths and not in the radio or visual regions of the spectrum.

Dyson Spheres – The Problems

Actually, Dyson did not really mean a sphere, as it might appear from the above quote. In a 2014 interview with Robert Wright,[659] he tries to correct the record:

> "Well it was really a joke which has been completely misunderstood … I suggested that people actually start looking in the sky with Infrared telescopes … unfortunately I added to the end of that the remark that what we're looking for in an Artificial biosphere *meaning by biosphere just a habitat something that could be in orbit around the neighboring star* (my emphasis) where the aliens might be living. So the word biosphere didn't imply any particular shape …"

This is a relief, because Dyson spheres are a plaguey nuisance. If they were ever to work, then they would need artificial gravity generators 🌀🌀🌀🌀🌀 plastered all over their surfaces. This would be for two reasons:

- If the gravitational pull of the star is to be countered by the rotation of the sphere (as the quote seems to suggest), then it can only be effective for the rotational equator of the sphere and the rotational velocity at that point would then equal the orbital velocity of a free-floating object. No other portion of the sphere, however, will have the star's gravitational force counteracted in this way by the centrifugal 'force'.[660]

  So, if the sphere is not to disintegrate into a zillion tiny fragments, then it will need the aforesaid gravity generators to balance the star's gravity and hence keep its structure entire.

- It may seem surprising, but a uniform hollow spherical shell does not exert any net gravitational forces (arising from the shell) on anything inside the shell[661]. So, everything inside the shell will need its own individual gravity generator to ensure that it remains comfortably positioned on the inner surface with its own appropriate weight. Otherwise, it could drift all the way into the star, go into an orbit around the star inside the sphere, be crushed

---

[658] This might sound adequate for almost any purpose, but, using Dyson's own figures, the human population density would be around 1,600 times the current human population density of the Earth.

[659] https://web.archive.org/web/20140109033551/http://meaningoflife.tv/transcript.php?speaker=dyson

[660] The inverted commas are there because centrifugal force is classed in physics as a pseudo-force – but I leave that for you to research for yourselves, should you so wish.

[661] Essentially, wheresoever the object inside the shell may be positioned, the gravitational pull from the shell's material on one side of the object is balanced by the gravitational pull from the shell's material on the other side.

into a flattened squidge on the inside surface of the sphere, or, if very lucky, find itself a small area on the inside surface of the sphere where it can be comfortable (but if it moves very far from that point, it will soon to find itself enduring one of the other three options).

Thus, the Dyson sphere problems (another is that the inhabitants would not be able to see the stars, but I guess the shell could have portholes here and there to solve that one) disappear because Dyson spheres themselves disappear, or as it was expressed earlier; they are 'un-buildable'.

DYSON SPHERES – THE SOLUTIONS – #1 DYSON SWARMS

What Dyson seems to have intended to describe is a structure, which *is* potentially buildable and which now goes by the name of a 'Dyson swarm'. A Dyson swarm bears a close resemblance to the discs of material forming proto-planets which condense around proto-stars, i.e., a large number of small(ish) solid bodies in independent orbits around the star. The difference from a proto-planetary disc would be that the swarm-bodies would be:

- individually large enough to form suitable environments for their inhabitants
- have orbital planes at many different angles and orientations (so that overall, the swarm-bodies have a spherical distribution centred on the star, not a disc-like distribution)

and

- that the gases and dust particles to be found in proto-planetary discs have been cleaned out so that only macroscopic bodies are involved.

Of course, the 'space-traffic' control would have to be perfect. Nevertheless, although the speeds are high, the distances involved are vast, so that there should be time for individual swarm-bodies to take avoiding action, if/when they get onto collision courses. Moreover, it would be impossible for all the star's radiation to be intercepted, all the time. There would always be crannies between adjacent swarm bodies through which the radiation could escape into the outer Universe. This problem, though, could be minimised by spreading the swarm-bodies' orbital radii over a wide range, then the more distant bodies could intercept the radiation missed by the inner bodies. This would also enable the ETIs to choose their own favourite climates; 'exo-polar bears' living on bodies in the larger orbits, say and the 'exo-thermophiles' living on the bodies closer to the star.

The formation of a Dyson swarm would clearly involve moving, at least moderately large asteroids, into new orbits. As we have seen above, moving Ceres within the Solar System so that it is near to the Earth would need several times 10^{29} J of energy. However, as we have also seen above, the dismantling and rearranging Jupiter might require 10^{37} J. Therefore, when we are thinking big, the amount of energy required is not automatically a limitation.

Thus, once we can move planets, we simply bring Mars in towards the warmer parts of the Solar System, whilst moving Venus into cooler regions and we will have then tripled our living areas (after a few centuries/millennia of settling-down). Following that, Uranus and Neptune should be kid's play to position where we want them to be and after *that*, manipulating Jupiter and Saturn will be easy-peasy.

Indeed, just like the Dyson swarm itself, the new system could be brought into being slowly and in a piecemeal fashion, so that the ETIs are not risking *everything* on a single throw of the dice (as with, for example, the Ring World concept below). All we need would be continuously emitting lasers, some million to a billion times more powerful than the current peak powers of our pulsed lasers and we could then arrange the whole Solar System to suit our convenience.

DYSON SPHERES – THE SOLUTIONS – #2 DISCWORLDS

The Discworld books by Terry Pratchett are intentionally comic stories (at least, I *think* it is intentional), not pretending to any scientific reality. Nonetheless a variant on Pratchett's Discworld could make at least as much sense as a Dyson swarm.

Pratchett's Discworld is a nonsense; a thin flat circular disc supported by four elephants, which in turn stand on the back of a turtle, which in turn stands … – well, that is where the sequence comes to an end.

However, if we take a Dyson sphere and flatten it to be a disc, then we have something that could be made to work. We need to cut the flattened sphere into narrow concentric rings and set each ring rotating about the star at the correct orbital speed for its distance from the star. We probably also need a high wall on the outer edge of the outermost ring to stop the atmosphere from being flung off into space and it probably also requires huge air pumps to blow the atmosphere back to the centre of the system. Furthermore, the inhabitants would have to be kangaroos-types in order to jump over the gaps between differently moving adjacent rings,[662] there are all sorts of stability and other problems still and we do have to find alternative homes for the elephants and turtle. But these are trivial concerns compared with inventing and manufacturing 10^{100} gravity generators needed for a Dyson sphere. At the end, we then have a viable(ish) megastructure with the main Dyson sphere problems at least partly solved:

- There is gravity – which pulls towards the disc (i.e., downwards, upwards if you are on the underside of the disc).
- Everything is orbiting the star at the orbital speed for its distance from the star and so is stable and safe
- You can see the stars.

There is a problem with centrifugal 'force' though; because of the rings' rotations you would feel a constant tug pulling you towards the outermost ring. If you were to slip you might tumble all the way to that ring, however, the atmosphere-retention wall would (probably) stop you from finding out what is on its other side.

DYSON SPHERES – THE SOLUTIONS – #3 RINGWORLDS

A Ringworld starts off as just one of the rings envisaged for the Discworld but with some changes. The main one is that it spins faster than the orbital speed for its distance from the star. This creates a centrifugal 'force' outwards to simulate gravity and puts the whole ring under a high tension. The ET's and ETI's living area is thus now the inner surface of the ring and not its top and bottom surfaces. Larry Niven has extensively explored the possibilities of a Ringworld in his *Ringworld* series of books and he estimates that the tensile strength of the material forming a Ringworld (which he calls 'Scrith') is "… on the order of the force which holds atomic nuclei together."[663] The 'force which holds atomic nuclei together' is called the 'Strong Nuclear Force' and it is about 100 times stronger than the electromagnetic forces holding atoms and molecules together. Scrith's tensile strength must thus be in the region of a hundred times that of normal high tensile material (actually I would go for '10,000 to 100,000 times stronger than normal high tensile material' before I felt that the safety factor was even beginning to be adequate).

Out of this list of imagined cosmic Mega-structures, the most likely to be constructed would probably be a Dyson swarm because it can be put together piecemeal; i.e., one, two or a few of the orbiting swarm-bodies would be installed at a time. So gradually building up to the full structure

---

[662] Think, getting on or off very fast escalators.
[663] *The Ringworld Engineers*, 1980, Victor Gollancz.

over a long period. A Ringworld by contrast is a single entity which would have to be constructed in one operation. Indeed, it is difficult to see how even very advanced ETIs could actually manage that: perhaps the Ringworld could be assembled well out into space and then moved to be centred on the star and spun up to its required rotation speed?[664]

SUPER ETIs (KARDASHEV SCALE)

OK – you've received a commission to build a Discworld around Vega. Everything is planned and designed, even the money is upfront, all you need to do now is to find a contractor to do the physical parts of the job. So where do you go to solicit quotations? You could probably do worse than to ask one of Nikolai Kardashev's pupils for some hints (Nicolai, himself, died in 2019).

Kardashev, when alive, thought BIG thoughts and one of the things he considered was the possibilities of life-forms capable of operating on galaxy-scale projects, so a mere Discworld, would be something they would do between elevenses and the lunch break. Moreover, Kardashev thought that some galaxy-scale operators might be bigger/more advanced than others. So, just in case it might be needed soon, he devised a scale for assessing the abilities of such civilisations. It is naturally called the 'Kardashev scale', and his article on it, called 'Transmission of Information by Extraterrestrial Civilizations', was published in the Astronomicheskii Zhurnal, 282, **41**, 1964. The details of the scale[665] are:

> "… it will prove convenient to classify technologically developed civilizations in three types;
> I – technological level close to the level presently attained on the earth, with energy consumption at ~ 4×10^{19} erg / sec[666].
> II – a civilization capable of harnessing the energy radiated by its own star (for example, the stage of successful construction of a 'Dyson sphere'); energy consumption at ~ 4×10^{33} erg / sec.
> III – a civilization in possession of energy on the scale of its own galaxy, with energy consumption at ~ 4×10^{44} erg / sec."

Kardashev, on the basis of a 1% per year growth rate, suggested that Humankind would be a Type II civilisation by the year 5,200 CE and Type III by the year 7,800 CE.

Other workers, since Kardashev, have added extra classification levels. Sagan for example added a level 0 for which the energy consumption was, in Kardashev's units, ~10^{13} erg/sec (= ~1 MW – i.e., ~Homo habilis level). A Type IV civilisation would use the entire energy of our presently visible Universe, whilst a Type V civilisation would find numerous Universes other than our own and utilise all their energy as well (though one does wonder what use they would have for that amount of energy).

---

[664] Maybe it would be easier to move the star??
[665] Taken from the translation at https://articles.adsabs.harvard.edu/pdf/1964SvA.....8..217K
[666] One joule = 10,000,000 erg. So, the current luminosity of the Sun in those units is ~ 3.9×10^{33} erg/sec.

Epilogue

With that last thought to resonate through your brain, I will bring this book to a close.

I hope that you have found it interesting and informative, but above all I hope that it will have stimulated you to keep in touch with how the various subjects covered herein develop and progress into the future.

It would be wonderful for me if even one reader were to be inspired actually to take up work in one or more of those subject areas and beyond my dreams if she/he were the one who in ??? years' time runs out of the telescope control room waving a printout over his/her head and shouting:

"BINGO!!!

They ARE there.

We've GOT THEM.

We are NOT alone!!!"

So long as the answer to the MHoQs remains 'Yes' it is likely that SETI in some or all of its manifestations will continue. Not until we *have* found any ET or ETI and the answer changes to 'No', is there any possibility of the hunt slackening, but such success is much more likely to push the hunt on towards even greater efforts and, with luck, to even greater future successes.

I hope that I may now be permitted to close (*not* end) this book with the following paraphrase of a well-known quote from Winston Churchill[667]:

Now this is not the end of our story.

It is not even the beginning of the end for, surely, searching the Universe for other life-forms is a task and a joy that can *never* end.

But we are, perhaps, near the end of its beginning.

Chris Kitchin
Hertford
March 2024

---

[667] From a speech delivered at the Mansion House, City of London, during the Lord Mayor's luncheon on the 10th November 1942. In original form (referring to the battle of El Alamein);"Now this is not the end. It is not even the beginning of the end. But it is, perhaps, the end of the beginning."

Appendices

APPENDIX A – INDEX NOTATION

The index notation, in so far as this book is concerned (mathematicians take it on to reach wondrous heights and complexities, but that is their affair), is just a convenient shorthand.

Essentially all it amounts to is if you have the same quantities being multiplied together several times, then instead of writing the multiplication out in full, you write the number of multiplications as a superscript, for example:

$$1.4142 \times 1.4142 \times 1.4142 \times 1.4142 \times 1.4142$$

Would be written in index form as

$$1.4142^5$$

and that is *it* for multiplication. When things are being divided, a minus sign is added, thus:

$$1/1.4142$$

would be written as

$$1.4142^{-1}$$

and

$$(1/1.4142) \times (1/1.4142) \times (1/1.4142) \times (1/1.4142) \times (1/1.4142)$$

would be written as

$$1.4142^{-5}$$

It has two main usages: in simplifying the representation of very large and very small numbers and in writing down complex measuring units.

UNITS

This usage is quickly dealt with. We see below that the basic measuring unit for speed is metres per second, which could be written as (and often is)

$$m/s$$

From what we have just seen though, this could equally well be written as

$$m \ s^{-1}$$

(and again, it often is). There is little time saved whichsoever version is used with such a simple type of unit. But acceleration has the units of [metres per second per second] and this can be written in index notation as

$$m \ s^{-2}$$

With really complex units, such as for the gravitational constant (G), the saving in time/space and, more importantly, in expressing *exactly* what is meant, becomes much more obvious. Thus, G is:

<u>force</u> multiplied by the square of the <u>distance</u> between the two gravitating objects and divided by the <u>masses</u> of the two gravitating objects.

So, its units are:

(Force x distance x distance) / (mass x mass)

which upon expansion becomes:

{[(kg x metres) / (seconds x seconds)] x (metres x metres)} / (kg x kg)

but this then simplifies down to

$$m^3 \ kg^{-1} \ s^{-2}$$

when we cancel things down and use the index format.

NUMBERS

A number like 1,000,000 results from the multiplication: 10 x 10 x 10 x 10 x 10 x 10. So, from what has already been said, this would be expressed in index form as:

$$10^6$$

Similarly, a number like 0.000001 results from 1/10 x 1/10 x 1/10 x 1/10 x 1/10 x 1/10. So again, from what has already been said, this would be expressed in index form as:

$$10^{-6}$$

That is really all that is involved: the index notation just shortens the writing-down of very large and very small numbers and also makes it easier to compare one number with another.

A COUPLE OF USEFUL EXTRAS

<u>Multiplication/Division of Index Numbers</u>

You just add or subtract the indices, i.e.,

$$1,000 \ x \ 10,000 = 10,000,000 = 10^7$$

and

$$10^3 \ x \ 10^4 = 10^7 = 10,000,000 - (\text{indices}; 3 + 4 = 7)$$

or

$$100,000 \ / \ 100 = 1,000 = 10^3$$

and

$$10^5 \ / \ 10^2 = 10^3 = 1,000 - (\text{indices}; 5 - 2 = 3)$$

<u>Square Roots</u>

Indices are useful for numbers other than 10 as well. Thus 2 x 2 = 4 or 2^2 = 4 in the index format. More often, though, this is used the other way round by saying that 2 is the square root of 4. Now we might often write this down as

$$\sqrt{4} = 2$$

Thus

$$\sqrt{4} \times \sqrt{4} = 4$$

In index form, the square root is symbolised by 0.5, because

$$4^{0.5} \times 4^{0.5} = 4^1 = 4 \text{ (since } 0.5 + 0.5 = 1)$$

APPENDIX B – THE SI SYSTEM AND SOME USEFUL CONSTANTS AND DATA FOR REFERENCE

THE SI SYSTEM

A very brief summary of the SI system was included in the Preparatory Notes at the start of this book. To refresh your memory, the list of the basic units of the system is repeated here:

- Time (SI unit; the second, symbol 's').
- Length (SI unit; the metre, symbol 'm').
- Mass (SI unit; the kilogram, symbol 'kg').
- Electric current (SI unit; the ampere, symbol 'A').
- Temperature (SI unit; the kelvin, symbol 'K').
- Amount of substance (SI unit; the mole, symbol 'mol').
- Luminous intensity (SI unit; the candela, symbol 'cd').

These are the fundamental quantities of the SI system which have to be defined in terms of something 'tangible'.

Thus, the unit of time, the second, is defined as the interval required for 9,192,631,770 wavelengths of the radiation emitted from the ground state hyperfine transition of the Caesium 133 atom to pass a fixed point when they are moving in a vacuum.[668] This radiation is an emission line in the radio region at a wavelength of ~326 mm and it is within the X-band spectral region used, *inter alia*, for radar work.

Most measuring units, though, are derived units and so are based upon one or more of these fundamental units. Accordingly, speed has units of metres per second and so utilises the fundamental units for length and time.

Another absolute unit, which is not a part of the SI system, though, is that which is used for measuring angles. In this book, the units of degrees (°) and seconds-of-arc (a.k.a.; " or as) are generally used (360° = a circle, 1° = 3,600"). Another commonly encountered angular unit, though, is the Radian (rad), and there are 2π rad to a complete circle (1 rad = 57.295 779 513 … ° = 206,264.806 …"). Despite what might appear to be an awkward definition for the rad, it is, in fact, a much more useful unit than the degree when it comes to undertaking calculations.

When dealing with very large or very small items, the basic SI units may be used in conjunction with the appropriate index number multipliers (see Appendix A). However, a system of prefixes is also used, and these are listed below in Table Appendix B.1 (non-standard prefixes like 'centi' and 'deca' are not included).

---

[668] If this means very little to you, then you are a member of the majority of the human race. However, since you know perfectly well what a second of time is, for almost all normal practical purposes, that will suffice.

Table Appendix B.1

Prefix	Symbol	Multiplier (see Appendix A)	Prefix	Symbol	Multiplier (see Appendix A)
Yotta	Y	10^{24}	milli	m	10^{-3}
Zetta	Z	10^{21}	micro	μ	10^{-6}
Exa	E	10^{18}	nano	n	10^{-9}
Peta	P	10^{15}	pico	p	10^{-12}
Tera	T	10^{12}	femto	f	10^{-15}
Giga	G	10^{9}	atto	a	10^{-18}
Mega	M	10^{6}	zepto	z	10^{-21}
Kilo	k[669]	10^{3}	yocto	y	10^{-24}
--- (No prefix)	---	1			

In consequence, a distance of 1,000,000,000 m can also be written as 1 Gm or 1 giga-metre or using index notation (Appendix A), as 10^9 m. All these forms will be found in this book, although I have a preference for the index notation when the multipliers are very large or small (because *I* can never remember what 'zetta' or 'yocto', etc., actually mean).

Useful Constants and Data for Reference

Quantity	Symbol	Currently accepted value
Astronomical unit	A.U.	$1.499\ 578\ 707 \times 10^{11}$ m (exact)
Earth mass	M_{Earth}	$5.972\ 2\ (\pm 0.000\ 6) \times 10^{24}$ kg
Earth radius at the equator	R_{Earth}	6,378,000 m (varies slightly)
Earth surface gravity	g	9.76 to 9.83 m s^{-2} (varies over the Earth's surface)
Electron mass (at rest)	m_e	$9.109\ 382\ 15\ (\pm 0.000\ 000\ 45) \times 10^{-31}$ kg
Gravitational Constant	G	$6.674\ 08\ (\pm 0.000\ 31) \times 10^{-11}$ m^3 kg^{-1} s^{-2}
Gravitational force from the Sun at the distance of the Earth		$\sim 6 \times 10^{-3}$ N kg^{-1}
Light year	ly	$9.460\ 730\ 472\ 580\ 800 \times 10^{15}$ m (exact)
Moon (lunar) mass	M_{Moon}	$7.340\ (\pm 0.0010) \times 10^{22}$ kg
Parsec	pc	$3.085\ 677\ 581 \times 10^{16}$ m (exact)
Proton mass (at rest)	m_p	$1.672\ 621\ 777\ (\pm 0.000\ 000\ 074) \times 10^{-27}$ kg
Radian	rad	1 rad = 57.295 8 ... °
Speeds of light and (probably) gravity	c	$2.997\ 924\ 58 \times 10^8$ m s^{-1} (exact)
Sun luminosity	L_{Sun}	3.828×10^{26} W
Sun mass	M_{Sun}	$1.988\ 5 \times 10^{30}$ kg
Sun radius	R_{Sun}	696,000,000 m (varies somewhat)

---

[669] This is lower case because 'K' is used for the Kelvin temperature scale. However, the usage of kilogram as non-multiplied-up basic unit is quite anomalous and confusing and a more appropriate name is needed (but any change seems unlikely at the moment).

APPENDIX C – THE GREEK ALPHABET

Greek letter, lower case	Greek letter, upper case	Name	Greek letter, lower case	Greek letter, upper case	Name
α	A	Alpha	ν	N	Nu
β	B	Beta	ξ	Ξ	Xi
γ	Γ	Gamma	o	O	Omicron
δ	Δ	Delta	π	Π	Pi
ε	E	Epsilon	ρ	P	Rho
ζ	Z	Zeta	σ	Σ	Sigma
η	H	Eta	τ	T	Tau
θ	Θ	Theta	υ	Υ	Upsilon
ι	I	Iota	φ	Φ	Phi
κ	K	Kappa	χ	X	Chi
λ	Λ	Lambda	ψ	Ψ	Psi
μ	M	Mu	ω	Ω	Omega

APPENDIX D – TIDAL-LOCKING

TIDES

The familiar tides in the sea which washed away our youthful sandcastles are a side effect of 'real' tides. Real tides are a force or forces which result from two (or more) differing gravitational forces acting simultaneously on an object or objects.

Taking, because of their familiarity, the Earth and Moon as examples, there is a mutual gravitation attraction between those two bodies (Newton's law of gravity – Box 3.2). Every 'bit' of the Earth experiences a force pulling it towards the Moon (and vice versa). But the magnitudes of those forces decrease or increase accordingly as the 'bits' of the Earth are further away from or closer to the Moon (Box 3.2). Thus, the 'bit' of the Earth which has the Moon directly overhead at any given time (the near-lunar 'bit') has a stronger pull towards the Moon (Figure Appendix D.1) than the 'bit' on the opposite side of the Earth, where the Moon is at the nadir (the far-lunar 'bit').

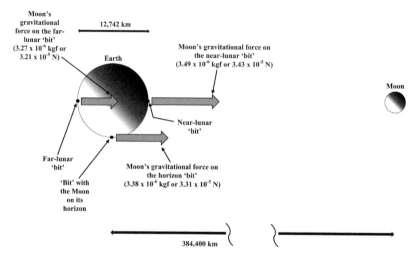

FIGURE APPENDIX D.1   The Moon's gravitational forces on the near-lunar and far-lunar 'bits' (not to scale).

If those 'bits' of the Earth were completely free to move, then they would both fall towards the Moon, with the near-lunar 'bit' accelerating faster and arriving at the Moon's surface sooner than the far-lunar 'bit'.

However, the 'bits' are not free to move; the Earth's much stronger gravitational force holds both 'bits' firmly in their places on the Earth's surface. Nonetheless the forces are still acting upon the 'bits' and these forces are part of the tidal effects of the Moon upon the Earth. If the 'bits' each have masses of exactly 1 kg each and, without the lunar forces, the Earth's gravity would be attracting them towards the centre of the Earth with weights[670] of exactly 1 kgf (9.80665 N) each.

The Moon's attractive force acting on the near-lunar 'bit' is 3.49×10^{-6} kgf (3.43×10^{-5} N), whilst that on the far-lunar 'bit' is 3.27×10^{-6} kgf (3.21×10^{-5} N).

Even though the 'bits' cannot move under these tidal forces, the weights of the 'bits' will change very slightly and the weights are, of course, the gravitational forces of the Earth upon the 'bits'.

A 'bit' on the Earth's surface which has the Moon on its horizon (Figure Appendix D.1) will have a lunar force acting on it of 3.38×10^{-6} kgf (3.31×10^{-5} N), and this is the average lunar gravitational force for all the 'bits' comprising the entire Earth. Considering the Earth as a whole, therefore, the far-lunar 'bit' has a force acting on it from the Moon of 3.27×10^{-6} kgf and this is 1.09×10^{-7} kgf *less* than the average lunar force of 3.38×10^{-6} kgf. The net force (i.e., the difference from the average) acting on the far-lunar 'bit' is thus 1.09×10^{-7} kgf (1.071×10^{-6} N) and it is acting *outwards* (away from the centre of the Earth).

Similarly, the near-lunar 'bit' has a force acting on it from the Moon which is 3.49×10^{-6} kgf and this is 1.15×10^{-7} kgf *more* than the average lunar force of 3.38×10^{-6} kgf. The net force acting on the far-lunar 'bit' is thus 1.15×10^{-7} kgf (1.13×10^{-6} N) and this is *also* acting *outwards* (away from the centre of the Earth).

Now 1.09×10^{-7} kgf (or even 1.15×10^{-7} kgf) is not a very large force. At the time of writing an ounce of gold is worth ~ \$1,800 (~£1,500, ~ €1,700, ~¥240,000) and 1.09×10^{-7} kgf is the weight of 0.000 004 7 ounces of gold. So, if that force is 'worth its weight in gold', then that 'worth' is just 0.8 ¢ (0.7 p, €0.008, ¥1.1). Nonetheless, since it is acting variously upon some 10^{24} kg of 'bits' of the Earth, it does have significant effects.

Those 'effects' are a stretching of the whole Earth along the Earth-Moon line; the net force on the far-lunar 'bit' and the rest of its hemisphere is away from the Moon, the net force on the near-lunar 'bit' and the rest of its hemisphere is towards from the Moon. The Earth therefore becomes elongated along the Earth-Moon line so that its shape resembles that of a Rugby or an American-football ball. A shape known correctly as a prolate-spheroid (Figure Appendix D.2). This effect is commonly described as 'the Earth having two bulges'. To me, though, this suggests it has something like enormous hills projecting out from the still spherical Earth. The Earth's stretched shape, however, remains a smooth geometrical solid, but one that is longer in one dimension than the other two.

---

[670] See the note on Units and Measurements in the Preparatory notes for the difference between mass and weight.

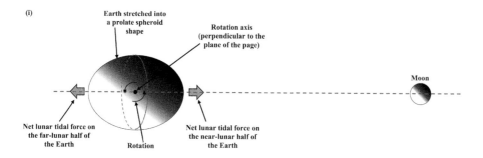

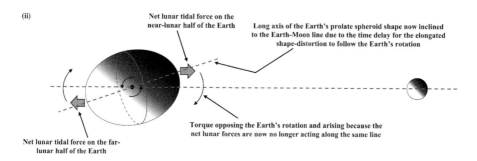

FIGURE APPENDIX D.2

(i)   The Earth's prolate spheroid shape produced by the Moon's tides (shape distortion exaggerated greatly for clarity).

(ii) How the delay in the Earth's distorted shape to follow the Earth's rotation leads to a torque acting upon the Earth, whose effect opposes the rotation and so slows it down.

The forces which have just been described are not, though, the ones which lead to your prize sand-castle disappearing under the waves. A related, but slightly different, process leads to that tidal effect.

The Earth and Moon are orbiting around their barycentre[78] and the centre of the Earth will be moving around its orbit at the correct speed for its distance from the barycentre. All the other 'bits' of the Earth will thus have to be orbiting at the same speed as the one at the centre, because the Earth's gravity is holding them rigidly in place. The near-lunar 'bit' though should, by Kepler's third law, be orbiting in a shorter time than the Earth as a whole. That near-lunar 'bit' is thus compelled to orbit more slowly than it 'wants' to do and so it must be experiencing a force, a tidal force, which slows it down. The far-lunar 'bit', via the same logic, experiences a force which speeds it up so that it is moving faster than *it* 'wants' to do.

In energy terms, the near-lunar 'bit' has too little kinetic energy for the orbit which it is forced to follow. If that 'bit' were free to move, it would drop down into a tighter orbit (closer to the bary-centre). Similarly, the far-lunar 'bit' would move into an orbit further out from the barycentre were it free to do so. This pair of tidal forces, acting parallel to the Earth's surface, will have little effect upon the solid parts of the Earth, but the liquid (water) parts are able to flow across the surface under their influence. This movement of masses of water across the surface of the Earth *is* the familiar sandcastle-washing-away tides. When the water movement is amplified by geographical features we can get huge tide ranges, such as those in the Bay of Fundy, where the high tide mark can be 16-m (53 feet) above the low tide mark.

Tidal-locking

How then does tidal-locking occur? Indeed, what is tidal-locking? The second question is quickly answered: the Moon is tidally locked onto the Earth; the Earth is not tidally locked onto the Moon. The difference being that the Moon is *not* rotating with respect to the Earth (we always see the same side and the same surface features, never the far side unless we're aboard a spacecraft orbiting the Moon) while the Earth *is* rotating with respect to the Moon (an astronaut on the Moon sees the whole of the Earth because it rotates). Thus, tidal-locking means that at least one of the components of a binary system has its rotational[671] and orbital periods of equal lengths so that it does not rotate with respect to the other component.

So how do tides produce tidal-locking? Well, no physical process is 100% efficient, nor can it occur instantaneously. Thus, because the Earth is rotating with respect to the Moon, the elongation of the Earth along the Earth-Moon line is moving around the Earth. This change in the position of the elongation, as just mentioned, cannot occur instantaneously and so it carries onward a little way beyond the position on the Earth where it *should* lie. The long axis of the Earth's shape is not therefore lying *exactly* on the Earth-Moon line (Figure Appendix D.2).

Now, a little bit earlier, we saw that the net force on the near-lunar 'bit' was 1.09 x 10^{-7} kgf (1.071 x 10^{-6} N) and that on the far-lunar 'bit' was .15 x 10^{-7} kgf (1.13 x 10^{-6} N), i.e., the two forces are unequal. Once the ellipsoid's long axis is at an angle to the Earth-Moon line, these two unbalanced forces produce a torque on the Earth which acts to slow down its rotation.

Given a long enough time, that torque will slow down the Earth's rotation until, with respect to the Moon, it has ceased to rotate – and, Hey Presto – the Earth will be tidally locked onto the Moon. Clearly that same process, acting on the Moon has, by today, had long enough to slow down the Moon's rotation so that it is already tidally locked onto the Earth.

Should you concerned that you might in the near future always (or never) have the Moon in your part of the sky, don't get too worried. It is estimated that it will take about four times the current age[672] of the Universe for that locking to occur. This tidal interaction, though, is also causing the distance (384,400 km[673]) between the Earth and Moon to increase[674] by about 38 mm every year. "Not a lot" you are now probably thinking and you would be right; the Moon's distance from us will only be about 50% larger than it is now in 5,000,000,000 years' time, when the Sun starts expanding and brings a halt to all our concerns by (perhaps) engulfing both Earth and Moon.

When an exoplanet is in an elliptical orbit, then both its orbital speed through space and its angular velocity around the host star are greater at periastron than at apoastron while its rotational angular speed is constant. In respect of tidal-locking, this situation has three possible outcomes:

- No effect, i.e., the exoplanet's rotation is much as it would be without the tidal forces on it (although there may be drastic effects upon its geological activity).
- Close to full tidal-locking: the exoplanet's rotation is equal to its orbital period, but during the fastest part of its orbit, the orbital speed 'gains' somewhat on the rotation and it loses against the rotation during the slowest part of the orbit.

   Although not very obvious, this is the situation for the Earth's Moon where it produces a phenomenon called 'Libration'.[675] Libration is seen from the Earth as being able to peek at small parts of the 'far' side of the Moon at times. It enables us to observe directly

---

[671] Viewed from anywhere other than from the component, the tidally locked component *is* rotating. When astronauts get to Mars, they will be able to see both the Earth and the Moon rotating with respect their viewpoint.

[672] The age of the Universe is ~ 13.7 billion years.

[673] This number is for the semi-major axis of the Moon's orbit. Its actual distance from the Earth can range between 362,600 km and 405,400 km.

[674] Technically, the tides are converting rotational angular momentum from the Earth into orbital angular momentum for the Moon.

[675] Actually, what is being discussed here is 'Libration in Longitude'. There are other, much smaller libration effects which need not concern us here.

about 59% of the total lunar surface. Libration arises because, when the orbital speed is 'gaining' on the rotation (around perigee), we observe a little further around *what we see from Earth*[676] as the Moon's *western* edge than is usual and, similarly, around apogee we see a little more of the Moon's observed eastern edge.

- Partial tidal-locking: when the exoplanet's orbit is significantly elliptical, the tidal forces near periastron are so much stronger than those near apoastron that they force the exoplanet's rotation to be tidally locked to the more rapid orbital speed around periastron. Since the exoplanet's orbital speed reduces as it moves away from periastron, the (constant) angular rotational speed decouples from the tidal-locking and the exoplanet resumes its rotation with respect to the host star. The tidal-locking will again come into effect towards the next perihelion passage, but almost certainly with the exoplanet's orientation with respect to the host star being different from the previous perihelion passage.

In the Solar System, Mercury is partially tidally locked to the Sun. Mercury's orbital period is 87.9691 days and its sidereal rotational period is 58.646 days. If you now get your calculator out and calculate ⅔ of the orbital period, it comes to 58.64607 days, a good enough match to the rotational period.

Thus Mercury, throughout one orbit of the Sun, will be rotating with respect to the Sun for 29.324 days and tidally locked for 58.646 days.[677] With the relationship between the orbital and rotational periods being a simple fraction, Mercury will actually have the Sun fixed overhead for 58.646 days at the same three points on its surface, separated by 120° of longitude, once each, every three orbits.

APPENDIX E – ARTIFICIAL INTELLIGENCES

We have mostly just considered life-forms produced 'naturally', i.e., by evolution from a First-Living-Molecule, throughout this book.

At the time of writing, though, the capabilities of what I will call 'Non-Organic Semi-Life-Forms'[678] (mostly computers with appropriate inputs and outputs) are advancing at a quite extraordinarily rapid pace. AI authoring of text material has now reached a level where it is often better than that produced by humans[679] and most major projects would now be impossible without the mind-blowing abilities of CAD-CAM systems.

So,

are Non-Organic Semi-Life-Forms real life yet? – No.

but

are Non-Organic Intelligences real yet? – Yes, or very nearly so.

A true Non-Organic Life-Form, to my way of thinking (please feel free to differ), in addition to all the other life-form requirements, *must* be able to reproduce itself, and that is not yet the case. However, it may be that at some time and somehow in the future ✿✿✿, a self-reproducing ability

---

[676] The rather clumsy language used here is because there can be confusion over what is defined as East and West for the Moon. The current official (I.A.U) definition is based upon the view that an astronaut on the surface of the Moon might take, i.e., that when he/she is facing North, the West is on his/her left side and East is on her/his right side. This definition places Mare Crisium on the Eastern side of the Moon for the astronaut. We on the Earth, though, see Mare Crisium as being to our West. Furthermore, until recently, many maps/drawings of the Moon *did/do* show Mare Crisium as being on the Moon's western side. This confusion is not helped by many maps of the Moon being drawn for the Moon as it is seen through an (image inverting) telescope.

[677] Actually, the transitions will occur smoothly over several days.

[678] Unfortunately, the older (and better) term 'Robot' nowadays brings up images of long lines of multi-armed, static machines each doing a bit of welding, putting in a nut and bolt or packing lettuces, etc., as part of a production line.

[679] This book, though, *has* been written by a human (with a good deal of help from the internet, but not from AI packages).

will be designed into at least a few devices by their (human) manufacturers.[680] Then, evolution of the devices becomes possible and we would have genuine Non-Organic Life-Forms, albeit with humankind midwives providing the First-Living-Molecule moment.

When it comes to considering Non-Organic Intelligences, this now has an extensive history. Automatons, able to walk or move in other ways and powered by clockwork, date back at least half a millennium. In a long series of short SF stories based around human-robot[681] interactions, Isaac Asimov examined many interesting potential situations and devised the 'Three Laws of Robotics'[682] (of which there are four) in the decade or so from 1940 onwards:

- Zeroth Law
    No machine may harm humanity; or, through inaction, allow humanity to come to harm.
- First Law
    A robot may not injure a human being or, through inaction, allow a human being to come to harm.
- Second Law
    A robot must obey the orders given it by human beings except where such orders would conflict with the First Law.
- Third Law
    A robot must protect its own existence as long as such protection does not conflict with the First or Second Law.

The first significant modern (theoretical) work though was by Alan Turing who suggested in 1950 a test which if an artificial intelligence was able to pass it, then that AI should be regarded as having the equivalent of human intelligence.

The Turing test[683] evaluates intelligent behaviour rather than pure intelligence and involves:

- The AI machine, a human and a (human) judge.
- The three participants are isolated from each other and use written communications.
- The judge is free to ask any questions of the other two participants.
- If, after a reasonable time, the judge cannot decide which of the two intelligences is the AI system and which is the human, then the AI system is judged to be of humankind-intelligence-level (at least).

Several AI systems have passed Turing tests, though until recently, not very convincingly. However, in 2014, a chatbot program given the name 'Eugene Goostman' succeeded in convincing a third of its panel of 30 judges that it *was* a human (a boy from Odessa). Whilst there is still controversy over this and other results, it is clear that *if* an AI system has *not yet* passed the Turing test, then one *will* do so in the fairly near future.

Thus, potentially within the foreseeable future, humankind may have produced both Non-Organic Life-Forms and Non-Organic Intelligences and will, almost certainly, soon have combined them ✿✿.

---

[680] Actually, commercial considerations will probably rule this out; would *you* go to the car dealer to buy a new car if the old car reproduced itself (say) every three years for free? However, if the Non-Organic Life-Form car was also a Non-Organic Intelligence, there would be significant ethical problems to be considered as well (Chapter 14).

[681] The mechanical-humanoid type in this instance.

[682] Asimov I. Introductory 'Quote' to *I, Robot*; Second story, *Runaround*, Gnome Press, 1950. Originally published in 1942 as a short story in *Astounding Science Fiction*. Asimov attributes the laws to *Handbook of Robotics*, 56th Edition, 2058 A.D.
    The zeroth law was described in *I, Robot*; Ninth story, *The Evitable Conflict*, Gnome Press, 1950. Originally published in 1950 as a short story in *Astounding Science Fiction*.

[683] The test has its critics, but it still suffices to give an idea of how an AI system might be classed as intelligent.

Well – if we can do that, then comparable ETIs should be able to do the same. Furthermore, their constructs may be better than ours and they may have been making them for billions of years. We must therefore hope, should we ever actually encounter them, that they are familiar with Asimov's laws.

Non-organic life-forms may, quite probably, be able to inhabit environments which are hostile to organic life-forms; surviving at low or high temperatures, in vacuums or at extremely high pressures, within poisonous (for organic life-forms) environments and so on. Even high levels of x-rays and γ rays/high energy particles/radioactivity might be tolerated by non-organic life-forms, although present-day terrestrial electronic equipment is still damaged by such emissions.

Quite possibly/probably and especially for environments inhospitable to organic life, when SETI finds life, it will be of the Non-Organic form (Figure 13.1 Left).

Bibliography

This book, covering, as it does, three of the currently hottest topics in science, could easily have a bibliography longer than the book itself – and most of it would be out of date by the time that the book was published.

A Google search just for '*book interstellar travel*' alone, at the time of writing, returns 14,700,000 results, whilst '*book extra terrestrial intelligences*' has 6,940,000 hits. However, just as the internet causes a problem in choosing a sensible list of references to list in this bibliography, it also provides the solution, which is to suggest that you conduct your own internet search for up-to-date and relevant sources of material on the specific topic in which you are interested, at the time when you are interested in it (I suspect most readers would do this anyway).

If you are a reader of this book without internet access, you will presumably have a book shop or library or another source of books within reach, otherwise it is difficult to see how you *could* be reading this book. Even if your book source does not have internet access, they must have a supplier who could enquire on your behalf for other items of interest to you.

This bibliography therefore lists just those books and other sources of reference of which mention has been made within the book itself, so that those sources, at least, may be found quickly when so desired. Academic journal references and some other similar types of reference are listed in the text, but are not separately listed here. Films/movies/videos/games, etc. are listed in the text, but are also not separately listed here, since they are probably difficult of access/no longer available by the time that this book is published. NB: the publishing details given below may not always be for the first edition but sometimes for the currently available edition(s).

BOOKS

10 Story Fantasy; Sentinel from Eternity, Clarke, A. C., Avon Periodicals, 1951, ISBN-10 1724409662, ISBN-13 978-1724409669.

Alice through the Looking Glass, Carroll, L., Macmillan, 1871, ISBN-10 0763642622, ISBN-13 978-0763642624.

Almagest, Ptolemy, C., ~150 CE. (Modern translation; ISBN-10 978-0691002606, ISBN-13 978-0691002606).

Amtor Series, Burroughs, E. R., Five books, first published in the Argosy magazine, 1934–1964.

Asimov's Mysteries; The Talking Stone, Asimov, I., 1968, Doubleday, ISBN-10 0449210758, ISBN-13 978-0449210758.

Astrophysical Techniques 7th Edition, Kitchin, C. R., Taylor & Francis, 2020, ISBN-10 1138591203, ISBN-13 978-1138591202.

Autour de la Lune (Around the Moon), Verne, J., Pierre- Jules Hetzel, 1869 (Modern edition, E. P. Dutton, 1970, ISBN-10 0525387048, ISBN-13 978-0525387046).

Barsoom Series, Burroughs, E. R., Eleven books, various publishers, 1912–1948.

Children of Time, Tchaikovsky, A., Pan Books, 2015, ISBN-10 1447273303, ISBN-13 978-1447273301.

Collected Works of C. G. Jung, Vol 8. The Structure and Dynamics of the Psyche, Jung C. G., Princeton University Press, 1972, ISBN-10 0691097747, ISBN-13 978-0691097749.

Contact, Sagan, C., Simon and Schuster, 1985, ISBN-10 0671434004, ISBN-13 978-0671434007.

De Magnete, Magneticisque Corporibus, et de Magno Magnete Tellure (On the Magnet and Magnetic Bodies, and on That Great Magnet the Earth), Gilbert, W., 1600, ISBN-10 1113856378, ISBN-13 978-1113856371.

Dragon Riders of Pern (Series), McCaffrey, A., Various Publishers, 1968–2022 (*Dragonflight*, 1968, ISBN-10 0552084530, ISBN-13 978-0552084536).

Entretiens sur la pluralité des mondes (Conversations on the Plurality of Worlds), de Fontanelle, B., 1686 (Modern edition, Tiger of the Stripe, 2008, ISBN-10 190479937X, ISBN-13 978-1904799375).

Exoplanets; Finding, Exploring and Understanding Alien Worlds, Kitchin, C. R., Springer, 2011, ISBN-10 1461406439, ISBN-13 978-1461406433.

Fibonacci's Liber Abaci: A Translation into Modern English of Leonardo Pisano's Book of Calculation, Sigler, L. E., Springer, 2002, ISBN-10 0387407375, ISBN-13 978-0387407371.

Fleet of Worlds, Niven, L., Lerner, E., Tor Books, 2007, ISBN-10 0765357836, ISBN-13 978-0765357830.

Frankenstein; or, The Modern Prometheus. Shelley, M., Lackington, Hughes, Harding, Mavor & Jones, 1818, ISBN-10 6257959624, ISBN-13 978-6257959629.

Goldilocks and the three Bears, Written down by Southey, R., Longman, Rees, 1837, ISBN-10 0721402690, ISBN-13 978-0721402697.

Guide to the Planets, Moore, P., Eyre and Spottiswoode, 1955, ISBN-10 0718817311, ISBN-13 978-0718817312.

Habitable Planets for Man, Dole, S. H., Blaisdell Publishing Co., 1964, ISBN-10 0833042270, ISBN-13 978-0833042279.

Handbook of Astrobiology, Kolb, V.M. (Ed), Taylor and Francis, 2019. ISBN-10 0367780488, ISBN-13 978-0367780487.

Hitchhikers Guide to the Galaxy, Adams, D., 1978 onwards, Complete set of books, Pan, 2020, ISBN-10 1529044197, ISBN-13 978-1529044195.

I, Robot, Asimov, I., Gnome Press, 1950, ISBN-10 0553294385, ISBN-13 978-0553294385

Juggler of Worlds, Niven, L., Lerner, E., Tor Books, 2008, ISBN-10 0765318261, ISBN-13 978-0765318268.

Just So Stories; The Elephant's Child, Kipling, R., Macmillan, 1902, ISBN-10 0099582589, ISBN-13 978-0099582588.

La Pluralité des Mondes Habités, (The Plurality of Inhabited Worlds), Flammarion, N. C., Librairie Académique Didier et C., 1862, (Modern edition, in French, Hachette Livre BNF, 2012, ISBN-10 2012563139, ISBN-13 978-2012563131).

L'Autre Monde: ou les États et Empires de la Lune (The Other World: or the States and Empires of the Moon) de Bergerac, C., 1657, (Modern edition in French, Mille et une Nuits, 1998, ISBN-10 2842053168, ISBN-13 978-2842053161).

Le Micromégas (The Small-Large), de Voltaire, M., 1752, (Modern edition, in French, Nathan, 2012, ISBN-10 2091884367, ISBN-13 978-2091884363).

Les États et Empires du Soleil (The States and Empires of the Sun), de Bergerac, C., 1662, (Modern edition in French, Flammarion, 2003, ISBN-10 2080711458, ISBN-13 978-2080711458).

Les Mondes Imaginaires et Les Mondes Réels, (Real and Imaginary Worlds), Flammarion, N. C., Librairie Académique Didier et C., 1864, (Modern edition, in French, Wentworth press, 2018, ISBN-10 0270808191, ISBN-13 978-0270808193).

Limits; Flare Time, Niven, L., Futura, 1985, ISBN-10 0708882013, ISBN-13 978-0708882016.

Magic Universe. A Grand Tour of Modern Science, Calder, N., Oxford University Press, 2002, ISBN-10 0192806696, ISBN-13 978-0192806697

Martian Chronicles, Bradbury, R., Doubleday, 1950, ISBN-10 096501746X, ISBN-13 978-0965017466.

Mémoires for the Instruction of the Dauphin, Louis XIV, 1661 to 1668. Translated by P. Sonnino, The Free Press, 1970, ISBN-10 0029301300, ISBN-13 978-0029301302.

On the Origin of Species by Means of Natural Selection, or the Preservation of Favoured Races in the Struggle for Life, Darwin, C., John Murray, 1859. (Facsimile, ISBN-10 1435393864, ISBN-13 978-1435393868).

Optical Astronomical Spectroscopy, Kitchin, C.R., Institute of Physics, 1995, ISBN-10 075030345X, ISBN-13 978-0750303453.

Optimism, Keller, H., T.Y. Crowell Co., 1903, ISBN-10 1594622086, ISBN-13 978-1594622083.

Orphans in the Sky, Heinlein, R., Victor Gollancz, 1963, ISBN-10 0671318454, ISBN-13 978-067131845.

Pale Blue Dot, Sagan, C., Random House, 1994, ISBN-10 0345376595, ISBN-13 978-0345376596

Poems chiefly in the Scottish Dialect; To a Mouse, on turning her up in her Nest, with the Plough, November, Kilmarnock volume, 1785, Burns, R., John Wilson, 1786, (Modern edition; Luath Press, 2014, ISBN-10 1910021512, ISBN-13 978-1910021514).

Poems in Two Volumes, Wordsworth, W. 1807, (Facsimile, ISBN-10 9333310576, ISBN-13 978-9333310574).

Red Planet, Heinlein, R., Scribner's, 1949, ISBN-10 0709068018, ISBN-13 978-0709068013.

Remote and Robotic Investigations of the Solar System, Kitchin, C. R., Taylor & Francis, 2018, ISBN-10 149870493X, ISBN-13 978-1498704939.

Rendezvous with Rama, Clarke, A. C., Victor Gollancz, 1973, ISBN-10 1399617176, ISBN-13 978-1399617178.

Ringworld Engineers, Niven, L., Holt, Rinehart and Winston, 1980, ISBN-10 0030213762, ISBN-13 978-0030213762.

Ringworld, Niven, L., Ballantine Books, 1970, ISBN-10 0575027916, ISBN-13 978-0575027916.

Sceptical Essays; On the Value of Scepticism, Russell, R., Geoge Allen and Unwin, 1928, ISBN-10 004104003, ISBN-13 978-0041040036.

Searching for Extraterrestrial Intelligence: SETI Past, Present, and Future, Shuch, P., Springer, 2011, ISBN-10 3642131956, ISBN-13 978-3642131950.

Solar Observing Techniques, Kitchin, C. R., Springer, 2002, ISBN-10 9781852330354, ISBN-13 978-1852330354.

Somnium, seu opus posthumum De astronomia lunari (A Dream, or a Posthumous Work on Lunar Astronomy), Kepler, J., 1608 – Published 1634.

Songe de Platon (Plato's dream), de Voltaire, M., 1756, (Included within Oeuvres Completes De Voltaire, Voltaire foundation, 1991, ISBN-10 0729404218 , ISBN-13 978-0729404211).

Stars, Nebulae and the Interstellar Medium, Kitchin, C. R., Adam Hilger, 1987, ISBN-10 085274580X, ISBN-13 978-0852745809.

Telescopes and Techniques 3rd Edition, Kitchin, C. R., Springer, 2013, ISBN-10 1461448905, ISBN-13 978-1461448907.

The Black Cloud, Hoyle, F., Penguin, 1970, ISBN-10 0141196408, ISBN-13 978-0141196404.

The Blind Watchmaker, Dawkins, R., Norton, 1986, ISBN-10 0393022161, ISBN-13 978-0393022162.

The Chrysalids, Wyndham, J., Michael Joseph, 1955, ISBN-10 071810062X, ISBN-13 978-0718100629.

The City and the Stars, Clarke, A. C., Corgi, 1975, ISBN-10 0552094730, ISBN-13 978-0552094733.

The Flight of the Horse, Niven, L., Futura Publications, 1975, ISBN-10 0345246926, ISBN-13 978-0345246929.

The Great Silence: The Science and Philosophy of Fermi's Paradox, Ćirković, M., Oxford University Press, 2018, ISBN-10 0199646309, ISBN-13 978-0199646302.

The Green Hills of Earth, Heinlein, R. A., 1947, The Saturday Evening Post.

The Island of Dr Moreau, Wells, H. G., Heineman, 1896, ISBN-10 1645940926, ISBN-13 978-1645940920.

The Left-hand of Darkness, Le Guin, U.K., MacDonald & Co. Ltd., 1969, ISBN-10 9781857230741, ISBN-13 978-1857230741.

The Long Arm of Gil Hamilton, Niven, L., Ballantine Books, 1976, ISBN-10 0345280385, ISBN-13 978-0345280381.

The Mote Around Murchison's Eye, Niven, L., Harper Collins, 1993, ISBN-10 0007350163, ISBN-13 978-0007350162.

The Mote in God's Eye', Niven, L., Weidenfeld and Nicolson, 1975, ISBN-10 0297770039, ISBN-13 978-0297770039.

The Naked Ape, Morris, D., Jonathan Cape, 1967, ISBN-10 0224612417, ISBN-13 978-0224612418.

The Origins of Life, Orgel, L., Wiley, 1973, ISBN-10 0471656933, ISBN-13 978-0471656937.

The Selfish Gene, Dawkins, R., Oxford University Press, 1976, ISBN-10 0192860925, ISBN-13 978-0192860927.

The Third Chimpanzee; The Evolution and Future of the Human Animal, Diamond, J., Harper Collins, 1992, ISBN-10 0060183071, ISBN-13 978-0060183073.

The War of the Worlds, Wells, H. G., Pearson's Magazine, 1897, ISBN-10 2874272159, ISBN-13 978-2874272158.

The Wind from the Sun; Crusade, Clarke, A.C., Corgi, 1974, ISBN-10 0552096547, ISBN-13 978-0552096546.

Understanding Gravitational Waves, Kitchin, C. R., Springer, 2021, ISBN-10 3030742067, ISBN-13 978-3030742065.

Universe, Life, Intelligence, Shklovskii, I. S., Eyre and Spottiswoode, 1962, (Printed in expanded form, in English with contributions from Carl Sagan, in 1966 as *Intelligent Life in the Universe*, Holden-Day, 1966, ISBN-10 0816279136, ISBN-13 978-0816279135).

INTERNET ITEMS

Centre de données astronomique de Strasbourg, https://cds.u-strasbg.fr/, [Astronomical data archive].

Exoplanet Exploration, https://exoplanets.nasa.gov/citizen-science/, [A medley of exoplanet topics].

Exoplanet Transit Database, http://var2.astro.cz/ETD/, [Exoplanets data archive].

Exoplanet Watch, https://exoplanets.nasa.gov/exoplanet-watch/, [Observing exoplanets].

List of Exoplanets, http://www.exoplanetkyoto.org/, [Exoplanets data archive].

NASA Astrobiology Institute – NAI, https://nai.nasa.gov/.

NASA Exoplanet Archive, https://exoplanetarchive.ipac.caltech.edu, [Exoplanets data archive].

NASA Exoplanet Exploration, https://exoplanets.nasa.gov/faq/6/how-many-exoplanets-are-there/ [General information on exoplanets].

Open Exoplanet Catalogue, https://www.openexoplanetcatalogue.com/, [Exoplanets data archive].

The Extrasolar Planets Encyclopaedia, http://exoplanet.eu/, [Exoplanets data archive].

Unistellar Citizen Science – UNITE, https://science.unistellaroptics.com), [Observing exoplanets].

Index

For Product Safety Concerns and Information please contact our
EU representative GPSR@taylorandfrancis.com Taylor & Francis
Verlag GmbH, Kaufingerstraße 24, 80331 München, Germany